**A to Z of South East Asian Orchid Species**

First Edition. September 2001



Published by
Orchid Society of South East Asia, Singapore
Robinson Road P.O. Box #2363, Singapore 904363
Fax (65) 467-9677
Email ossea@pacific.net.sg
Website: www.mediav.com.sg/ossea


For book orders please contact the above addresses and fax numbers.

Colour separations by Unique Colour Separation Pte Ltd
Printed by KHL Printing Co Pte Ltd

ISBN 981-04-4479-6

Cover: A lowland form of *Bulbophyllum lobbii* from Sumatra. *B. lobbii* is widespread and common in many parts of S.E. Asia.

# A to Z of South East Asian Orchid Species

Peter O'Byrne

Orchid Society of South East Asia/ Singapore

*Malleola insectifera*

## *Foreword*

IT is a delight to write the foreword to a book which fulfills the only basic requirement of orchid literature. Peter O'Byrne is an extraordinary man who provides that elusive essential, that rare commodity, the antithesis of banal anecdote and horticultural mythology, the material that makes a book really useful but is often overlooked, the stuff that gives the reader knowledge and confidence - beautiful, indispensable, closely observed and succinctly expressed - facts. In an era when amateur orchid growers in the West are returning to their nineteenth century roots, growing more and more species, raised by the hundreds of thousands in laboratories and nurseries all over the world, cultural information is desperately needed.

Those of you who have read his 'Lowland Orchids of Papua New Guinea', or his articles in The Orchid Review, will know that he is not only an author, but an authority. There are neither generalisations nor gardening folklore to be found in his writing, and herein Peter O'Byrne gives us information on nearly 400 species of orchid that grow in their varied habitats in South East Asia, where he has studied them for the past 14 years. Each plant has a colour photograph - all but two taken by the author - and a brief account of the plant and its distribution. A good photograph can inspire as well as educate, but his photograph of *Coelogyne rochussenii* with 2,500 flowers on 63 spikes is, additionally, breath-taking. Where this book excels, is in the distribution maps and easy to read pictograms of plant form and flowering period; habitat rainfall, sunshine and temperature month by month; and growing media. A quick glance tells one that, for instance, *Calanthe rosea* is warm growing, receives heavy rain in its habitat only in September and August, and is best cultivated in a pot with a terrestrial compost. These are the essential facts that every grower needs.

Ninety four genera are covered in this book, and many are little known in cultivation outside South East Asia. They are all fascinating, intriguing and tempting and by making them more widely known, and by giving us the information to grow them successfully, the author guides us to do our part, as ex situ conservationists and orchid lovers, with confidence.

*DR. HENRY OAKELEY*
Chairman, Orchid Committee of the Royal Horticultural Society
United Kingdom

*Coelogyne celebensis* growing on a dead tree trunk, Sulawesi.

*Liparis minahassae* in full bloom on a forked tree-trunk in Sulawesi.

*Oberonia* species growing on a tree in Sulawesi after a forest fire. Almost everything else on the hillside was burned to a cinder, but these trees were protected by surrounding rocks, so they (and their cargo of orchids) survived.

Masses of *Cymbidium aloifolium* growing on a palm tree in Luang Prabang, Laos.

Wild-collected specimens of *Dendrobium chrysotoxum* (above) being transported to the market on a lorry (right) in Phonsavan, Laos, where they will be sold to collectors.

# Introduction

THIS book was produced by the Orchid Society of South East Asia (OSSEA) to coincide with the 17th World Orchid Conference 2002 in Kuala Lumpur, Malaysia. The WOC attracts visitors from all over the world who congregate to enjoy the profusion of orchids (and orchid-related items) on display during the week. Many of these visitors know little about the local orchid species that they see at the Conference or on trips around the area, but want to find out as much as they can while they have the opportunity.

This book was written for people who want a simple, convenient, inexpensive guide to the orchids of South East Asia. There are other books available, but the most comprehensive ones are too large and expensive, while the smaller ones are either too general or cover too wide a geographic area. This book was designed to avoid these problems.

This book is not a complete guide to the orchid species of the region. The number of species in South East Asia could amount to more than ten thousand. Here we illustrate and discuss 401 species in 94 genera; this is a small selection. We have attempted to include many of the more common species, many of the showier species, and a fair number of extremely rare (or little-known) species. We hope there is something for everyone!

In making the choice of species, we have tried to position Kuala Lumpur at the centre of the region, so there are relatively many Malaysian species included, but relatively few from Vietnam or the Philippines. The further away from Malaysia you travel, the less useful this book becomes.

All the funds from sales of this book will go to support the Orchid Society of South East Asia. OSSEA, the original orchid society in South East Asia, meets on the first Sunday of each month in Singapore, and provides a forum for discussions, exchange of views, general orchid-related chat, and (of course) an opportunity to display your best plants. Visitors are always welcome. More information can be found at: www.mediav.com.sg/ossea

I would like to thank the Singapore Botanic Gardens and members of OSSEA for allowing me to photograph their orchid collections. This book would not have been possible without their assistance and input.

Finally; the names of orchid species are often a source of contention. The names used in this book reflect the author's opinion. You are welcome to disagree.

*PETER O'BYRNE*
UWCSEA, Pasir Panjang P.O.Box 15, Singapore 911121.
e-mail: oberonia@singnet.com.sg

## *The Orchid Society of South East Asia*
SINGAPORE

THE Orchid Society of South East Asia, Singapore (OSSEA) is one of the few non-profit institutions that have not only survived but remained relevant ever since its founding in 1928, a period in Singapore's history which saw the remarkable rise of the rubber industry in Malaysia and renewed interest in many things horticultural in this part of the world. Then known as the Malayan Orchid Society, OSSEA was established by three great names in orchidology, viz. R.E. Holttum, a well known writer and Director of the Singapore Botanic Gardens and whose special interest was seed germination, John Laycock, an established lawyer and an avid collector of species, and Emil Galistan, a city councillor and an expert gardener.

OSSEA's objectives had been and continue to be the promotion of orchid conservation, hybridization and cultivation, and the society works for the interests of hobbyists, botanists and orchid growers in the region as well as worldwide. The society's early days saw great strides in cultivation and germination and witnessed the tentative beginnings of a cut flower industry. But it was the society's participation in the Second World Orchid Conference in Hawaii in 1957 that raised its international image. Another historical occasion that put Singapore firmly in the world's orchid map was the award of an FCC (First Class Certificate) by the Royal Horticultural Society to *Vanda* Tan Chay Yan at the Chelsea Flower Show in 1954. These events culminated in what must be OSSEA's defining moment when Singapore was accorded the honour of hosting the Fourth World Orchid Conference in 1963, a truly spectacular affair in the annals of the region's rich botanical history.

Besides the World Orchid Congress, OSSEA today participates in regional shows that include the Asia Pacific Orchid Conference, the Japan Grand Prix International Orchid Festival in Tokyo and the Shah Alam Orchid Show in Kuala Lumpur. The society also organizes trips to nurseries and members' homes in Singapore and Malaysia, and holds a local orchid show each year in conjunction with the Singapore Botanic Gardens. Amongst its many activities are awards given by the society to plants deserving of FCCs, AMs and HCCs, which are subsequently published in the Review. Other awards include the OSSEA Fellowship Award which is presented to an individual or individuals who have made substantial contributions to the aims of the society, and the Eric Holttum Gold Medal to the breeder with the most outstanding local hybrids or species plant exhibited during the year in question.

The society has a vast store of knowledge in the collective histories of the region's more important growers and hybridisers, and contributes articles and journals to support such efforts. A comprehensive collection of photographs and other archival material is currently being recorded,

In 1931, OSSEA published the first Malayan Orchid Review, which featured mainly species. Today, the Review has grown to well over 100 pages, with virtually every page in full colour, and contains information concerning new hybrids, awards, reviews, shows and articles related to orchid discoveries, scientific research projects and breeding trends. Amongst the society's publications are "Orchid Growing in the Tropics" and a fully illustrated Orchid Hybrids of Singapore, the latter of which will be published in late 2002. To help members stay in touch, OSSEA also produces a monthly Bulletin and maintains a website.

Orchid enthusiasts worldwide are invited to join the society as ordinary members for a nominal fee of S$40 per year (excluding postage), or S$50 if they are Singapore residents. The subscription includes a free copy of the Malayan Orchid Review, and discounted prices for the purchase of the society's publications. The society holds a meeting once a month, where orchidists are invited to give a talk in their area of specialty and a panel gives a discourse on plants brought to the meeting by members for judging. Besides tours and excursions, members also get together for an annual dinner.

For more information, please contact the Secretary at:
Fax (65) 467-9677 Email: ossea@pacific.net.sg
Website: www.mediav.com.sg/ossea
Or write to: Orchid Society of South East Asia
Robinson Road P.O. Box #2363, Singapore 904363

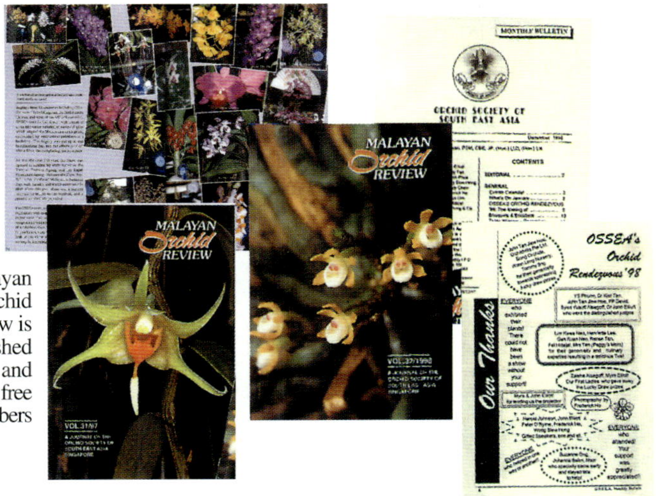


The Malayan Orchid Review is published annually and distributed free to all members

*Dendrobium senile*
**- a Thai species, common in cultivation**

# A

## Acampe

This genus of at least 4 rather large coarse species is distributed from Africa to China and Indochina, with one species occurring as far south as Malaysia. The species have fairly short branched inflorescences with up to 12 rather crowded flowers that are all open at the same time. *A. rigida* and *A. papillosa* are quite commonly seen in orchid collections.

### ***Acampe ochracea*** (Lindl.) Hochr.

In some parts of Myanmar this large tough species is easy to spot as it is present in the crowns of almost all the roadside trees, scrambling up the trunks and along the branches.

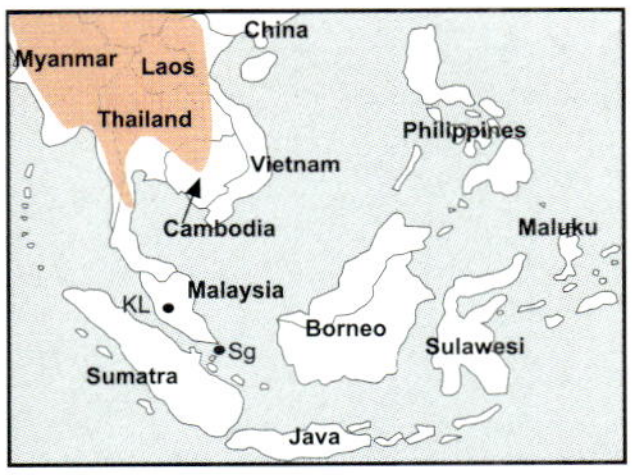


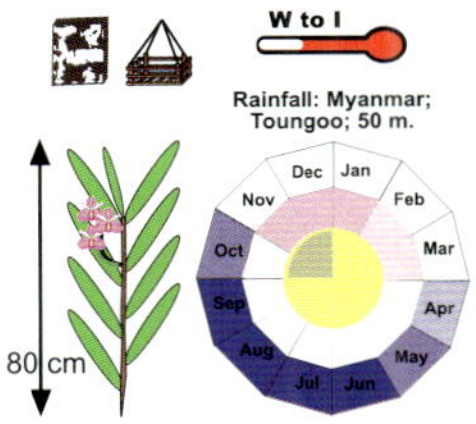


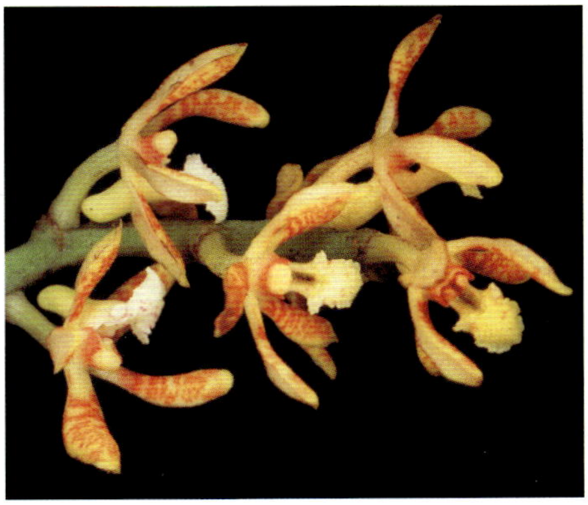

## Acriopsis

This genus of six epiphytic species (and five varieties) is distributed from Sikkim through S. E. Asia to Australia and the Solomon Isles. *Acriopsis* species are often grown in village gardens throughout the region and are easily viewed in the wild. They can be recognised by their crowded egg-shaped pseudobulbs surrounded by a 'nest' of erect white catch-roots. Their cross-shaped flowers are unique; the lateral sepals are fused together and the lip-base is joined to the column margins to form a nectary.

### ***Acriopsis densiflora*** Lindley

This rather uncommon species is the most attractive in the genus due in part to it's compact growth habit. It has egg-shaped pseudobulbs that are 2-3 cm tall, and 12 to 15 cm long unbranched inflorescences each with up to 30 small but brightly coloured flowers. There are two varieties; var. *densiflora* (illustrated here) occurs in the hill country of Malaysia, west Sumatra and northern Borneo. Var. *borneensis* (from Borneo) is sweetly scented and has rectangular keels with a flattened tip.

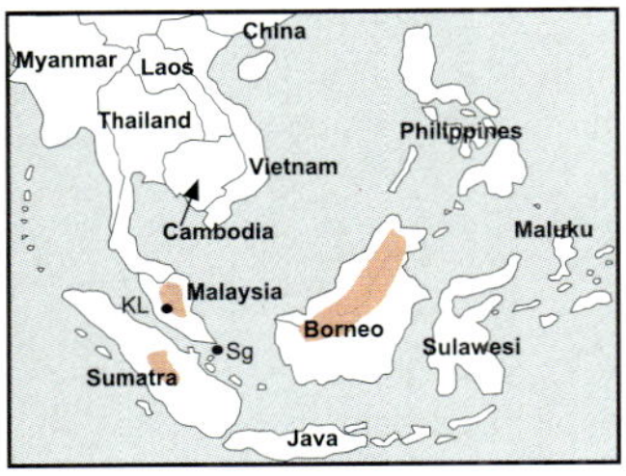


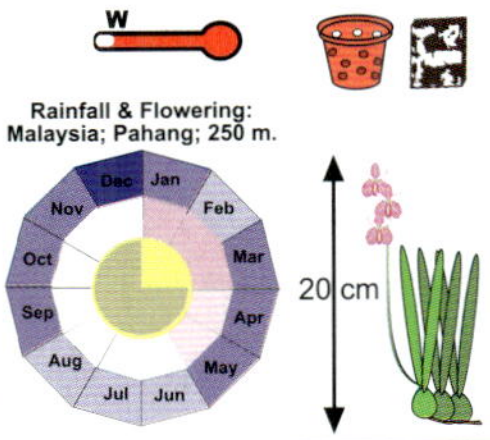

# A

## *Acriopsis liliifolia* (Koen.) Ormerod

This very common species, more widely known by the name *A. javanica*, is present throughout most of S. E. Asia from sea-level to at least 1500 metres altitude, in all lowland forest types as well as trees in plantations, villages and parkland. It is one of the few orchids that grows on bamboo plants. It is often associated with ants, which nest in the roots. In Malaysia the roots and leaves have been used to treat fevers. It has a long branching inflorescence with up to 200 tiny cross-shaped flowers.

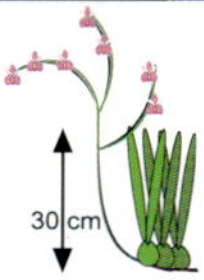

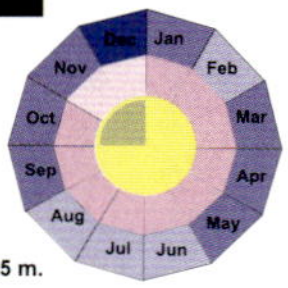


Rainfall: Malaysia; Ipoh; 35 m.

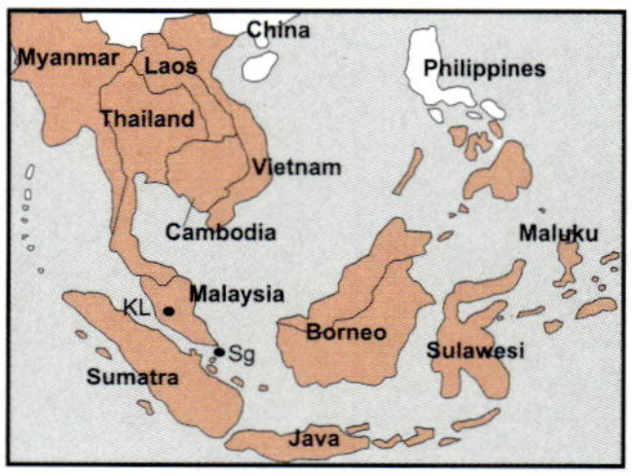


## Adenoncos

This genus of about 16 rather small species is distributed from Thailand and Indochina to New Guinea. The plants have a very characteristic habit, with short narrow fleshy leaves (which are often hard and sharp-tipped) arranged in two close rows. The short inflorescences have 1-5 green (or yellowish flowers) that have a large lip with a white or green keel. *Adenoncos* species are of no commercial value and are only occasionally seen in orchid collections.

## *Adenoncos sumatrana* J.J.Sm.

This small epiphytic species is quite common in the forests of Malaysia and Sumatra, where it grows from sea-level to at least 1200 metres altitude. It has short stems which branch frequently, sometimes resulting in a quite large plant. The 3 cm long leaves are very thick and fleshy and have a sharp-pointed tip. The 1 cm fragrant flowers are easily overlooked as they are produced on a very short inflorescence and are usually held underneath the stem.

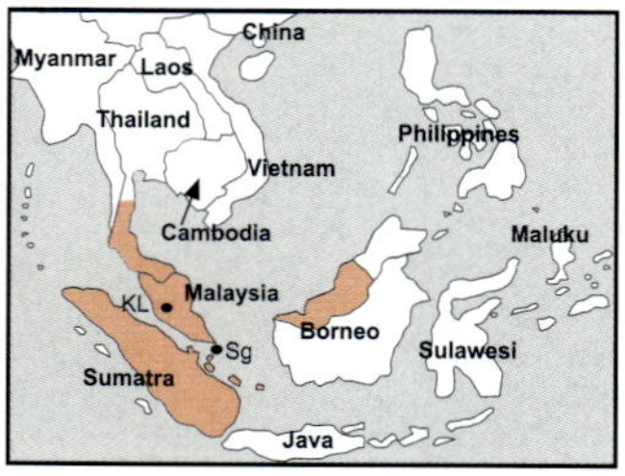


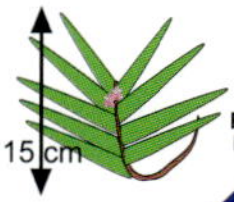


Rainfall: Sumatra; Pekanbaru; 35 m.

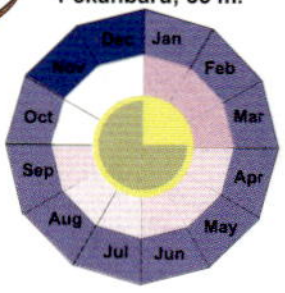

# A

## Aërides

This genus is sought-after by orchid enthusiasts around the world and is grown in gardens and orchid collections throughout Asia. There are some 17 to 20 species, found from India to the Philippines. Although the Philippines is usually considered to have the best *Aërides* species, other parts of S.E.Asia have very attractive members of this showy genus. The species are all medium to large epiphytes or lithophytes and have strap-shaped leaves and an arching / descending inflorescence of close-set showy flowers.

### ***Aërides flabellata*** Rolfe ex Downie

Less well known than several other *Aërides* species, *A. flabellata* is found in the forested lowlands and hills of northern Thailand and Indochina, where it is often seen being grown in village gardens. It is relatively small-growing (for the genus) and carries up to 15 scentless flowers on the inflorescence. The flower are about 2 cm wide and last for 10-14 days. Unlike many *Aërides* species, the colour is not particularly variable.

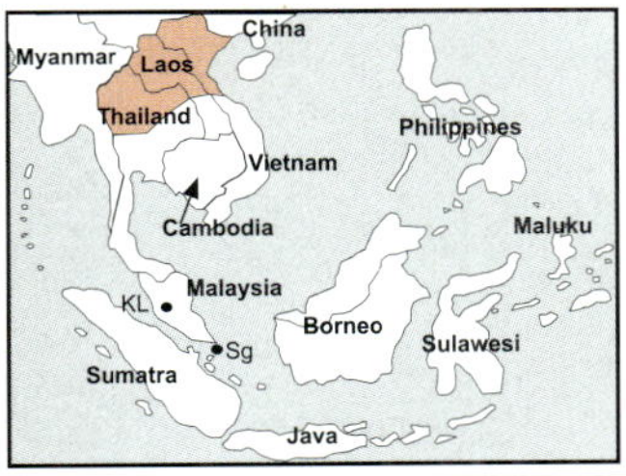

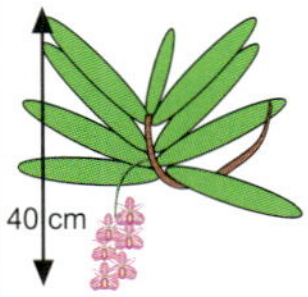


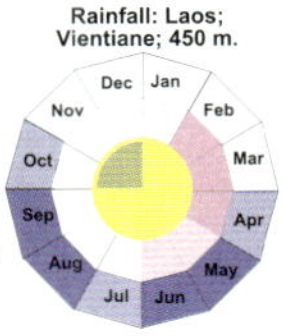

### ***Aërides houlletiana*** Rchb.f.

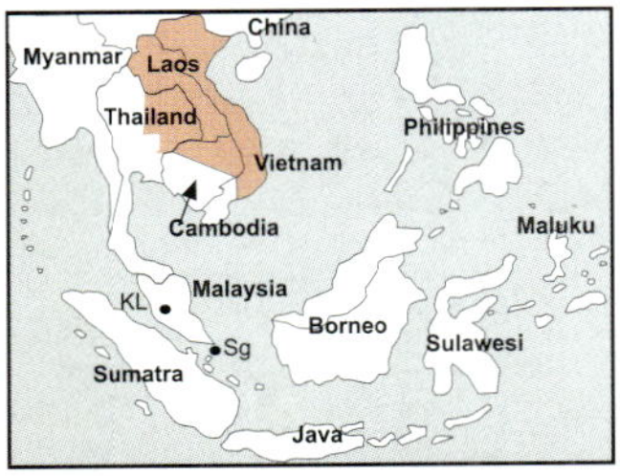


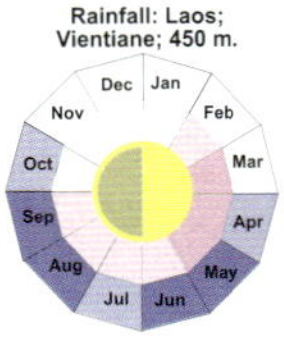


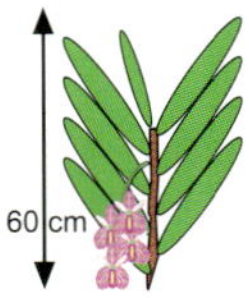

Some authorities consider this fragrant species to be a variety of *Aërides falcata*, which occurs from Assam (N.E. India) and Myanmar to Vietnam. *A. houlletiana* likes more shade than most *Aërides* species. The sepal and petal colour varies from yellow to yellow-green to orange or pale brown; the lip always has a lavender apex.

# A

## *Aërides inflexa* T. & B.

This very attractive species has long been overlooked by orchid growers, although it is common throughout much of southern Sulawesi where it is widely grown as a  garden plant and often becomes quite large. The inflorescence has up to 20 flowers, each of which is 3 cm high. The flowers last for 3-4 weeks; much longer than most other *Aërides* species.

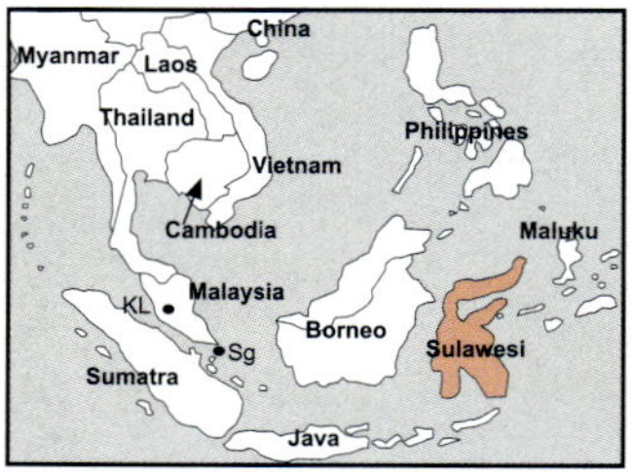


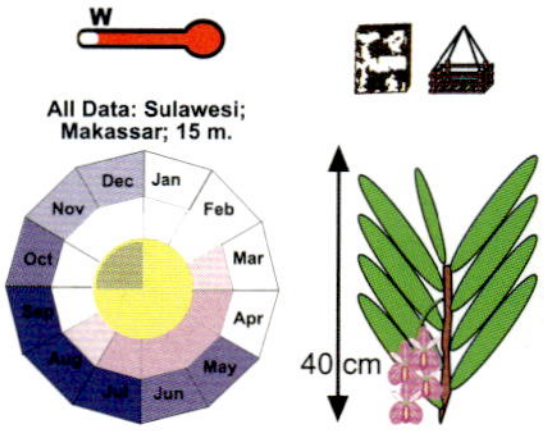

## *Aërides multiflora* Roxb.

This attractive species is well-known in Thailand, where it is commonly grown in gardens and is readily available from nurseries. The illustrated specimen has unusual coloration; most plants have white sepals and petals with purple spots and marks towards the tips, and a purple lip with white markings and lines at the base.

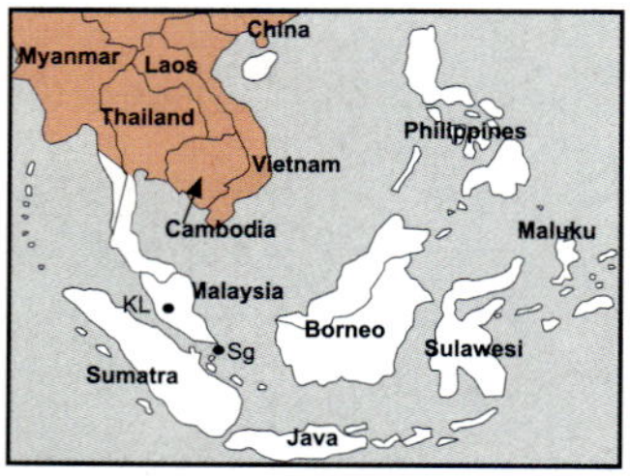


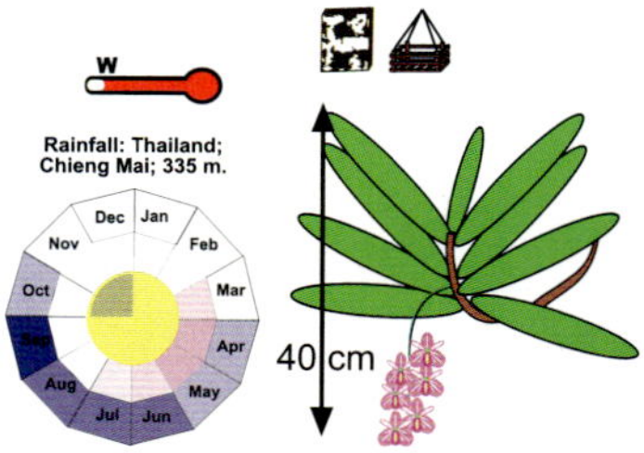

## *Aërides odorata* Lour.

This widespread species can be found through much of the region from sea-level to 500 metres altitude. It is highly variable in flower shape, size and colour; 3 of the main colour forms are illustrated. *A. odorata* is strongly scented, although not everyone finds the lemon-honey scent attractive. Some specimens from Borneo and the Philippines are sweet-scented.

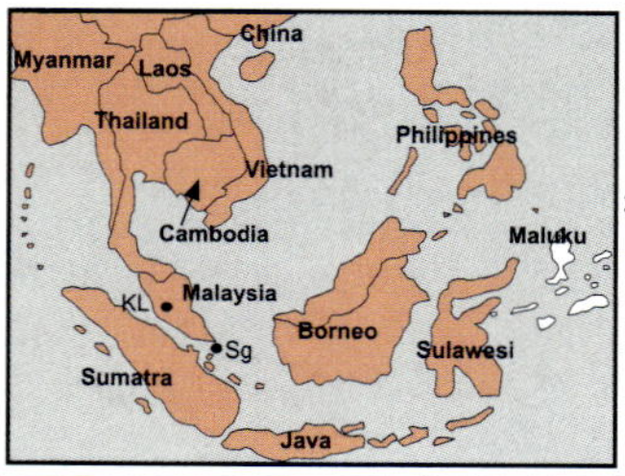


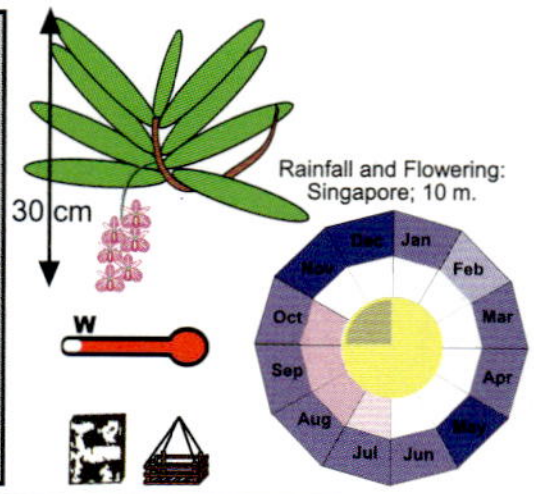

# A

## ***Aërides odorata*** Lour. **var. *lawrenceae*** Rchb.f.

Most orchid growers know this plant as *A. lawrenceae*, although it differs from the normal form only in scent and colour. This variety, which is much sought-after by orchid growers for its beautifully fragrant flowers, grows into quite large plants and comes in a variety of colours, including an *alba* form.

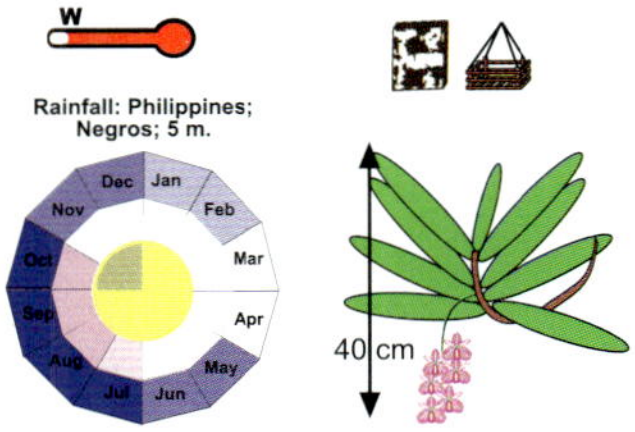


## ***Aërides odorata*** Lour. **var. *quinquevulnera*** Lindl.

Widely known by the synonym *A. quinquevulnera*, this variety differs from the normal form only in scent and colour. A number of different colour forms have been described. The most common form (illustrated) has white flowers with pink-tipped tepals. Plants from Sulawesi have larger flowers but are often scentless.

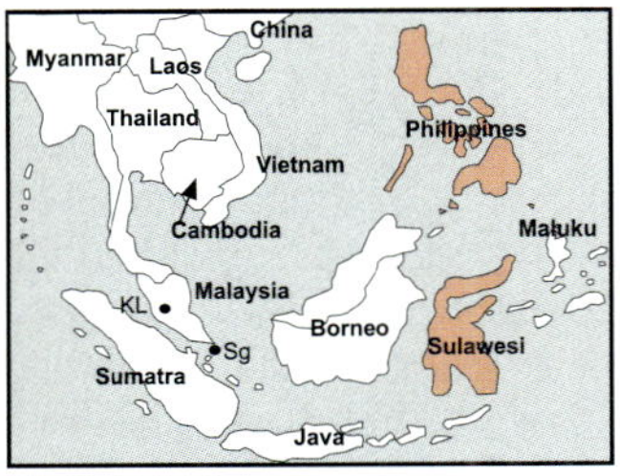


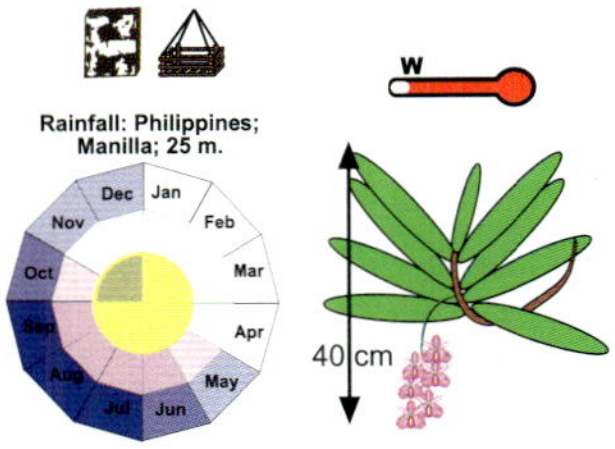


## ***Aërides rosea*** Loddiges ex Lindl. & Paxton

Widely known by the synonym *Aërides fieldingii*, this beautiful species grows in regions with a long dry season. The flower colour varies from rosy-pink to lilac, becoming paler towards the base of the sepals and petals. The flowers are usually produced towards the end of the dry season and last about 2 weeks.

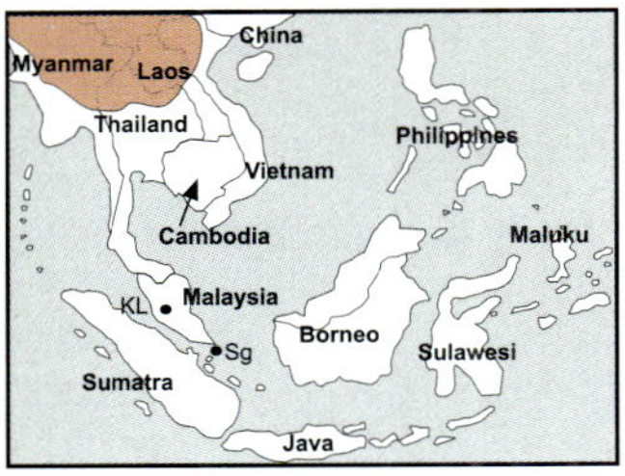


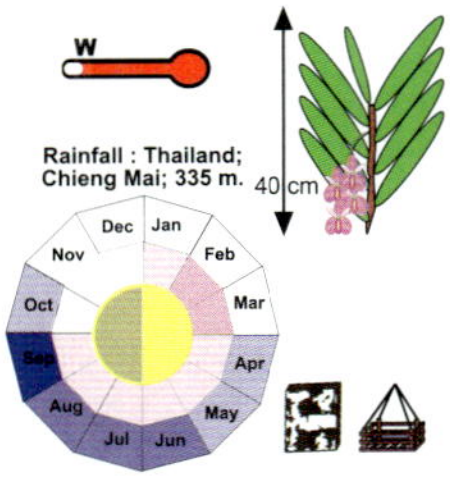

# A

## Agrostophyllum

This genus of 50-60 tropical species is found from the Seychelles to the Pacific Islands with the majority of species being found in New Guinea. Most species are epiphytes, but some occasionally grow on rocks. Most *Agrostophyllum* species have slow-growing upright stems that terminate in a tight ball of inflorescences, but some are fast-growing creeping species with branching stems. None are particularly attractive, and despite being quite common, they are seldom seen in collections.

### ***Agrostophyllum glumaceum*** Hook.f.

This species is quite common in parts of Malaysia and Borneo, growing up to 1100 metres altitude. It is very distinctive because it has only 3-5 leaves on each side of the stem, to which they are attached by long loose sheaths. It forms a 3 cm wide ball of inflorescences, each of which can produce up to 5 off-white flowers.

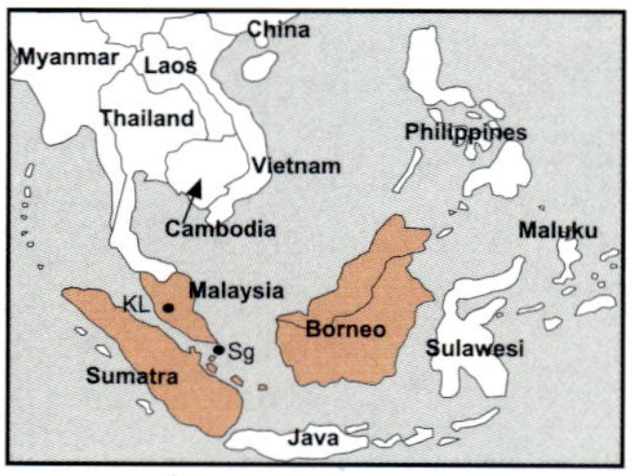


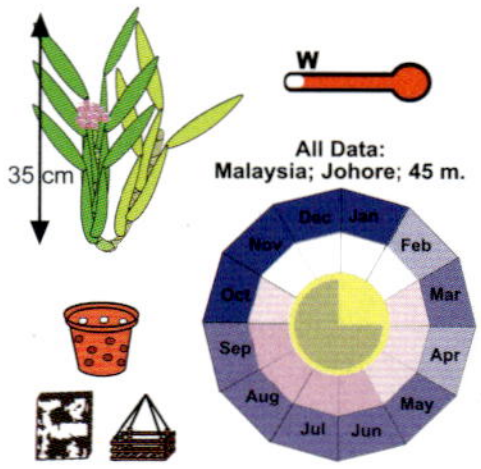

## Anoectochilus

This genus of some 40 small terrestrial species is distributed from India and Sri Lanka to Japan and the Pacific Islands. Although most species have green leaves, some have a beautiful velvety sheen or colourful silvery, red or pink veins and are cultivated as "Jewel Orchids" (the name includes other terrestrials with variegated leaves). *Anoectochilus* species have a creeping rhizome which produces erect shoots at intervals. The flowers have distinctive lips with a bilobed apex behind which are rows of lateral filaments.

### ***Anoectochilus setaceous*** Bl.

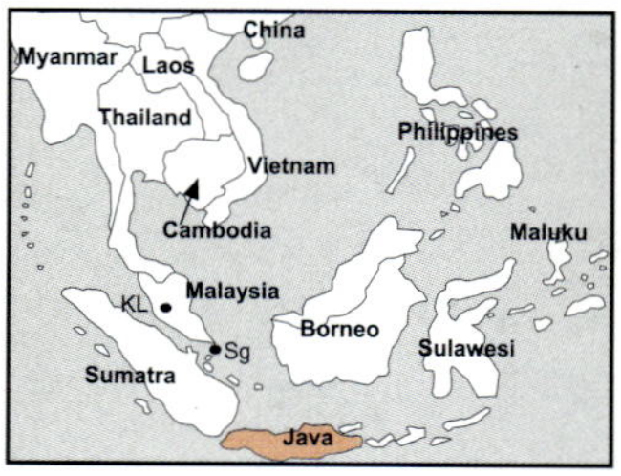

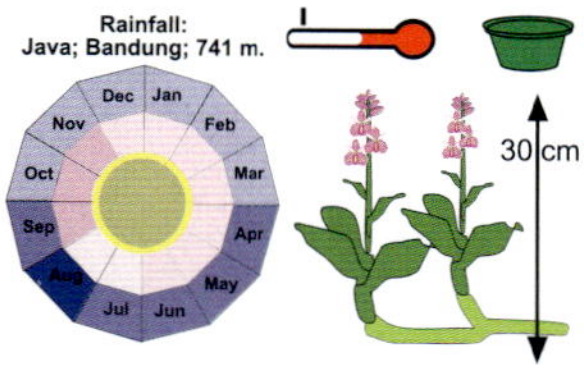


Found growing in humus on the forest floor from 1000-1800 metres altitude, this attractive terrestrial is occasionally seen for sale in nurseries and garden centres. It is easy to grow but requires deep shade with no direct sunlight. Most plants have white veins but yellow or pink-veined plants have been recorded.

# A

## Appendicula

This genus of 60-70 tropical species is found from the Himalayas to the Pacific Islands with the majority of species being found in Sumatra; none are of any commercial value. They are sympodial epiphytes or lithophytes with thin (sometimes woody) stems that carry 2 rows of small leaves which are often twisted at the base to lie in one plane. The inflorescences can be terminal or lateral, and the flowers are always small.

### *Appendicula anceps* Bl.

This species is very common in many parts of lowland Malaysia and Sumatra, where it is easily spotted low down on trunks of trees alongside rivers, or on mossy rocks next to waterfalls. In Java it grows on trunks of trees from 750-1600 metres altitude. The stems are usually 20-40 cm long, although much bigger specimens are occasionally seen. The stems are slightly flattened and have sharp edges (hence the name *anceps*) and carry two rows of leaves that are twisted at the base so they all lie in one plane. The descending inflorescences are up to 4 cm long and can form up to 30 flowers, although only 2-4 are open at one time.

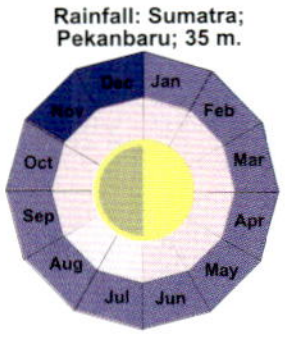


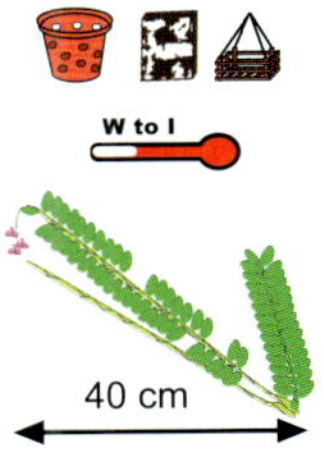


### *Appendicula undulata* Bl. var. *longicalcarata* (Rolfe) Ames

This normal form of this species is found from Sumatra and Java to the Philippines, and has a medium-length spur. Var. *longicalcarata* is a large plant that grows high up in lowland forest trees, usually in forks or on large branches. It has pendulous stems that are usually about 80 cm long. The stems are dorso-ventrally slightly flattened and carry two rows of small leaves that are twisted at the base so they all lie in one plane, and are often a dark green-purple colour. The inflorescences are up to 2 cm long and produce a series of flowers that are so thin they look like bits of white thread.

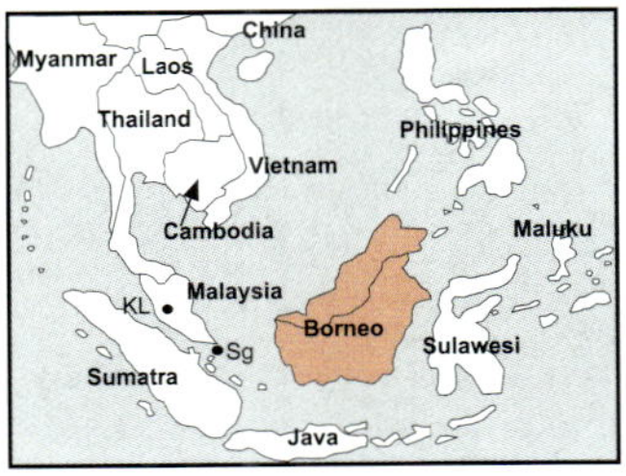


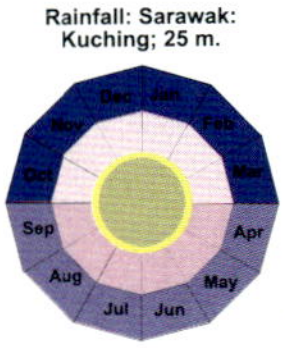


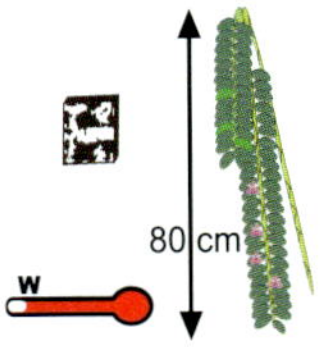

# A

## Arachnis

Widely known as the "Scorpion Orchids", this well-known genus was once common in gardens throughout Asia, but today the species have largely been displaced in orchid collections by hybrids such as the ever popular *Arachnis* Maggie Oei. *Arachnis* contains 13 species (and one natural hybrid) and occurs from Sikkim to the Solomon Islands. Most of these plants are epiphytes with scrambling stems that may be several metres long. The large fragrant flowers often have bold brown or green markings.

### ***Arachnis flosaeris*** (Lindl.) Rchb.f. **var. *gracilis*** Holttum

This tall species is usually seen scrambling up through trees until it reaches the upper branches although it is occasionally found growing over rocks or up cliff faces. The flowers have a musky odour. Variety *gracilis* (illustrated), has pinkish red blotches and a sweet scent; this variety has been recorded from Thailand, Malaysia and Borneo.

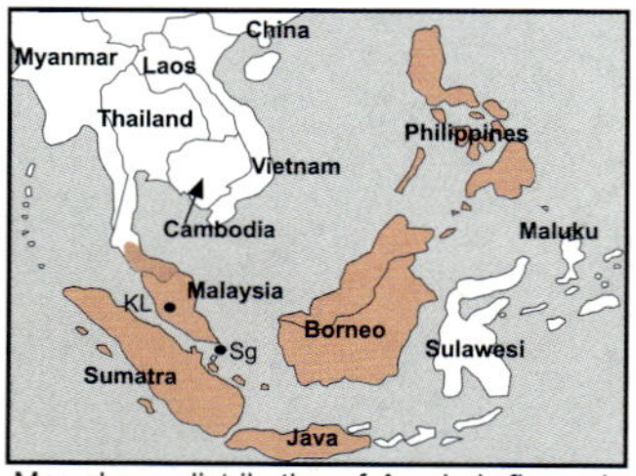


Map shows distribution of *Arachnis flosaeris*

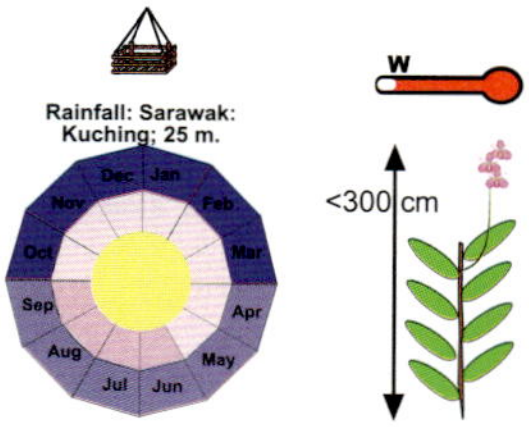


## Arundina

This genus of 2 to 5 terrestrial species is distributed from India and Sri Lanka to Sulawesi and China. The only significant species in S. E. Asia is *A. graminifolia*, which occurs throughout the region and has become naturalised in the Pacific Islands.

### ***Arundina graminifolia*** (D.Don.) Hochr.

This terrestrial orchid is found from sea-level to at least 1500 metres altitude and is usually seen on roadside embankments and cuttings. It is commonly grown as a garden plant and I have seen it used as a windbreak around a swimming pool in a Malaysian hotel. A Highland form (illustrated) has white sepals and petals, while the lowland form has larger flowers with pinkish sepals and petals.

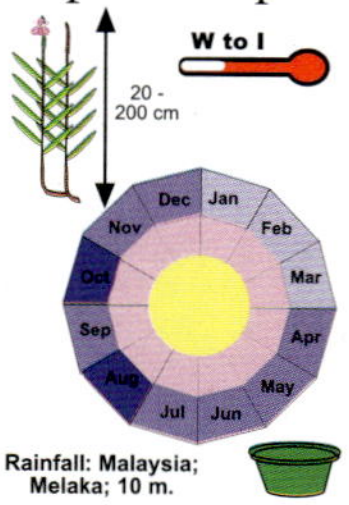

# A

## Ascocentrum

This genus of 8-10 epiphytic species is well known to orchid enthusiasts around the world and is valued for its brilliantly coloured flowers borne on small compact plants. The best-known species is *Ascocentrum miniatum* (some consider the Thai plants to be *A. garayii*) with its dense inflorescence of red-orange flowers, but the genus contains less well known plants of other colours.

### ***Ascocentrum ampullaceum*** (Lindl.) Schltr.

This is the northernmost representative of the genus, being found from Bhutan to China and extending as far south as the Andaman Islands. Plants are rather variable; some have spotted leaves (possibly caused by exposure to the sun), and the colour varies from the normal red through red-purple to violet-purple. A red-orange colour form from Munipore (India) has been called *var. aurantiacum*; this should not be confused with *Ascocentrum aurantiacum*, which is a different species. *A. ampullaceum* often produces a spectacular display, as each stem can form several inflorescences of densely packed flowers.

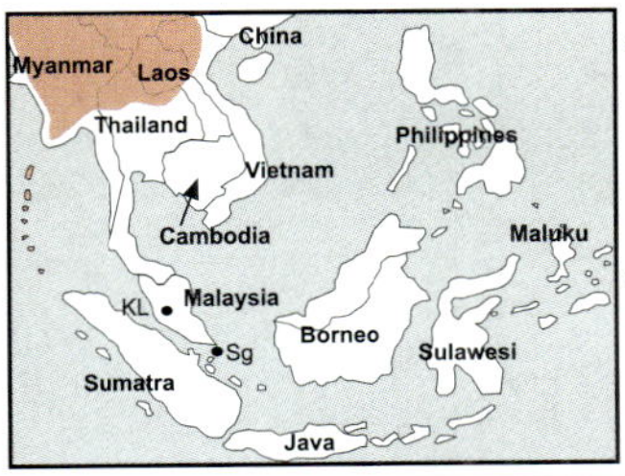


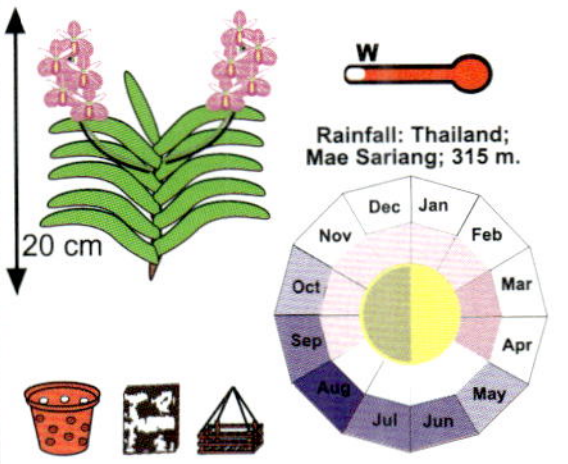

### ***Ascocentrum aurantiacum*** Schltr.

This species was originally described from Sulawesi, where the flowers are a glowing orange with a paler column and spur (left). Philippines plants, recently given subspecies status by Christenson, are yellow-gold with a red-orange area around the spur aperture (right), and appear to be identical to the Sumatran form of *A. miniatum*.

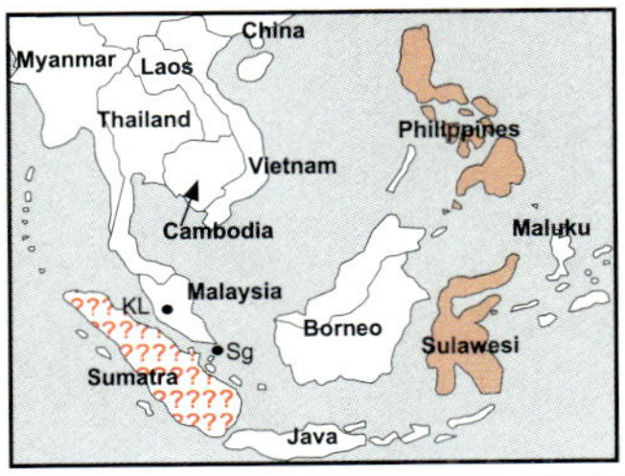


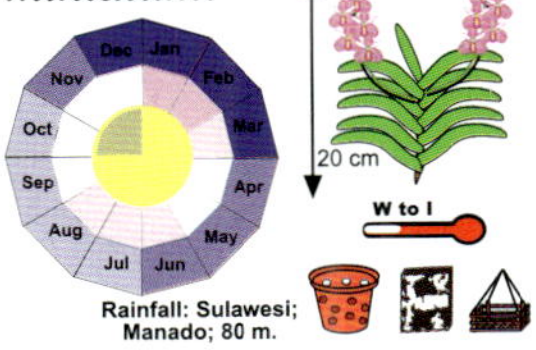

# A

## *Ascocentrum aureum* J.J.Smith

Little is known about this extremely rare species. It was first described in 1917 from specimens collected in the Sula Islands, and since then has been lost to the outside world, although there are rumours of it growing on the eastern 'arms' of Sulawesi. The plant in the photo, which was growing in the Singapore Botanic Gardens collection, perfectly matches the description of this long-lost species. The plants are rather small, with a few slender leaves and a short uncrowded inflorescence of small but brilliantly coloured flowers in which the sepals and petals are rather cupped and the short spur curves forward.

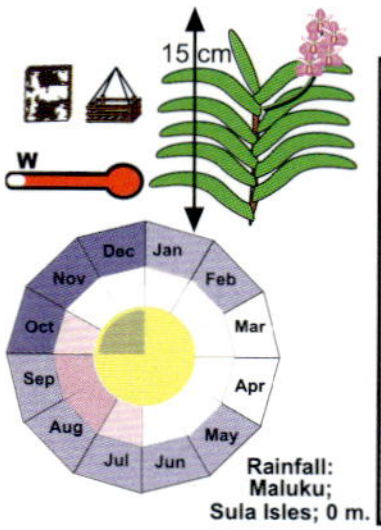


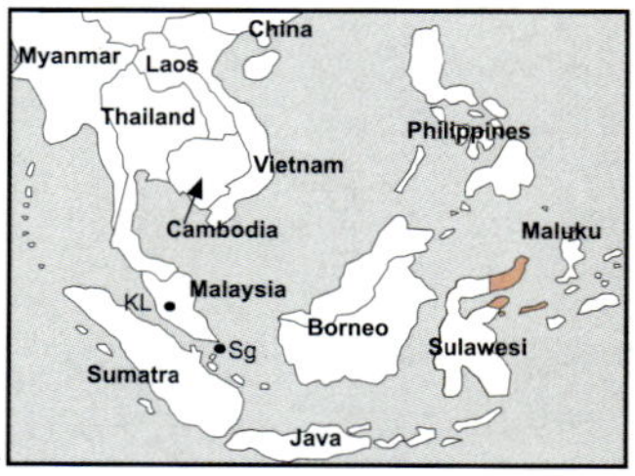


## *Ascocentrum christensonianum* J.R. Haager

This delicately-coloured species was discovered as recently as 1993 and is currently known only in a small region of Vietnam, where it grows on the branches of forest trees. In the last few years specimens of this species have been offered for sale in some orchid nurseries in S. E. Asia. *A. christensonianum* differs from other *Ascocentrum* species by having an elongated growth habit and short inflorescences and the flowers have a long slender spur that has a swollen tip. It's pale lilac flowers are a welcome extension to the colour range of this genus.

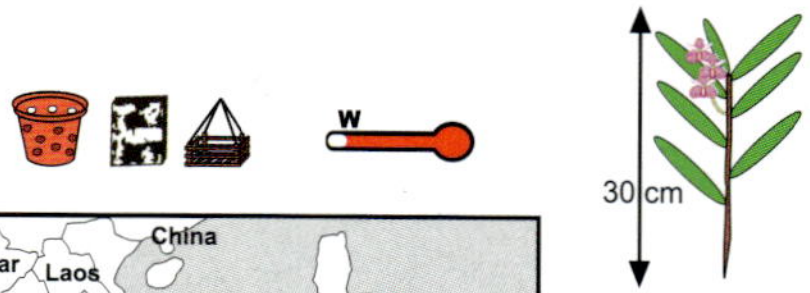


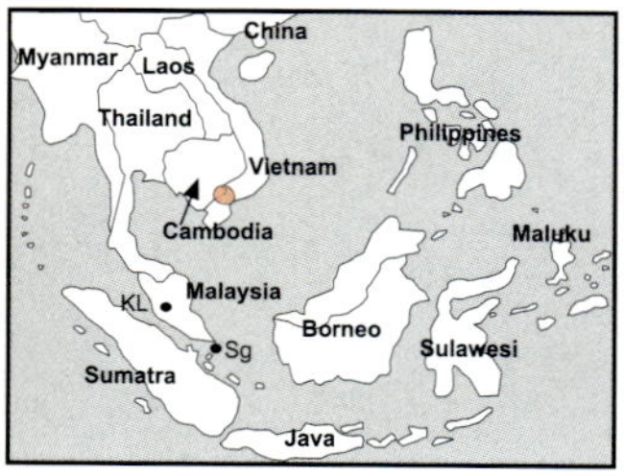


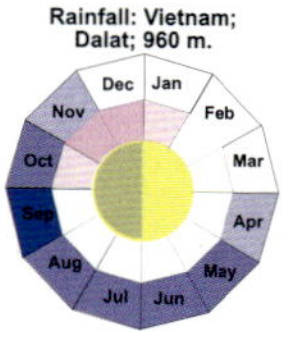

# A

## *Ascocentrum miniatum* (Lindl.) Schltr.

This spectacular species is known to occur in Java, Malaysia and peninsular Thailand, although the exact distribution is unclear. In Java it is quite common, growing from 0-1200 metres altitude in forests and plantations, preferring bright airy situations. According to Christenson, the plants of north Thai origin that are in cultivation worldwide under the name *A. miniatum* are actually *A. garayii*. It is unclear if the plants from Sumatra are *A. miniatum* or *A. aurantiacum*. *A. miniatum* forms clumps of 20-30 cm tall plants with close-set thick leathery leaves. It typically produces several 20-25 cm inflorescences at a time, each with about 40 crowded 12 mm wide red-orange flowers.

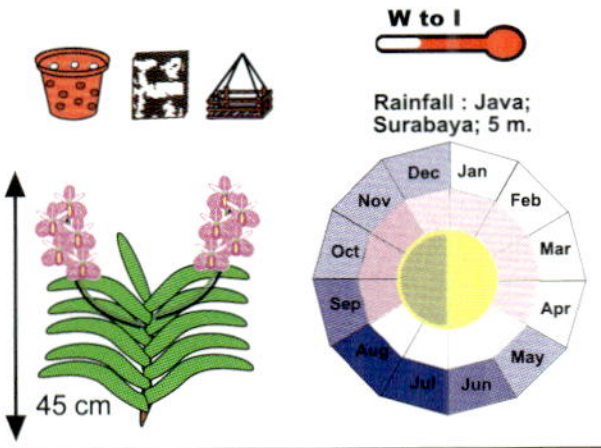


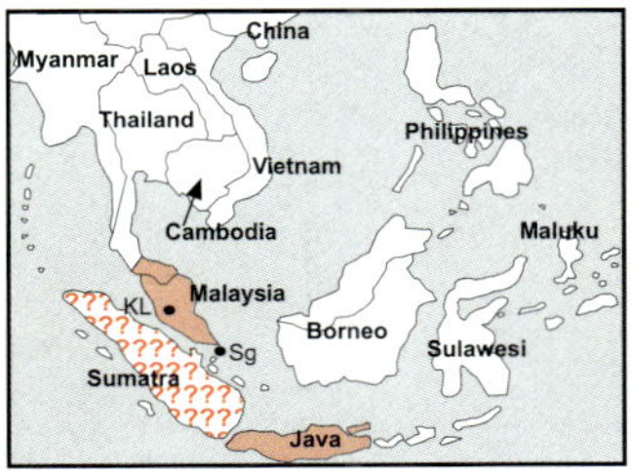

## Ascochilopsis

This genus contains only 2 species; *A. myosurus* (below) and *A. lobata* which was described in 1994 and is only known from Sabah in Borneo. Both species are very small epiphytes with branching stems and relatively large close-set fleshy leaves. Several inflorescences are produced at the same time; these are club-shaped with a narrow base and a swollen upper part, and produce flowers in succession over a period of several months. The tiny flowers only last 1 day.

## *Ascochilopsis myosurus* (Ridl.) Carr

This small species grows on twigs and small branches in moist lowland forests. The inflorescences are 4 cm long, and produce up to 4 tiny flowers at a time. Each inflorescence continues to flower at irregular intervals over a period of several months.

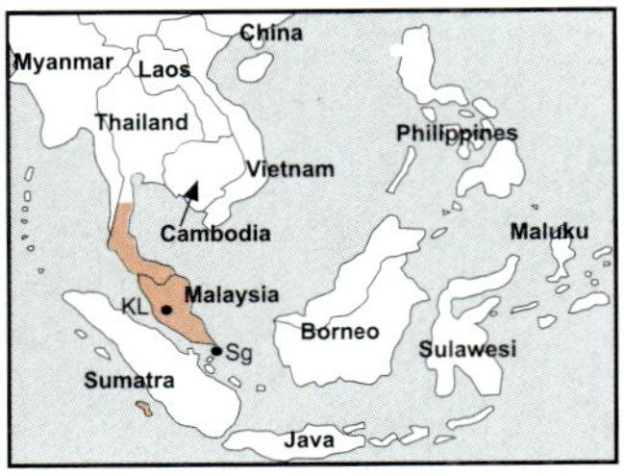


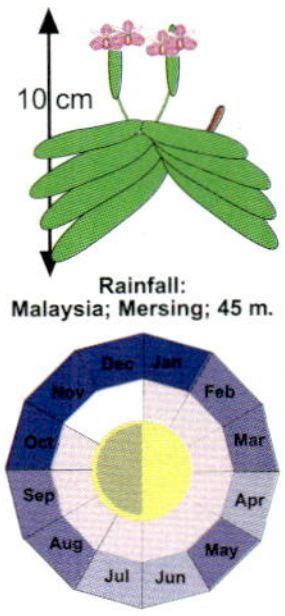

# A

## Ascochilus

This genus contains about 5 to 10 species; it is hard to be precise because botanists disagree about whether several species belong here or in the closely-related *Pteroceras*. *Ascochilus* species are very small epiphytes with close-set leaves. The inflorescences are narrow with a swollen tip from which the flowers emerge. Flowering occurs in successive bursts over a period of several months. The small flowers only last 1 day.

### ***Ascochilus emarginata*** (Bl.) Schuitemann

This species has been known as *Pteroceras emarginatum* and *Grosourdya emarginata*. It is a very small twig epiphyte that is common on mangroves along the coast of north Borneo; elsewhere it has been found up to 1400 metres altitude. Each plant produces several inflorescences that are active for several months, producing up to 6 flowers at a time at irregular intervals.

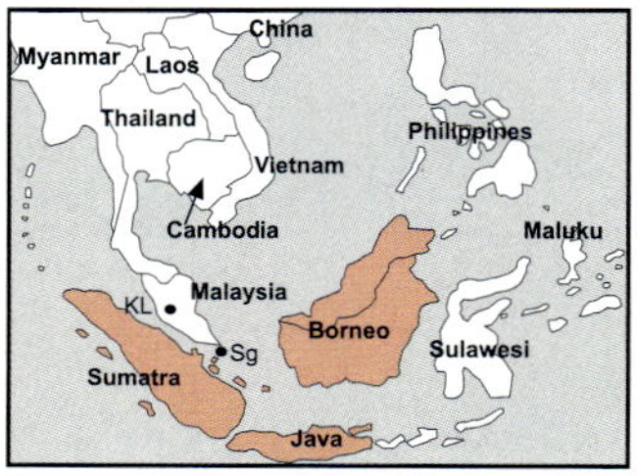


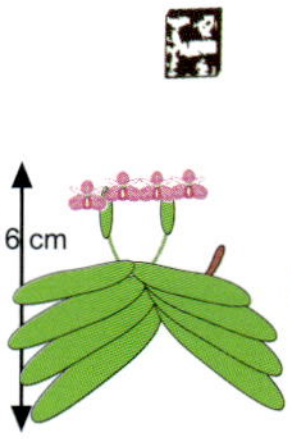


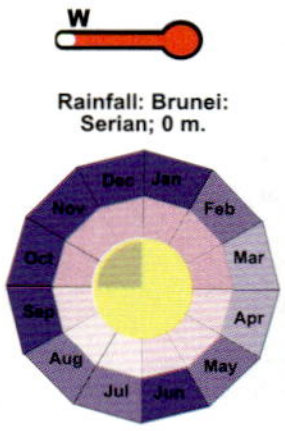

## Ascoglossum

This genus contains only 1 species.

### ***Ascoglossum calopterum*** (Rchb.f.) Schltr.

This species has two different growth habits; plants from Sulawesi and Maluku (previously known as *A. purpureum*) are tall coarse specimens that grow well in full sun, while plants from New Guinea prefer light shade and are seldom more than 30-40 cm long and have shorter inflorescences. Both forms have identical flowers.

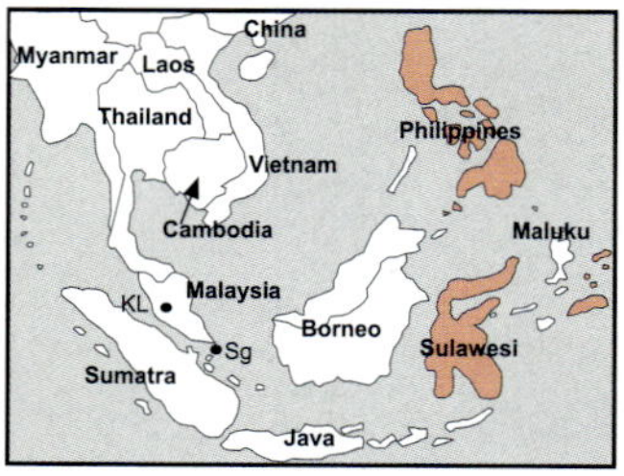


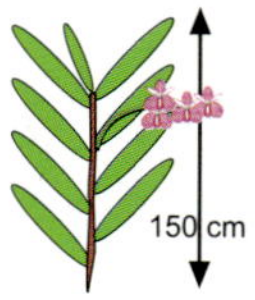


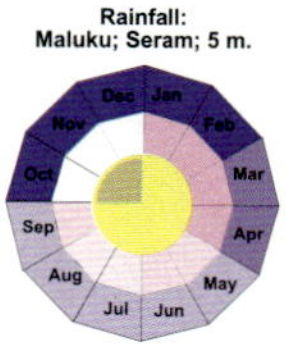

# B

## Brachypeza

This small genus of 6 or 7 epiphytic species is distributed from Laos to Indonesia, the Philippines, New Guinea and the Pacific Islands. One species (below) is common and widespread. They have a growth habit similar to *Phalaenopsis* (with which they are easily confused) but differ in having a swollen rachis that produces a succession of short-lived flowers. The sidelobes of the lip are either absent or much reduced. *Brachypeza* have no commercial value (although they have been sold to unsuspecting orchid-purchasing tourists as *Phalaenopsis* species) and are seldom seen in orchid collections.

### ***Brachypeza indusiata*** (Rchb.f.) Garay

This species is quite common in some of the moist forests in the area, growing on tree trunks and branches in deep shade from sea-level to 300 metres altitude. It has a short stem with 5-7 rather limp leaves that are often flushed purple. One or two flowers are open at a time; the colour varies from white to yellow, with pink spots.

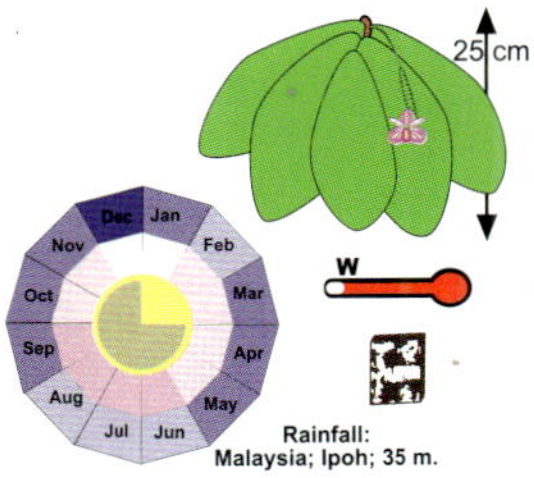


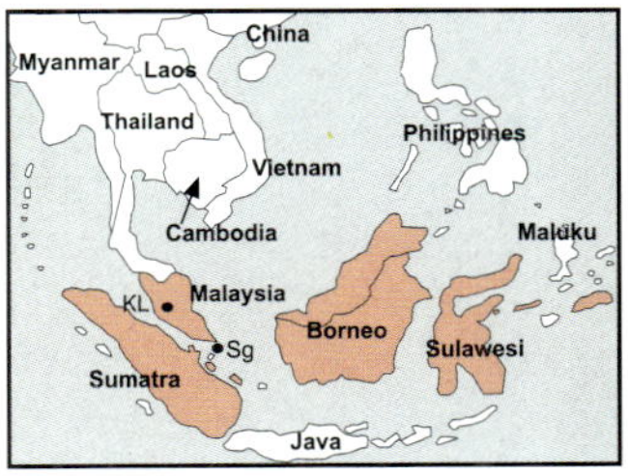

## Bromheadia

This genus of 26 species and 2 varieties is distributed from Sri Lanka across S. E. Asia to New Guinea and Australia. It is divided into two sections; section *Bromheadia* are tall terrestrial (occasionally epiphytic) plants with cane-like stems; section *Aporodes* are epiphytic (rarely terrestrial) plants with a tufted habit, short stems and compressed or terete leaves. The inflorescence has 2 rows of flattened bracts and produces a succession of short-lived flowers. Only *B. finlaysoniana* is common in cultivation.

### ***Bromheadia finlaysoniana*** (Lindl.) Rchb.f.

This terrestrial species is common in many parts of the region, growing in peaty, sandy or swampy soils, along footpaths or in openings in the forest from sea-level to 1100 metres altitude. It is often grown in full sun in village gardens. Each inflorescence produces up to 75 large showy fragrant flowers, with 1-4 open at the same time. The size of the plants and flowers can vary considerably.

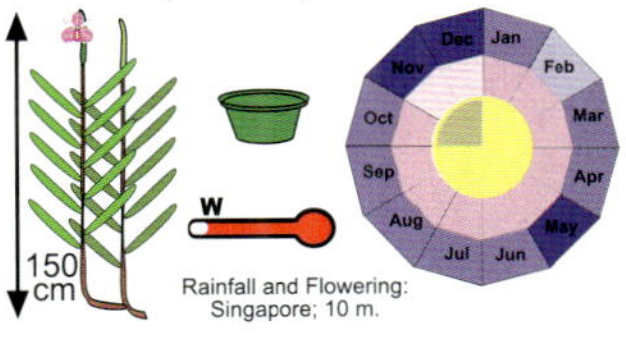


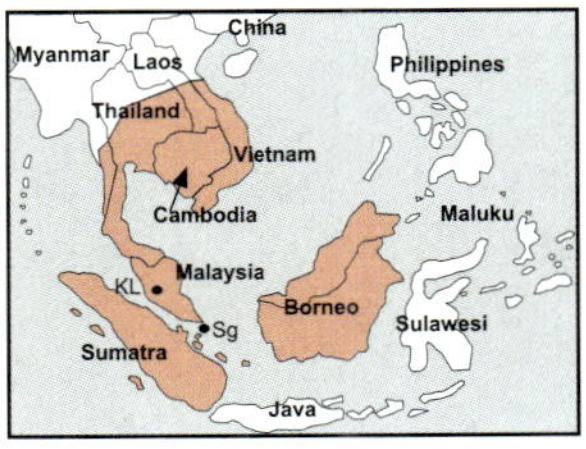

# B

## *Bromheadia scirpoidea* Ridl.

This small epiphytic species grows on low branches and in forks of forest trees from sea-level to 1500 metres altitude. It forms dense tufts, with each compressed stem having 2 or 3 rather soft recurving flattened leaves on each side. The short inflorescence extends a bit beyond the bracts and can have up to 4 branches. It forms up to 40 delicate translucent flowers, but only 1-2 are open at a time, and they only last 1 day.

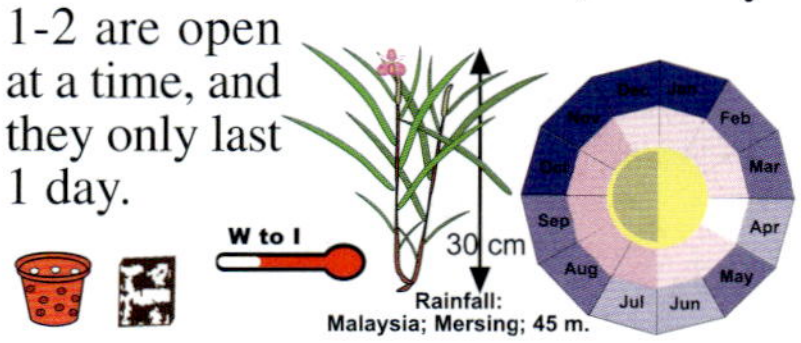


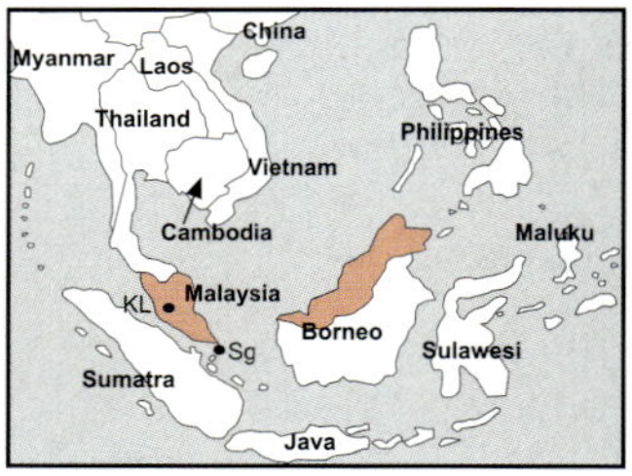

## *Bulbophyllum*

This huge genus of over 1200 mostly epiphytic species occurs in tropical regions around the world, with Africa, S.E.Asia and New Guinea being particularly rich in species. In order to make the genus easier to handle, it is often subdivided into sections; some botanists consider some of these (such as section *Cirrhopetalum*) to be genera in their own right. In S. E. Asia *Bulbophyllum* probably has about 400-500 species, only a few of which are commonly cultivated, although a steady rise in their popularity over the last decade has seen an increase in the number of species available at nurseries. Although these species are generally quite easy to grow, their cultivation requirements can vary enormously, so it is particularly important to find out about the climatic conditions that prevail in the species' natural habitat.

## *Bulbophyllum acuminatum* (Ridl.) Ridl. Section: *Cirrhopetalum*

This species is very common in parts of Malaysia and Borneo, where it grows as an epiphyte in forests from sea-level to at least 600 metres altitude. It has a creeping rhizome with ovoid pseudobulbs at 2-6 cm intervals. The slender inflorescences emerge from the rhizome below the pseudobulb and carry 5 to 8 flowers in a half-circle. The flower colour varies considerably. It is occasionally seen in orchid collections, where it is often confused with *B. pulchrum* or other similar species.

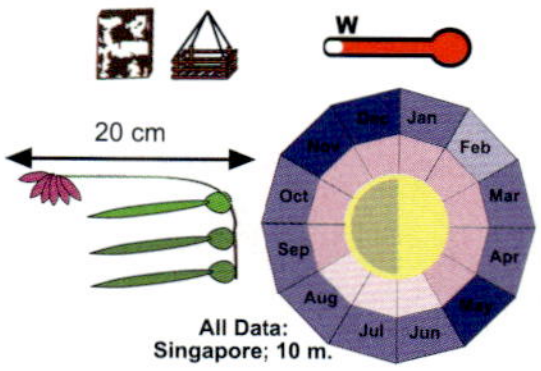


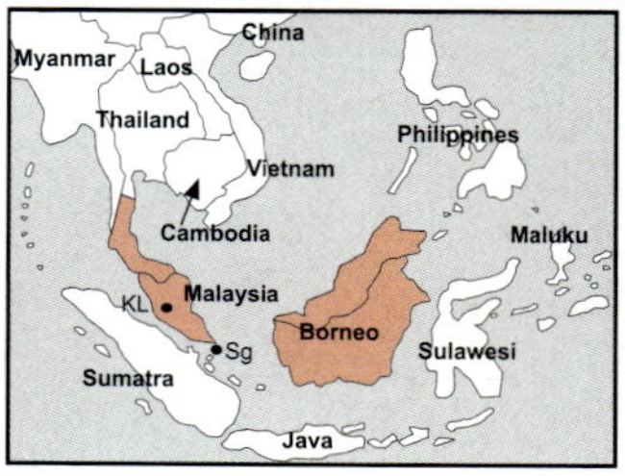

# B

## *Bulbophyllum amplebracteatum* T.&.B. Section: *Intervallatae*

This dramatic large-flowered species was seldom seen until recently, when a few nurseries in S. E. Asia started offering it for sale. It has a large ovoid pseudobulb and a large thick oblong leaf. Once the long arching inflorescence has grown to about the height of the leaf, it starts to produce flowers, which are formed one at a time. After a flower has fallen the next starts to form; the inflorescence may continue flowering for a 2-3 month period and produces a total of 6 to 12 flowers. Each 7 cm long flower lasts for 1-2 weeks.

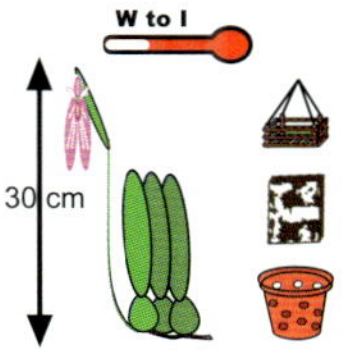


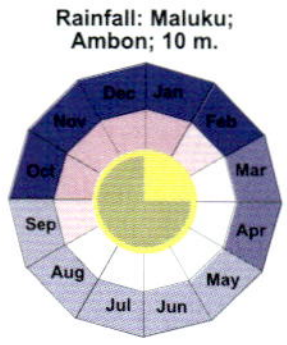


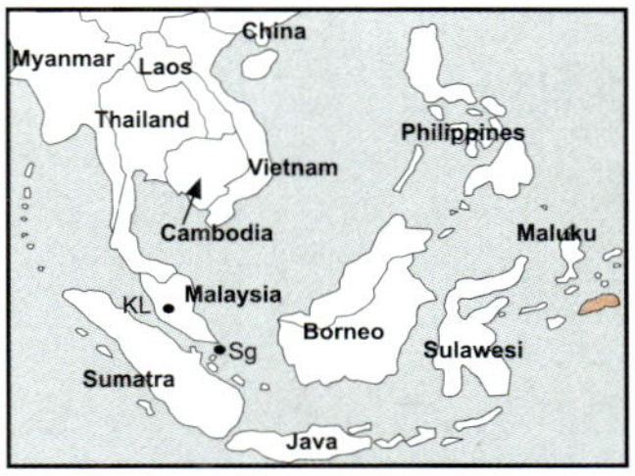

## *Bulbophyllum angustifolium* (Bl.) Lindl. Section: *Desmosanthes*

This species is a common epiphyte in montane forests from 1000 - 1800 metres altitude, where it forms mats that hang down from branches of exposed trees. The pseudobulbs are about 3-10 cm apart. The 2-10 cm long inflorescences have 3-7 flowers near the end. Flowers are usually cream to yellow with orange to red tips; Sumatran plants (illustrated) are usually yellow with darker yellow tips.

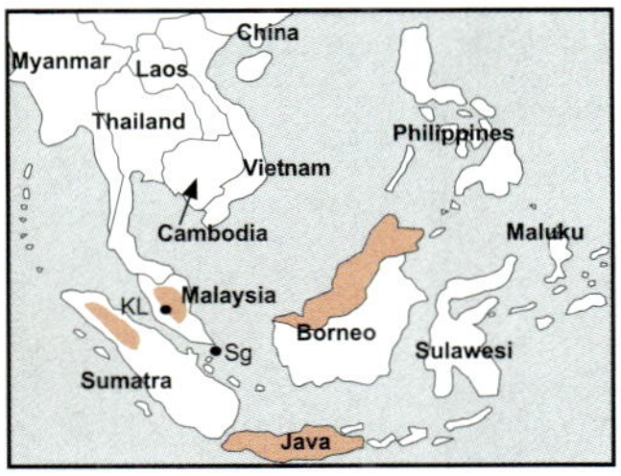


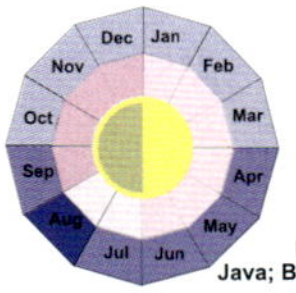


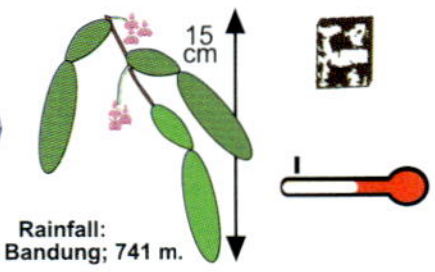


## *Bulbophyllum annandalei* Ridl. Section: *Cirrhopetalum*

This rare large-flowered epiphytic species is known from only a few specimens that were found in hilly country from 600-1000 metres altitude. The plant has a creeping rhizome with angular pseudobulbs at 2-5 cm intervals. The inflorescence has up to 8 flowers that only last a few days; there is some variation in the flower colour. Some flowers have the lateral sepals joined in a "*Cirrhopetalum*" manner, while other flowers open wide.

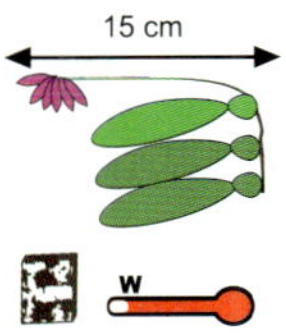


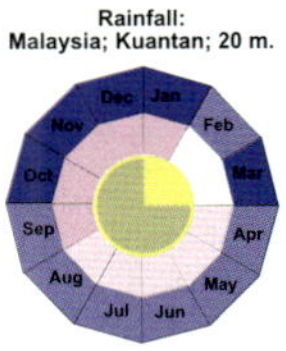


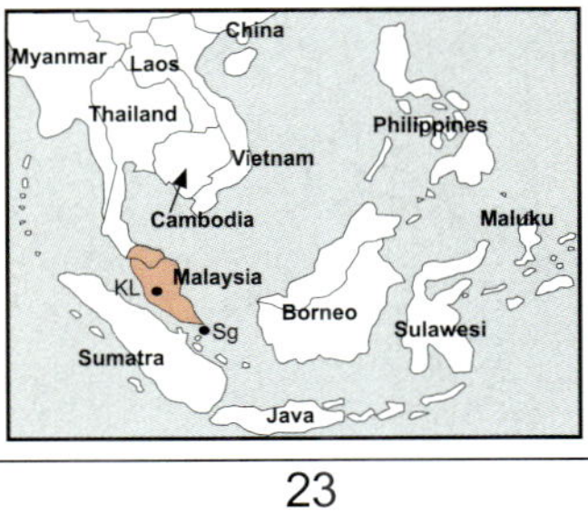

# B

## *Bulbophyllum atratum* J.J.Sm. Section: *Hirtula*

This extremely rare species is known from only a few specimens that were found growing on the branches of forest trees in the mountains from 600-1200 metres altitude. It forms clumps of pea-sized pseudobulbs that have a large leathery leaf on a distinct petiole. The pendulous inflorescence is up to 50 cm long; the final half is swollen and fleshy, and bears many 5 mm long striped flowers that are covered with minute hairs.

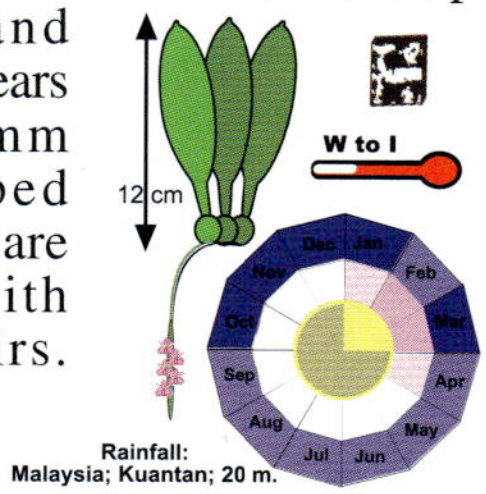


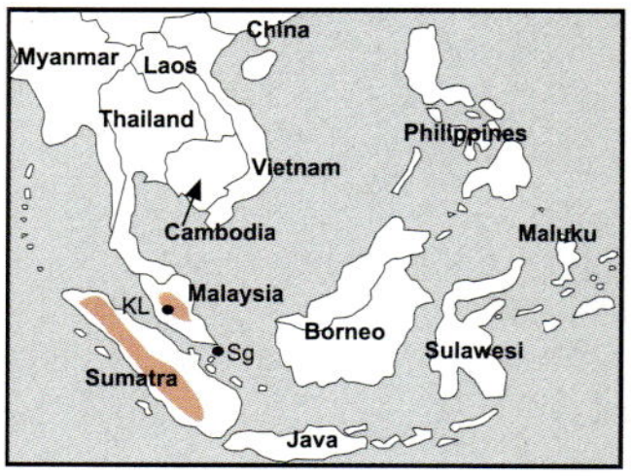


## *Bulbophyllum auratum* (Lindl.) Rchb.f. Section: *Cirrhopetalum*

This common species is well-known to orchid enthusiasts, who often call it by the old name *Bulbophyllum campanulatum*. It grows in forests from sea-level to at least 1100 metres altitude. The developing inflorescence is usually covered in a thick layer of mucous that disappears as the flowers open. The flowers usually have pink lateral sepals, but the rarer *alba* form is more popular with orchid growers.

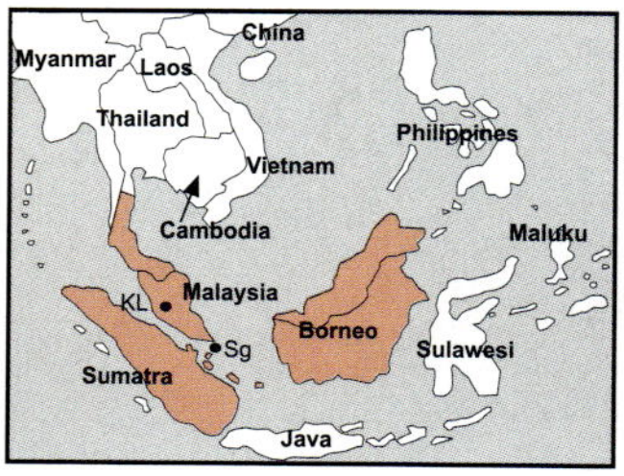


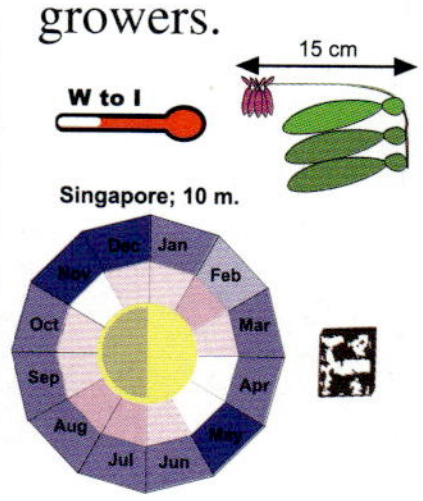


## *Bulbophyllum auricomum* Lindl. Section: *Pleiophyllus*

*B. auricomum* belongs to a small group of deciduous *Bulbophyllum* species which have two leaves that fall at the onset of the dry season, just before flowering commences. It only grows where there is a distinct dry season, so it is quite common in southern Myanmar and has been recorded from one area in Sumatra. In parts of Myanmar this attractive species is traded in local markets; it is grown over doorways and on the walls of houses. It has a long arching inflorescence with flowers that last about 10 days.

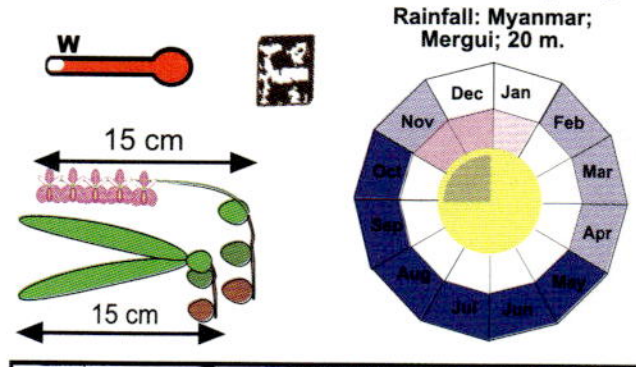


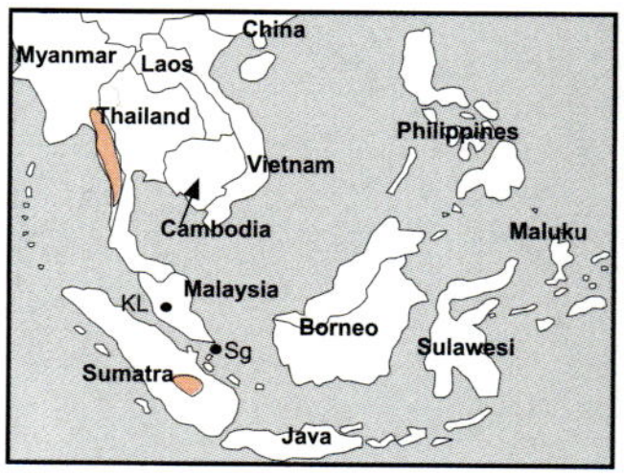

# B

## *Bulbophyllum blepharistes* Rchb.f. Section: *Tripudianthes*

This tough 2-leafed species is quite common in woodlands and forests in areas with a pronounced dry season, growing from sea-level to at least 800 metres altitude. It has large leathery 4-angled pseudobulbs with stiff leathery leaves that can withstand the long dry period. The long slender inflorescence produces a succession of buds at the tip; 1-3 flowers are open at a time and each lasts for a few days.

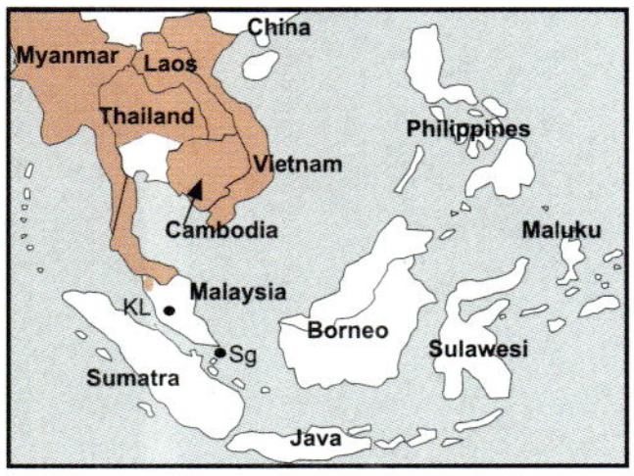


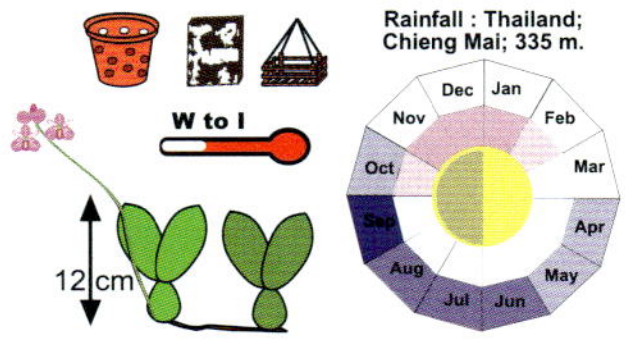


## *Bulbophyllum blumei* (Lindl.) J.J.Sm.

Section: *Ephippium*

This widespread species is found in a range of lowland forest types, growing on trees from sea-level to at least 600 metres altitude. In some areas it is quite common. It is well-known to orchid enthusiasts, often by the names *Bulbophyllum masdevalliaceum* (in New Guinea and Australia), *Bulbophyllum maxillare* or *Cirrhopetalum ciliatum*. The plant forms large masses, with 2-4 cm tall conical pseudobulbs and produces many slender inflorescences that have a single 6 cm tall flower. A white flowered form exists, but is very rare.

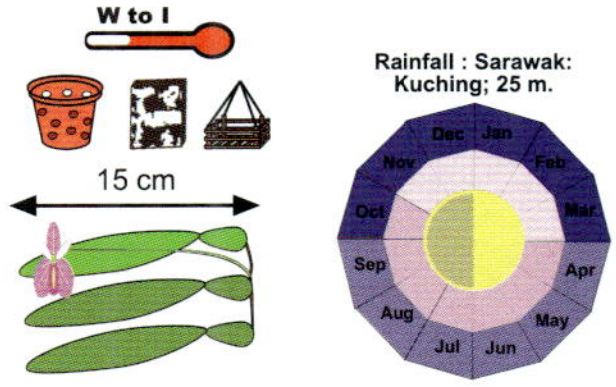


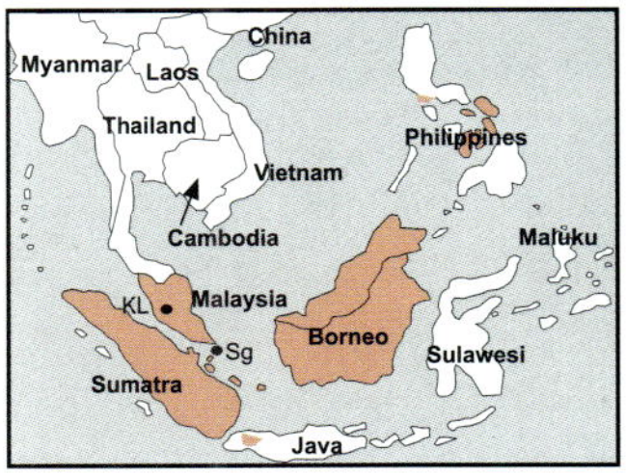

## *Bulbophyllum brienianum* (Rolfe) Ames Section: *Cirrhopetalum*

This species is often confused with *B. makoyanum*, although it has different colours. It grows on trees in lowland forests, and is quite common in swamp forests in southern Malaysia. It forms a long slender brittle inflorescence with up to ten 3 cm flowers in a fan.

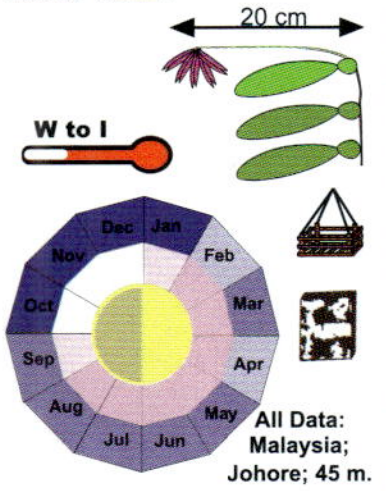


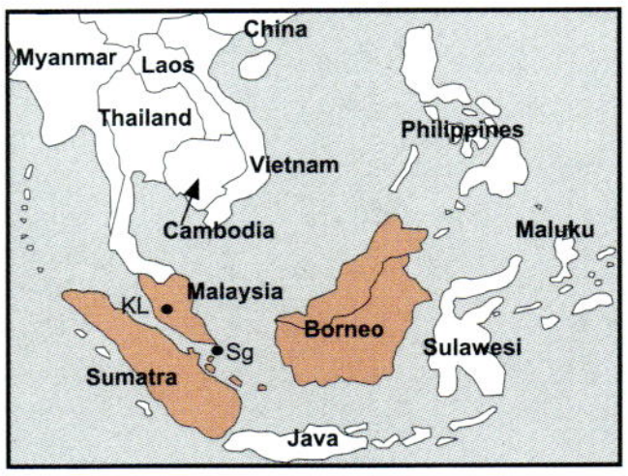

# B

### ***Bulbophyllum burfordiense*** Hort. ex Garay et al. Section: *Hyalosema*

This spectacular species, which is known to orchid enthusiasts around the world, is often confused with *B. grandiflorum*, which has 15 cm long flowers. *B. burfordiense* has single flowers that are 7-10 cm long, and tiny petals and lip. A rare *alba* form is illustrated; most flowers have brown, yellow-brown or green-brown sepals with white window patches.

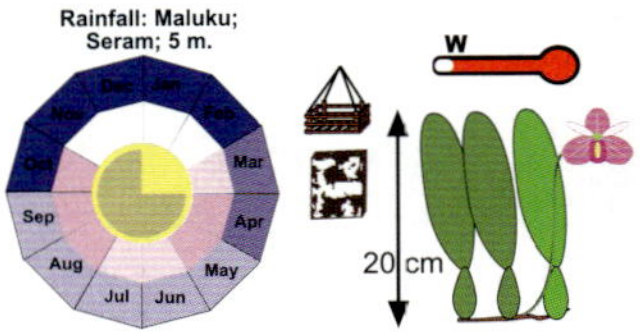


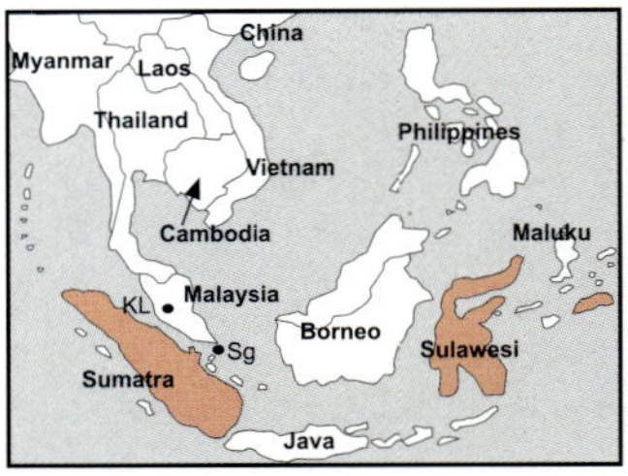

### ***Bulbophyllum coniferum*** Ridl. Section: *Globiceps*

This distinctive species is found in mossy montane forests from 1000-2200 metres altitude. It has very small pseudobulbs, large broad leaves, and produces an unmistakable inflorescence with unpleasant-smelling flowers packed into a cone. The sepal colour varies from pale green to dark purple, usually with brown to purple stripes or spots.

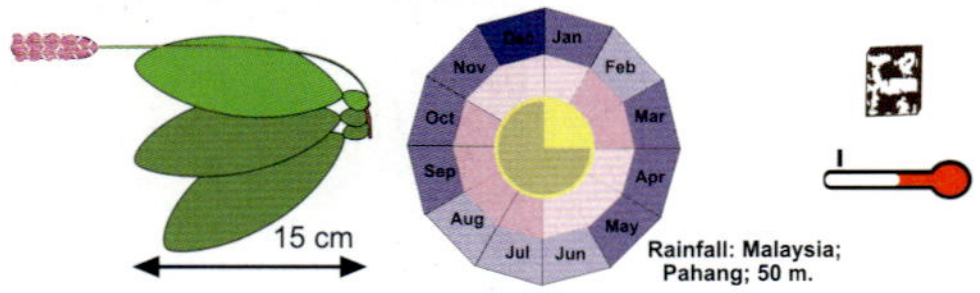


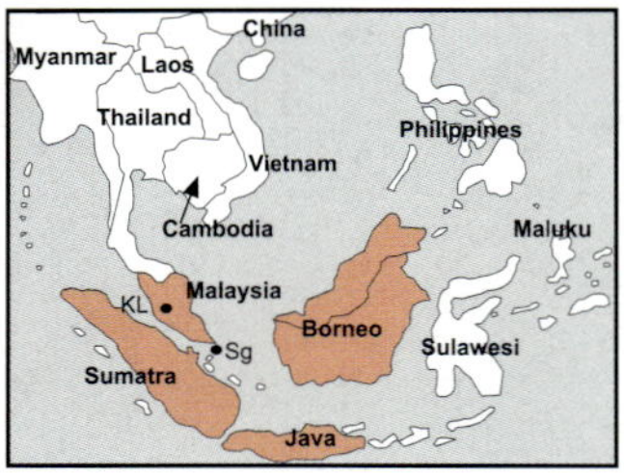


### ***Bulbophyllum dearei*** (Hort.) Rchb.f. Section: *Sestochilus*

This species is well known to orchid enthusiasts, although it is often confused with *B. lobbii*, which lacks the large backward-pointing sidelobes on the lip. *B. dearei* is a highland species, found in hill-forest and montane-forest from 700-1200 metres altitude, often growing on large bare tree-trunks. It has rather large ovoid pseudobulbs, a thick leathery leaf, and produces many single-flowered inflorescences. The 4-5 cm long flowers are often fragrant.

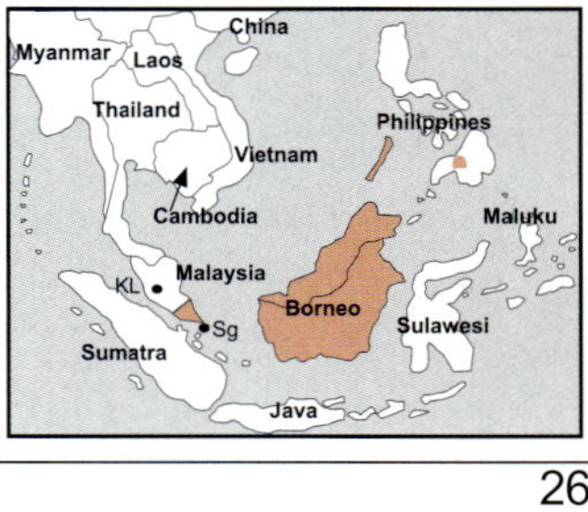

# B

## ***Bulbophyllum dentiferum*** Ridl. Section: *Cirrhopetalum*

This lowland epiphytic species is quite common in Peninsular Thailand. It produces more flowers than most in this section; the 12-30 flowers often form a full circle, sometimes a double whorl is formed. The amount of pink spotting on the sepals varies considerably.

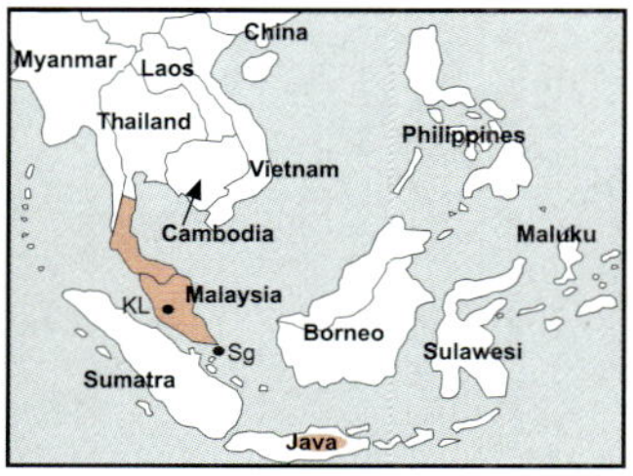


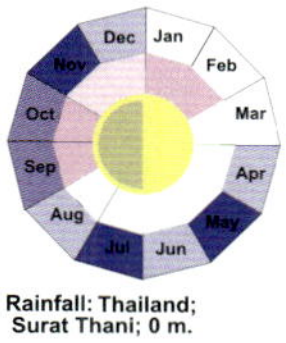


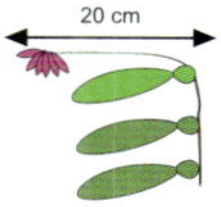

## ***Bulbophyllum echinolabium*** J.J.Sm. Section: *Intervallatae*

This is the world's largest-flowered *Bulbophyllum* species, with flowers up to 35 cm (14 inches) long. It has a very limited distribution in the wild as it is known only from the lower montane forests of one range of hills in Sulawesi. It is now endangered due to over-collecting for sale in the USA and Europe. It has a long slowly-extending inflorescence that bears up to 10 flowers which open one at a time over a 2-3 month period. Each flower lasts about 1 week.

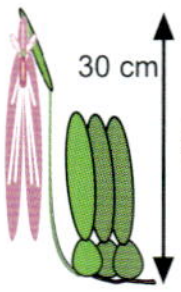


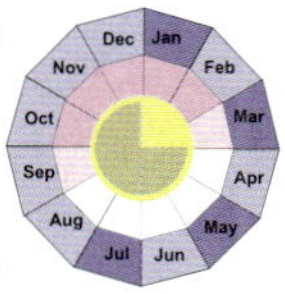


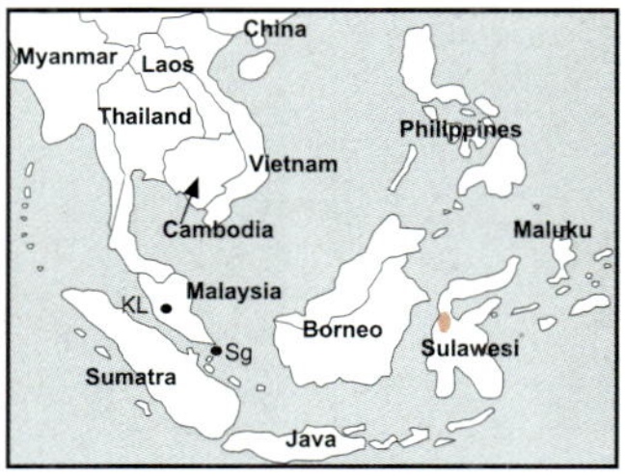

## ***Bulbophyllum flavofimbriatum*** J.J.Sm. Section: *Epicrianthes*

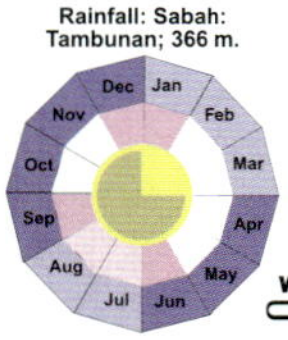

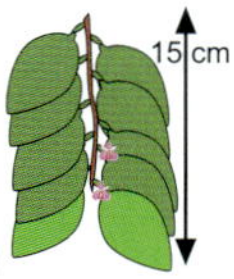


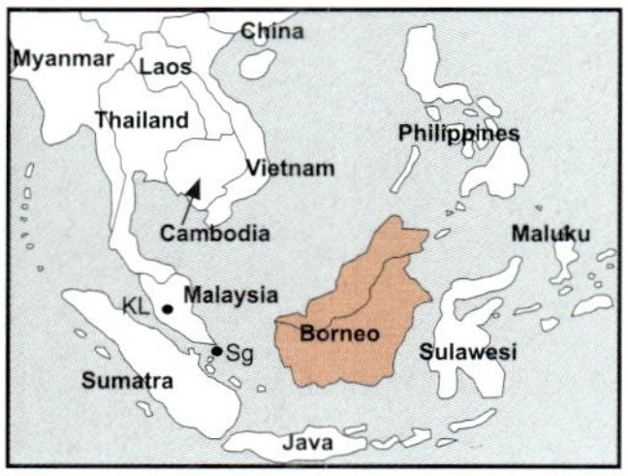


Although most species in this section are rare, *B. flavofimbriatum* is an exception, being moderately common in montane forest from 800-1500 metres altitude, where it grows on the trunks and branches of trees and shrubs. The hanging rhizome grows to well over 1 metre long. The single-flowered inflorescences are hidden under the leaves. The petals each have 6-10 rather immobile appendages.

# B

## *Bulbophyllum gibbosum* (Bl.) Lindley Section: *Aphanobulbon*

This large variable species has been found growing at sea-level, but is usually found in montane forests where it can be very common from 1000-2500 metres altitude, growing on tree branches and in forks on tree trunks. The rhizome has tiny pseudobulbs at 2 cm intervals. The species is easy to identify because the lower half of the tall inflorescence is wrapped in large loose bracts.

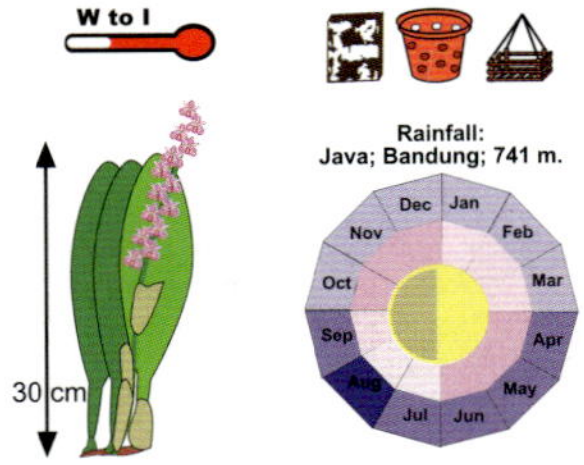


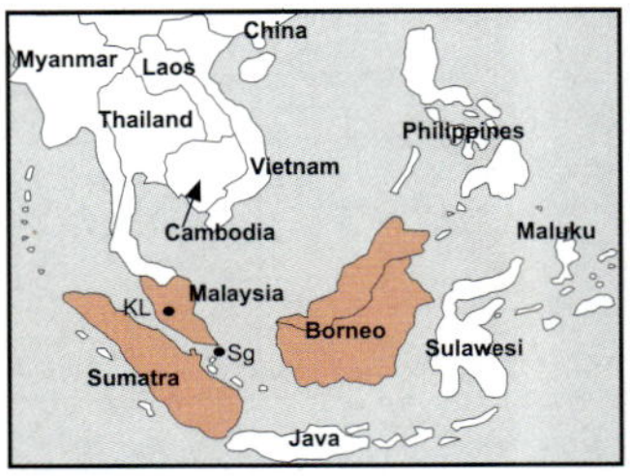


## *Bulbophyllum hodgsonii* Henderson Section: *Monilibulbus*

This uncommon miniature jewel is found in the mountains of Malaysia where it forms a chain of 10 mm long pseudobulbs on tree branches at about 1600 metres altitude. During the flowering season it produces 9 cm tall single-flowered inflorescences. The 3 cm wide flowers vary from orange to reddish-brown and last for several days.

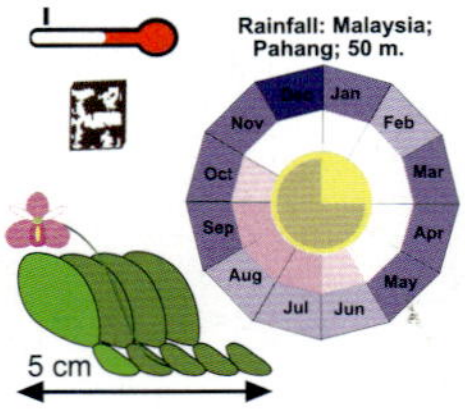


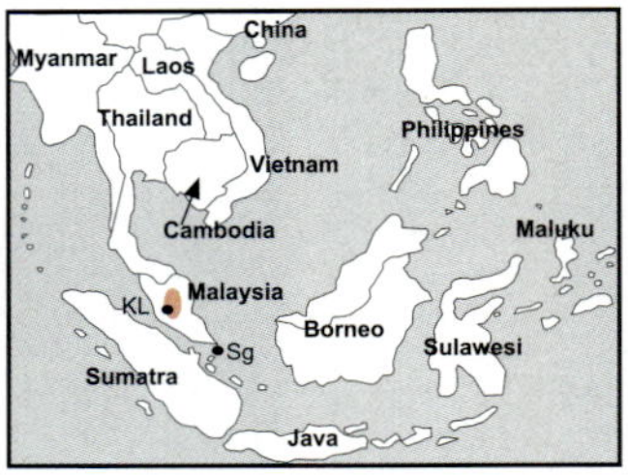

## *Bulbophyllum klabatense* Schltr. Section: *Intervallatae*

Quantities of this species have recently been shipped to the USA, where they were sold as "*B. echinolabium*". The fraud became apparent as soon as the plants started to flower. *B. klabatense* has large pseudobulbs and leaves, and a very long arching inflorescence that produces a succession of 3 cm high flowers, one at a time, over a period of up to 11 months.

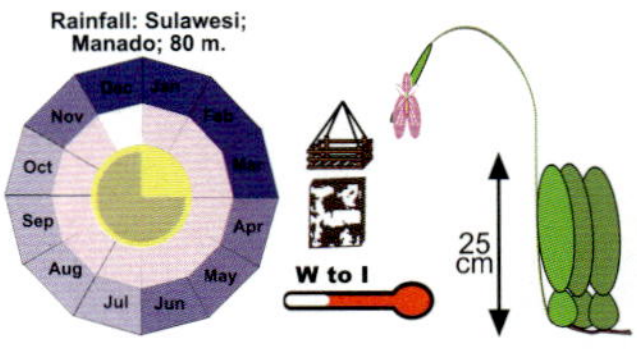


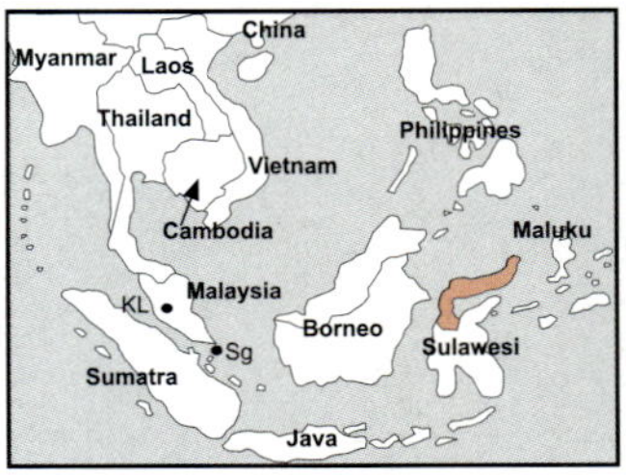

# B

## ***Bulbophyllum korinchense*** Ridl. Section: *Aphanobulbon*

This pretty species is found in the mountains of Sumatra where it grows on tree branches and partly-rotted logs from 900-1700 metres altitude. It has a long creeping rhizome with minute pseudobulbs and tall narrow leaves. The 30-40 cm tall inflorescences have up to 20 rather distant 10 mm wide flowers. Although it has two named varieties, *B. korinchense* may turn out to be synonymous with plants from other parts of the region.

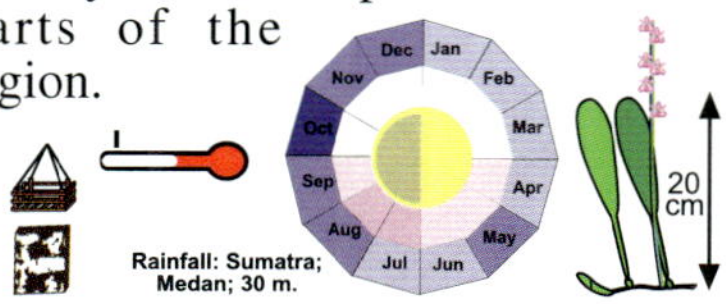


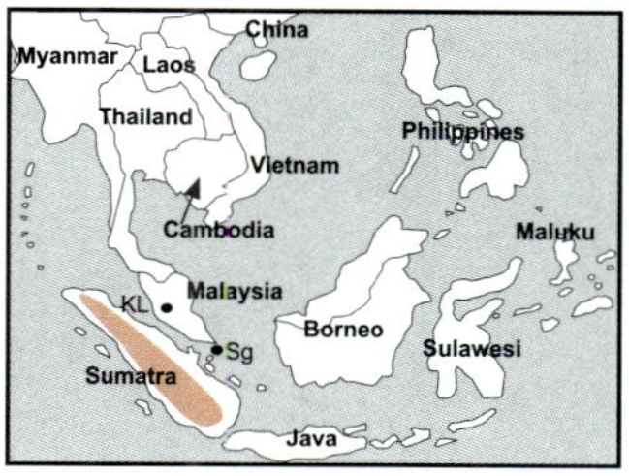

## ***Bulbophyllum lasianthum*** Lindl. Section: *Sestochilus*

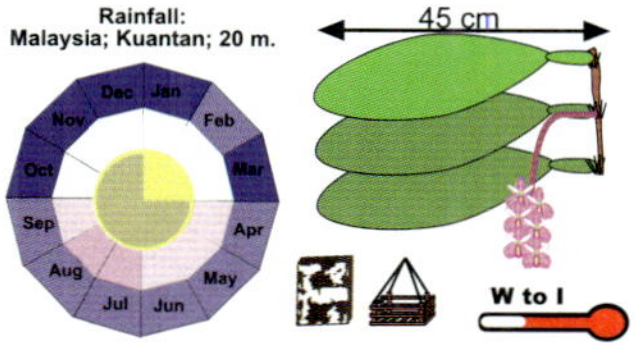


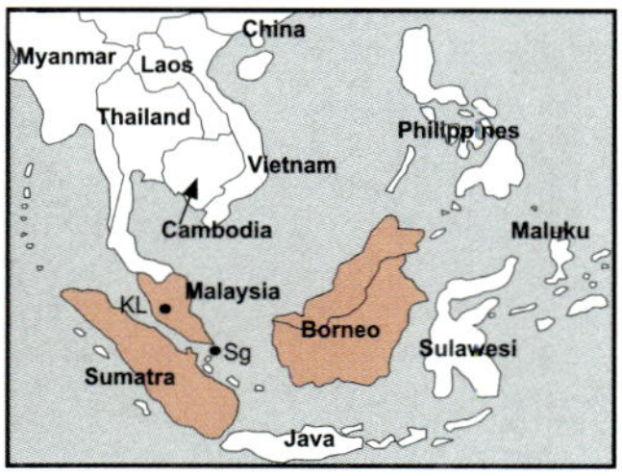


This species forms very large plants that are spectacular when in full bloom, although the dark-coloured foul-smelling bristle-clad flowers could hardly be called beautiful. Although it is widespread in the region and can be found on rocks and trees from the lowlands to 1200 metres altitude, *B. lasianthum* is uncommon and is usually seen in hilly country at about 600 m. The large pseudobulbs are carried 5-25 cm apart on a stout rhizome, and bear a tough 40 cm long broad leaf. The inflorescence can be up to 45 cm long, and the 4 cm long flowers are either dark wine-red or yellowish-green with a violet lip.

## ***Bulbophyllum lepidum*** (Bl.) J.J.Sm. Section: *Cirrhopetalum*

This species is grown in orchid collections around the world, although it is often confused with *B. pulchrum* and other species. *B. lepidum* is a common lowland species which grows from sea-level to 900 metres altitude throughout most of the region. Although most plants have white to cream lateral sepals with pink to purple markings extending from the base, there is considerable variation in the colour of the rest of the flower. This species can be identified by the presence of round lateral flaps alongside the anther at the end of the column.

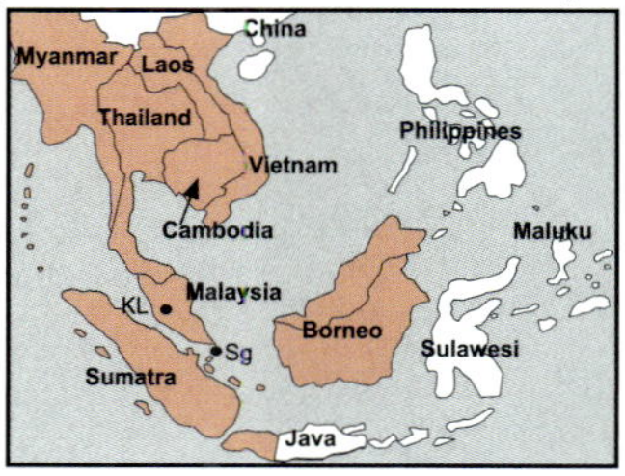


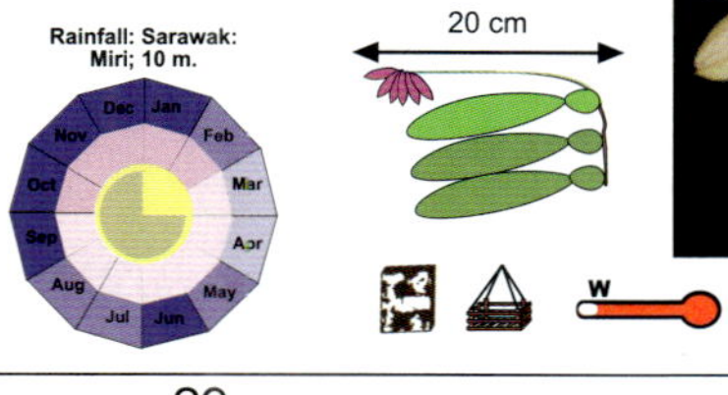

# B

## ***Bulbophyllum lobbii*** Lindl. Section: *Sestochilus*

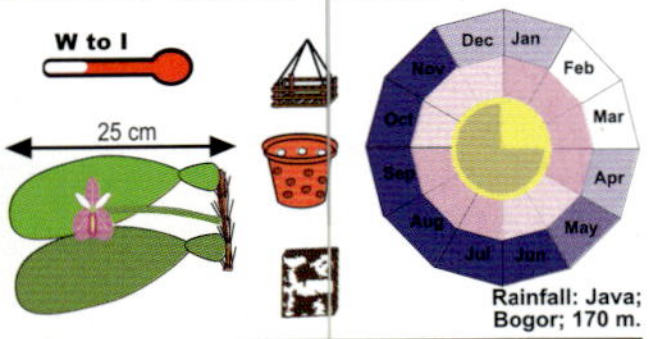


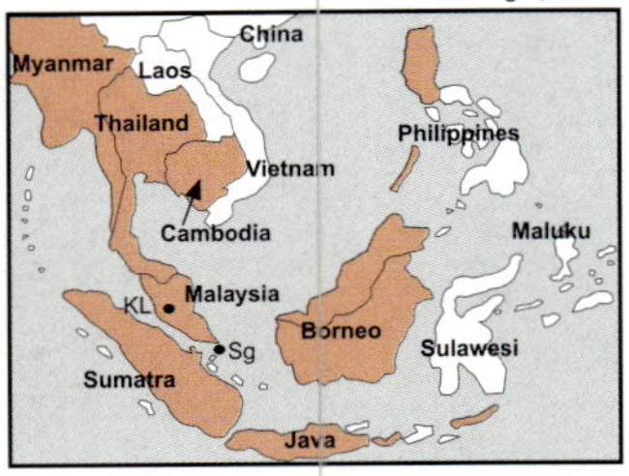


This well-known species is usually found growing on tree trunks and branches in the forest from 700-1200 metres altitude, but in some places the plant is found down to sea-level or as high as 2000 m. The 3-7 cm wide flowers, which usually have a spicy scent, show considerable variation in shape, colour and markings.

## ***Bulbophyllum macranthum*** Lindl. Section: *Sestochilus*

This well-known species is common on old mangroves and in lowland rainforest up to about 700 metres altitude, but in some places has been found in montane forests up to 1500 m. It has a stout bristle-covered rhizome with 2-4 cm high pseudobulbs at 3-10 cm intervals. The 3-5 cm high flowers, which have white to creamy-yellow sepals with purple spots, have a thick spicy odour. If the anther is dislodged the flowers respond rapidly, changing their shape by folding up their lateral sepals.

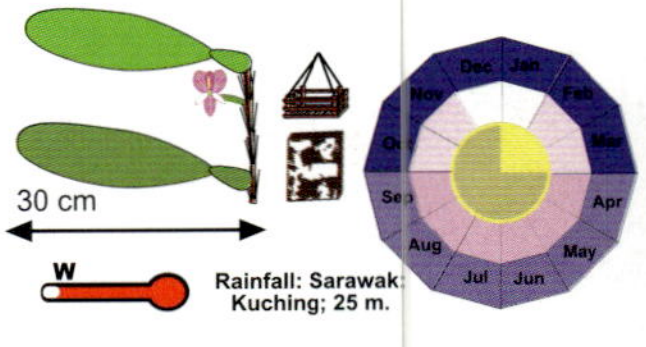


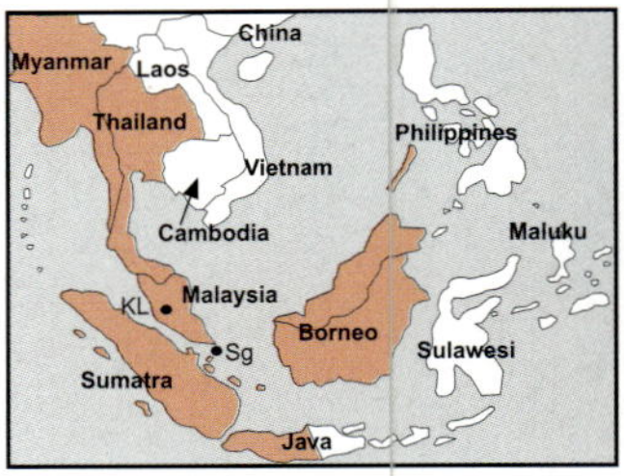

## ***Bulbophyllum makoyanum*** (Rchb.f.) Ridl. Section: *Cirrhopetalum*

This lowland species is common in parts of the Philippines and is occasionally found in Borneo, growing from 100-200 metres altitude. It was once found growing in Singapore. Although it is often confused with *B. brienianum* and other species with narrow sepals, it can be identified by the many small orange spots on the sepals.

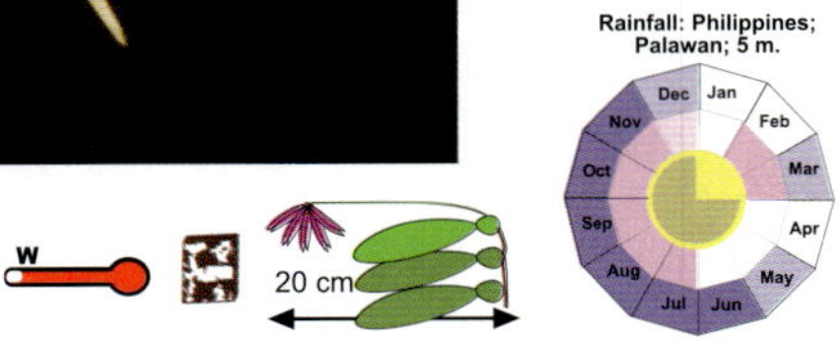


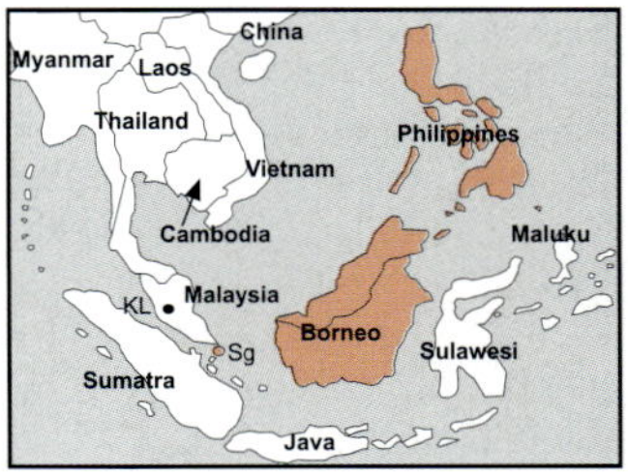

# B

## ***Bulbophyllum medusae*** (Lindl.) Rchb.f. Section: *Desmosanthes*

This species is well-known to orchid enthusiasts worldwide. It can be found from sea-level to at least 400 metres altitude, preferring fairly open sites such as tree branches that overhang rivers. It flowers several times a year, often producing large numbers of white mop-like inflorescences that may be 12-15 cm across.

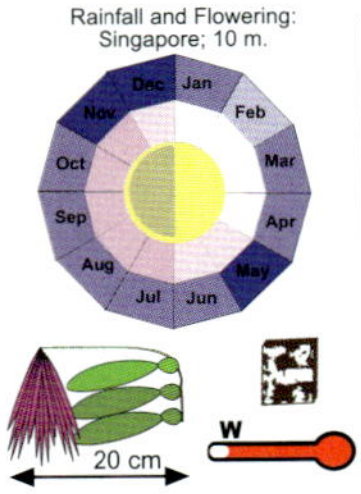


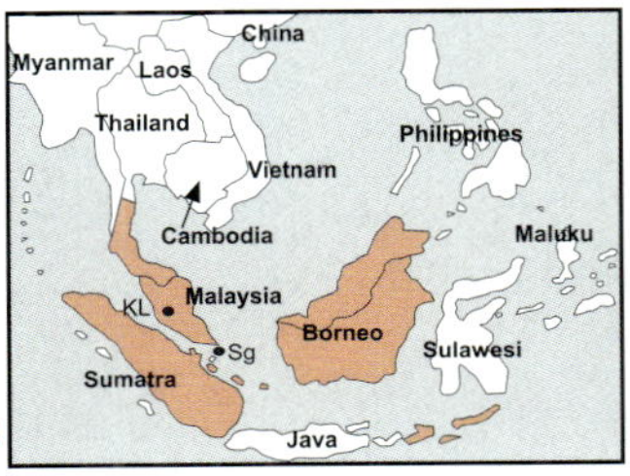

## ***Bulbophyllum mutabile*** (Bl.) Lindl. Section: *Aphanobulbon*

This epiphytic species can be found from 400-2700 metres altitude. It is highly variable in plant form and flower size and colour. The rhizome, which varies from erect to pendulous, is usually under 20 cm long with several branches, tiny pseudobulbs and thick leaves. Several miniature 'alpine' forms exist; one is illustrated here.

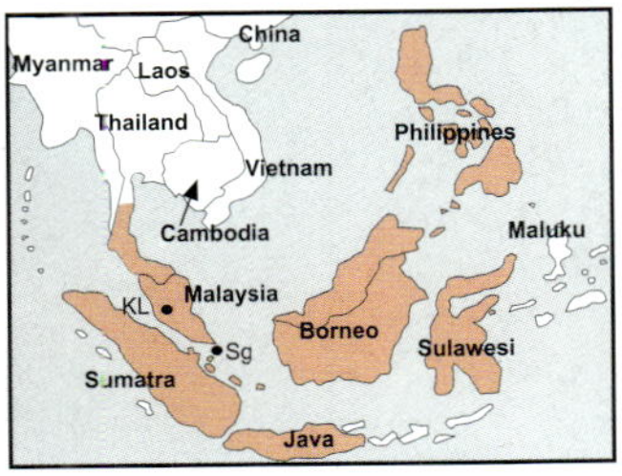


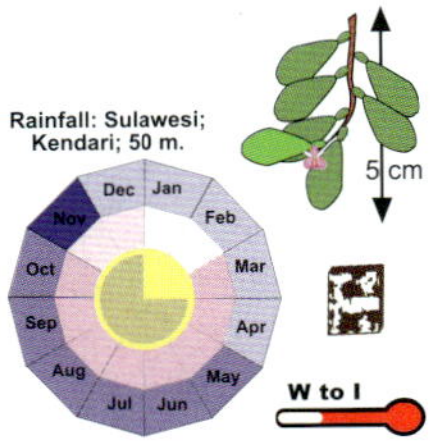


## ***Bulbophyllum ovalifolium*** (Bl.) Ridl. Section: *Monilibulbus*

This very small epiphytic species is common in the mountains from 900-2000 metres altitude, where it creeps along mossy tree branches. Each pseudobulb appears to be joined to the predecessor, without a visible rhizome. It produces numerous 5-6 cm long inflorescences which bear a single rather delicate 2 cm wide flower. The sepal colour may be cream, yellow, orange or red; the lip varies in colour and wartyness.

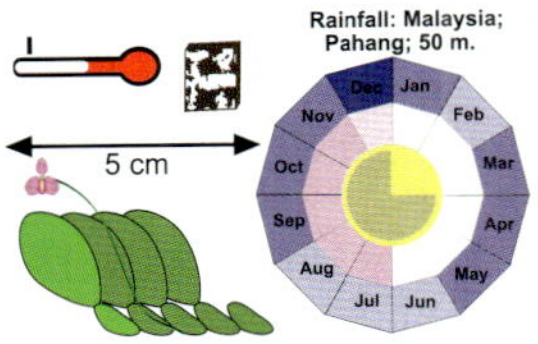


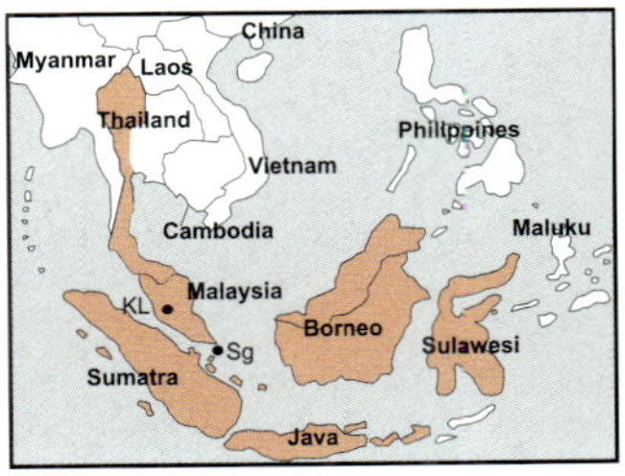

# B

## *Bulbophyllum patens* King ex Hk.f. Section: *Sestochilus*

This species is usually found creeping up tree trunks in swamp forests, and can also be found in other lowland forest types from sea-level to 200 metres altitude. The single lightly-scented flowers have a straight lip that is highly mobile.

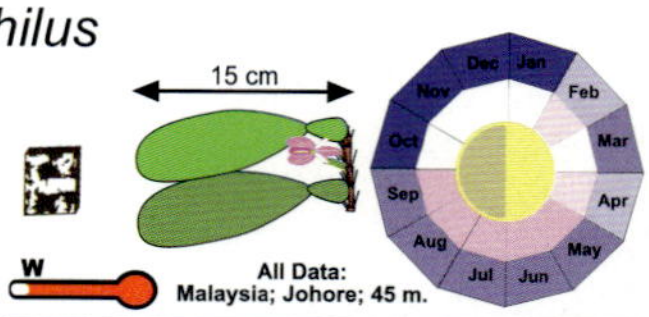


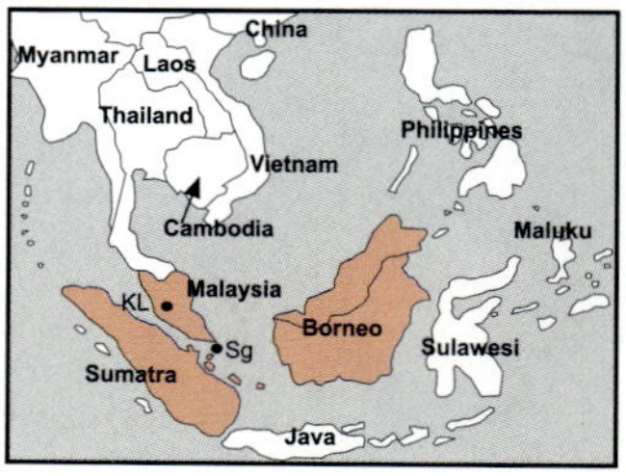


## *Bulbophyllum perpendiculare* Schltr. Section: *Aphanobulbon*

Little is known about this extremely rare epiphytic species, which is known from only a few locations in the mountains of Sulawesi, where it grows at about 1000-1200 metres altitude. It has rope-like rhizomes that hang freely from tree branches and grow slowly to about 80 cm long. The tiny pseudobulbs are hidden by brown bracts, and bear a single sharp-pointed lanceolate greyish-green leaf. Many short inflorescences are formed at a time, each with 3-6 flowers that do not open widely and are about 5 mm long. The flowers are pale yellow with a darker lip.

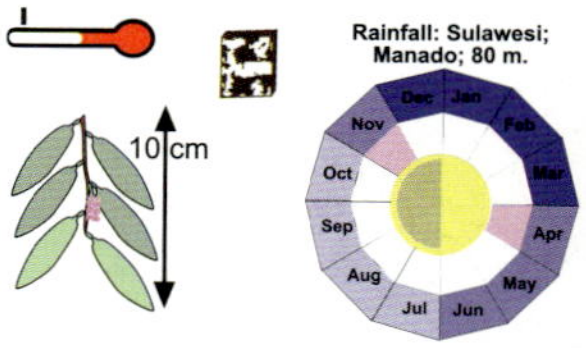


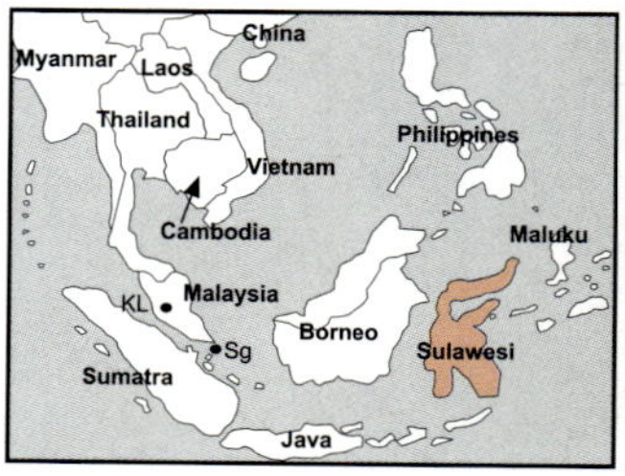

## *Bulbophyllum pileatum* Lindl. Section: *Sestochilus*

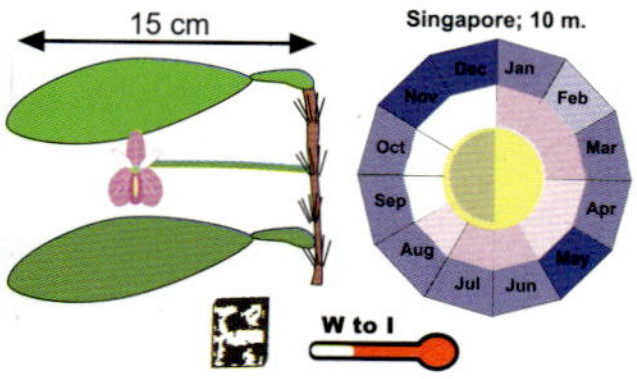


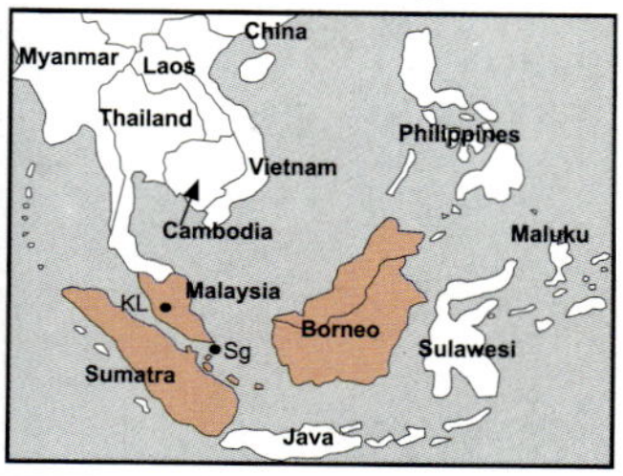


This little-known epiphytic species occurs at around 1000 metres altitude in the mountains of Borneo, but in Malaysia it is found from sea-level to about 200 metres altitude. In Johor it is common in some swamp forests, while in the Riau Islands it is found on mature mangroves. The single 4 cm flowers are dull yellow to pale orange and have a large straight lip with a deep groove running from the base to the middle.

# B

## *Bulbophyllum purpurascens* T. & B. Section: *Cirrhopetalum*

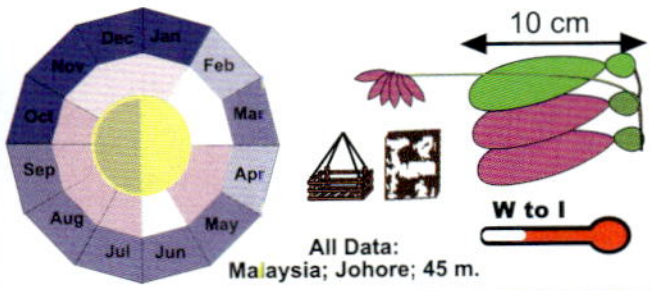


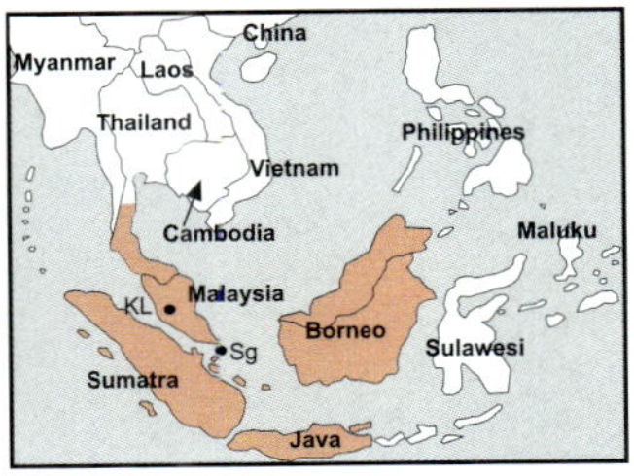


This common epiphytic species is usually found in lowland forests, often on tree branches overhanging rivers. It has a creeping rhizome with 1 cm pseudobulbs and leaves that are often purple on one or both sides. The inflorescence is up to 18 cm long, with a tight cluster of 1-2 cm long pale yellow or white flowers.

## *Bulbophyllum putidum* (T.& B.) J.J.Sm.

Section: *Cirrhopetalum*

This species, shown on the right in the photograph, is in cultivation around the world, often under the old name *B. fascinator*. The species on the left in the photo, which is also in cultivation and is often labelled *B. putidum*, is actually *B. appendiculatum* (Rolfe) J.J.Sm. *B. putidum* is found growing in deciduous woodlands in hilly country from India to northern Indochina. It has a tough creeping rhizome with stout 2 cm tall pseudobulbs with a tough leathery leaf. The inflorescence has a single 5-8 cm flower.

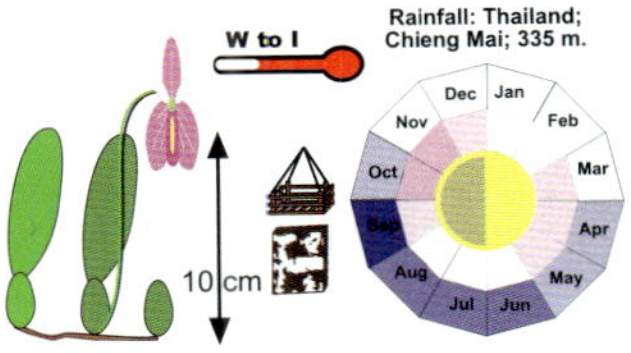


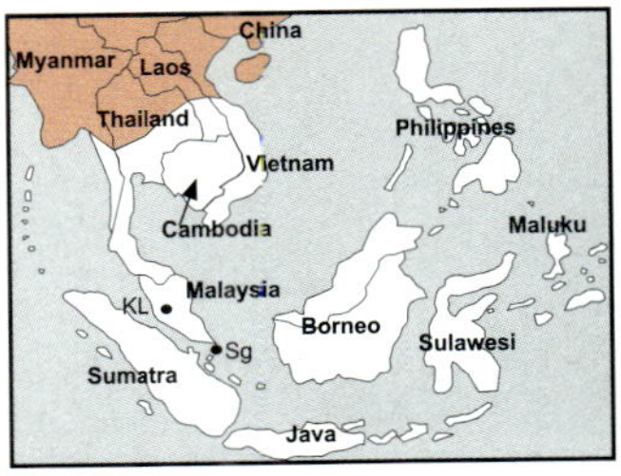

## *Bulbophyllum reticulatum* Batem ex Hook.f. Section: *Sestochilus*

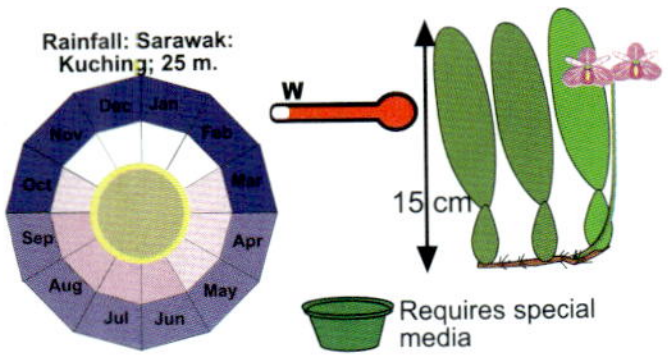


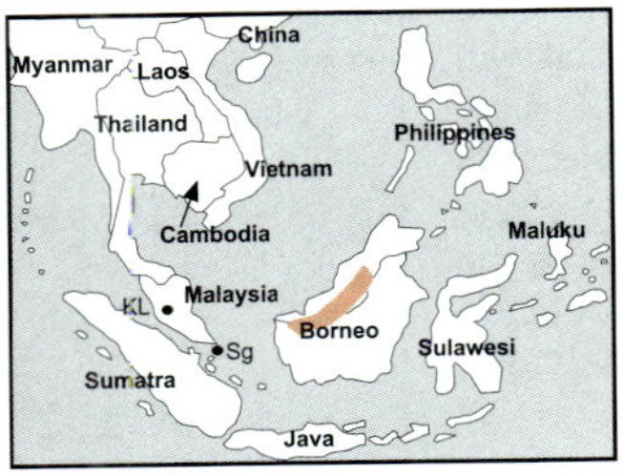


This species grows in limestone regions of Borneo, scrambling over decayed limestone rocks and rooting into a thin covering of humus. Well established plants often scramble up into nearby trees for long distances, although they seldom flower when growing as epiphytes.

# B

## *Bulbophyllum rufinum* Rchb.f. Section: *Careyana*

This tough epiphytic species is usually found on trees in deciduous woodlands in regions with a long dry season. There are a number of similar species in Thailand and Indochina, and accurate identification can be a problem. *B. rufinum* has large 4-angled pseudobulbs with a tall narrow leaf, and a 30-40 cm long inflorescence with many 1 cm long yellow flowers.

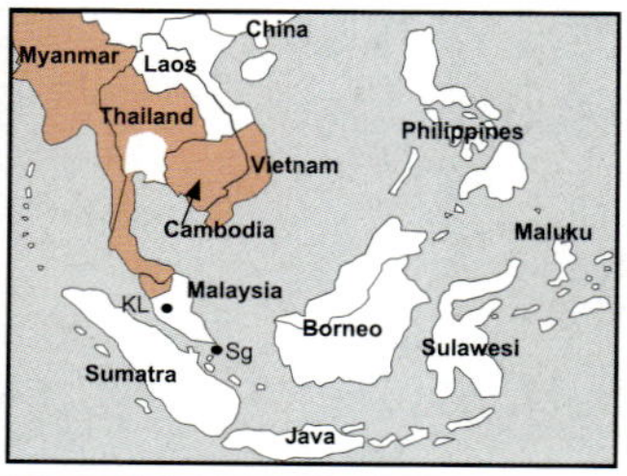


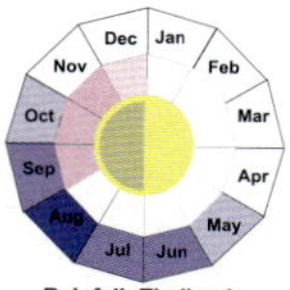


Rainfall: Thailand; Mae Sariang; 315 m.

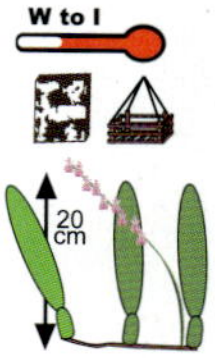


## *Bulbophyllum sopoetanense* Schltr. Section: *Aphanobulbon*

This species grows in moss pads on tree branches in montane forest from 1000-1500 metres altitude. The creeping rhizome has tiny pseudobulbs at about 4 cm intervals. The inflorescence is 10-15 cm long with up to 14 flowers that are 16 mm wide.

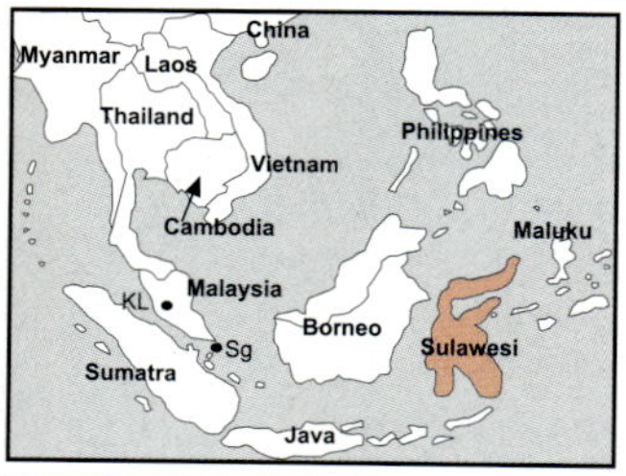


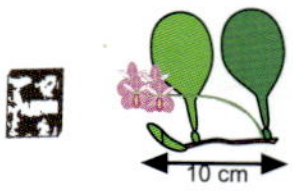


Rainfall: Sulawesi; Kendari; 50 m.

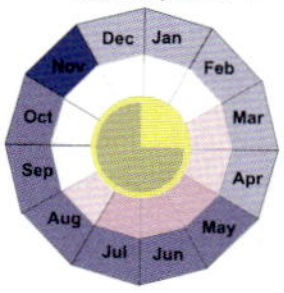


## *Bulbophyllum spissum* Vermeulen Section: *Leptopus*

This diminutive recently-discovered species is known only from two small mountainous islands where it grows on the trunks and branches of trees in dense primary rainforest on steep hillsides from 250-1000 metres altitude. It has tight clumps of small egg-shaped pseudobulbs, each with a single thin lanceolate leaf. Many single-flowered inflorescences are produced during the long flowering season, each about as high as the leaves. The 1 cm long delicate flowers do not open widely and last for a few days.

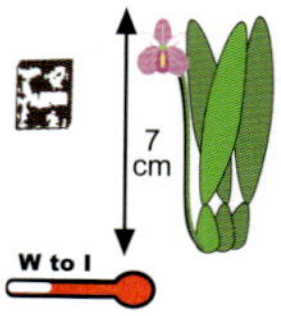


Rainfall and Flowering: Singapore; 10 m.

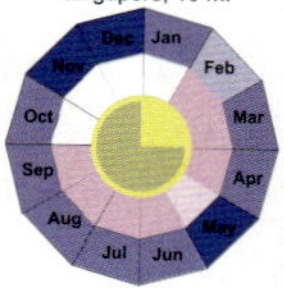


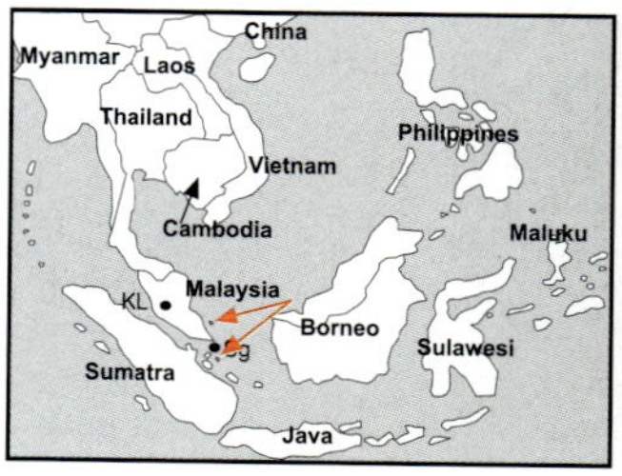

# B

## ***Bulbophyllum stormii*** J.J.Sm. Section: *Monilibulbus*

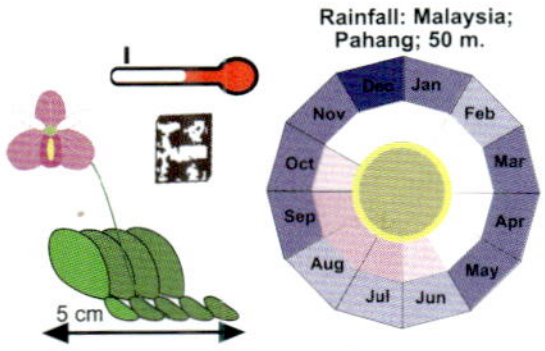

This beautiful miniature species is found in the mountains from 1300-2000 metres altitude, where it forms creeping chains of pseudobulbs on understorey trees or shrubs in deep shade. Many 6-9 cm tall single-flowered inflorescences are produced during the flowering season. The flowers vary from 4-7 cm across, and the sepal colour varies from yellow to orange, sometimes with white tips and darker veins.

## ***Bulbophyllum subumbellatum*** Ridl.

Section: *Sestochilus*

This distinctive species is usually found in lowland swamp forests from 0-250 metres altitude, where it forms large untidy masses that creep along and hang from tree trunks and branches. It has rather narrow pseudobulbs about 5-12 cm apart on a thin creeping and branching rhizome. The inflorescences each have 1-3 flowers that are 3 cm high and are easily identified by the long tail at the tip of each petal.

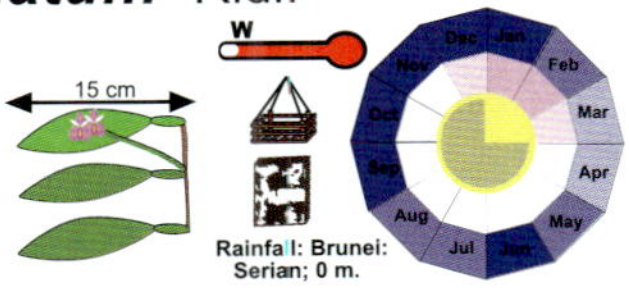


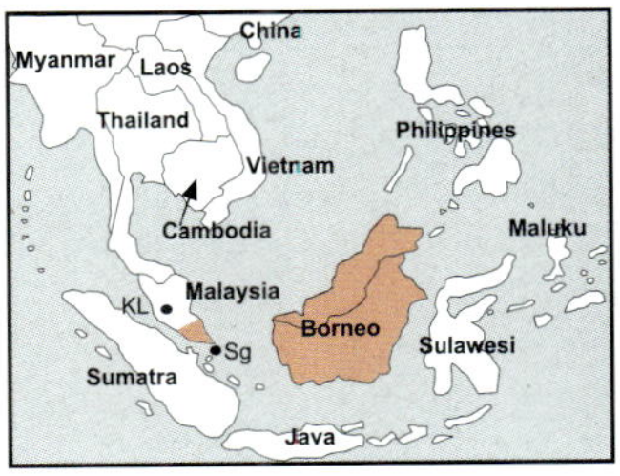

## ***Bulbophyllum sulawesii*** Garay, Hamer & Seigerist Section: *Intervallatae*

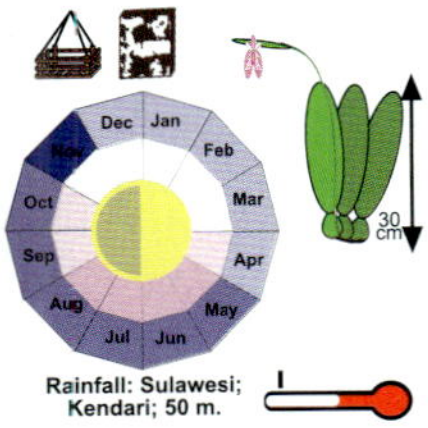


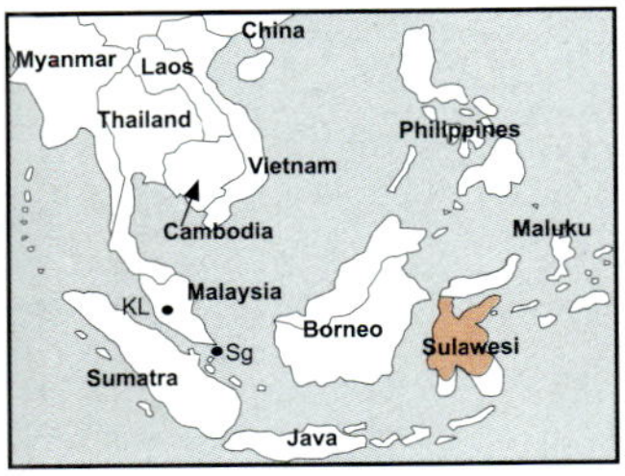


This large showy species is common in Central Sulawesi from 1200-1500 metres altitude, where it grows on horizontal branches of trees in rather stunted montane forest. It has clusters of large ovoid pseudobulbs. The long arching inflorescence extends slowly over a 3-5 month period and produces up to 10 flowers that open one at a time and are 8-9 cm tall.

# B

## *Bulbophyllum taeniophyllum* Par. & Rchb.f. Section: *Cirrhopetalum*

This misunderstood species has 4-5 cm long flowers with smooth-edged narrow sepals. It is found from 500-1500 metres altitude in the hills and mountains of the northern part of the region and is usually called *B. helenae* (O. Kuntze) J.J. Sm. Plants with purple-spotted white flowers from the southern part of the region, are actually *B. fenestratum* J.J.Sm., although they are usually called *B. taeniophyllum*.

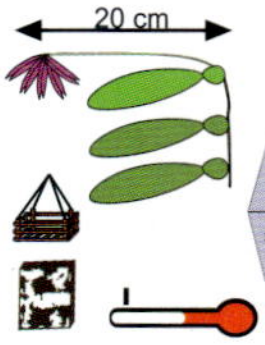


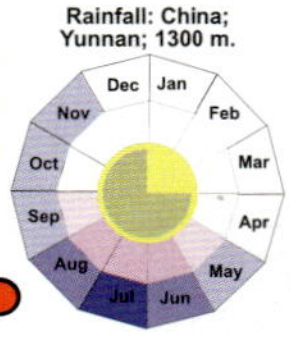


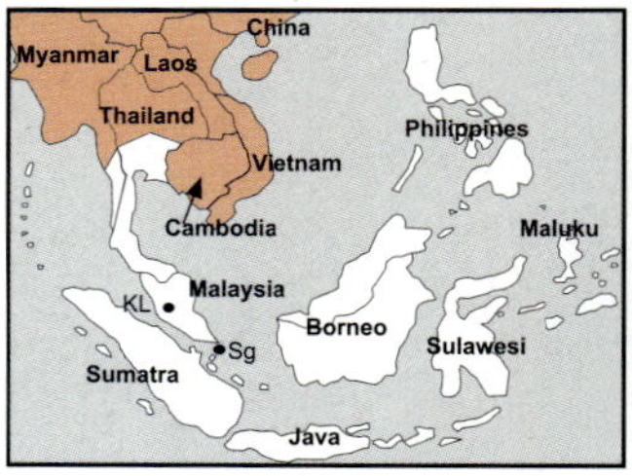

## *Bulbophyllum taeter* Vermeulen Section: *Sestochilus*

This extremely rare species is known only from a small area of Borneo, where it grows low down on tree trunks in primary rainforest at about 25 metres altitude. It has a creeping rhizome with 4-7 cm tall egg-shaped pseudobulbs at about 12 cm intervals. The large leathery leaf is very dark green and often suffused purple on the underside. The 50 cm long arching inflorescence has about 15 flowers in a crowded raceme. The 2 cm long flowers have a sharp offensive smell that, in combination with the red and yellow colours, is thought to be attractive to carrion-feeding flies.

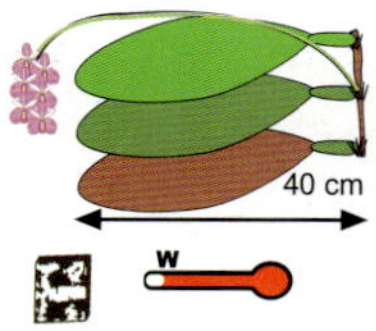


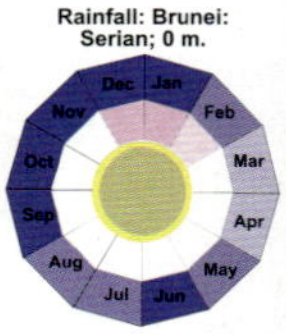


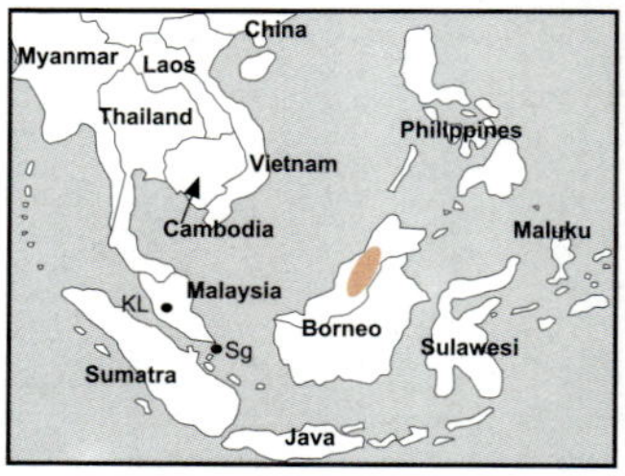

## *Bulbophyllum uvaeflorum* O'Byrne Section: *Osyricera*

This recently-discovered species was found growing in a moss-pad on a tree branch in montane forest at about 1350 metres altitude. The colour of both the soft cylindrical pseudobulbs and the leaves varies from maroon-green to brown to olive. The outer half of the inflorescence is swollen and carries a number of 6 mm long flowers that look rather like split grapes.

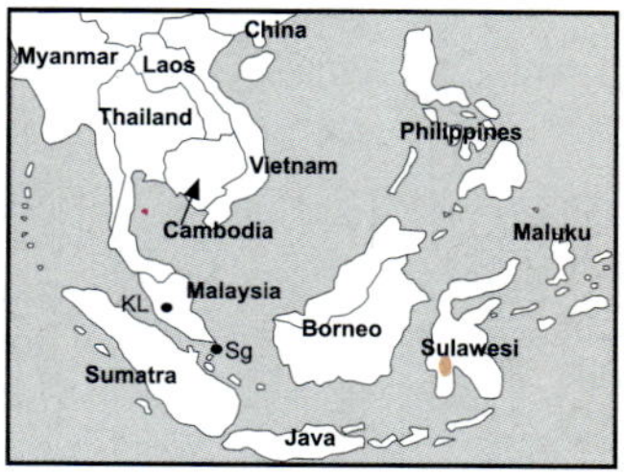


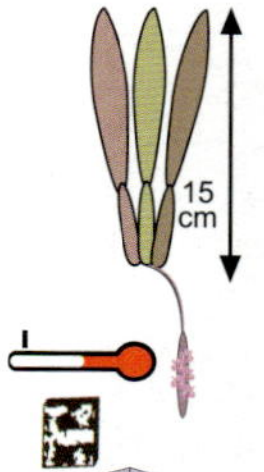


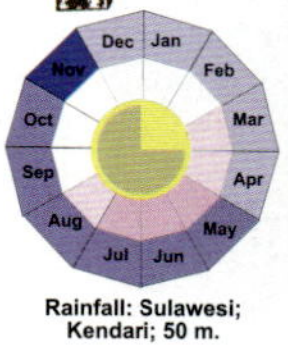

# B

## ***Bulbophyllum vaginatum*** (Lindl.) Rchb.f. Section: *Cirrhopetalum*

This lowland species forms colonies on large trees and is a common sight in cities, towns, villages, farmland areas and on roadside trees. It has two to three main flowering periods a year, when every specimen in an area flowers simultaneously in a mass display that lasts 2-3 days.

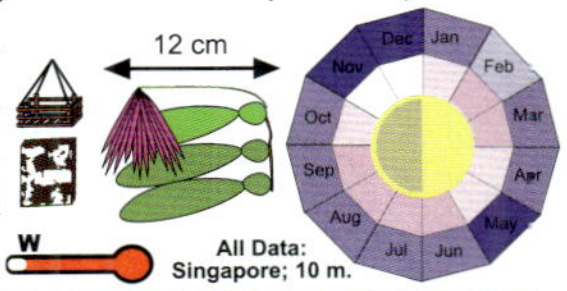


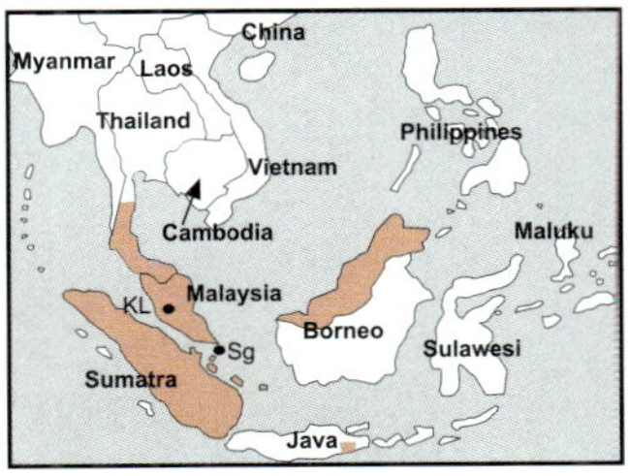

## ***Bulbophyllum vermiculare*** Hook.f. Section: *Aphanobulbon*

This tiny species grows on the branches of trees in mangroves and lowland forest from 0-600 metres altitude. It has a thin creeping rhizome (the name *vermiculare* means 'wormlike') that bears tiny pseudobulbs at 1-4 cm intervals. The inflorescence is 2-5 cm long and carries up to 6 flowers that are 3-4 mm across. The lip is narrow with stiff hairs along the margins.

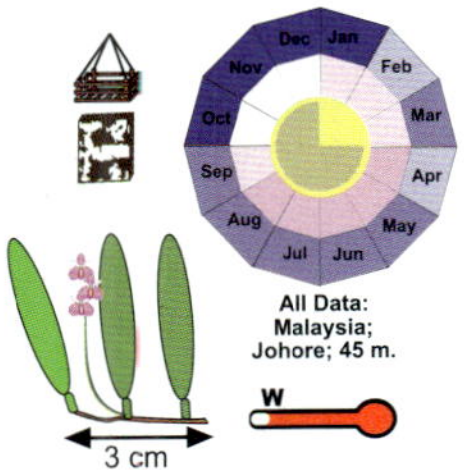


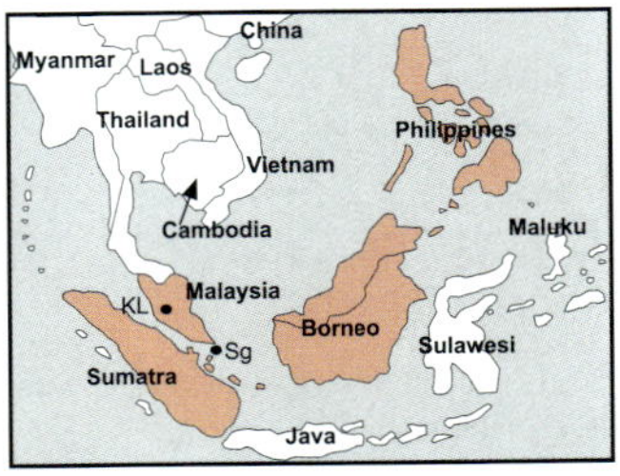


## ***Bulbophyllum woelfliae*** Garay, Senghas & Lemcke Section: *Megaloglossum*

This very rare large-flowered species grows in montane forest at about 1400 metres altitude. It has a creeping rhizome with rather squat pseudobulbs, and forms many long single-flowered inflorescences. The flowers are 8-12 cm across.

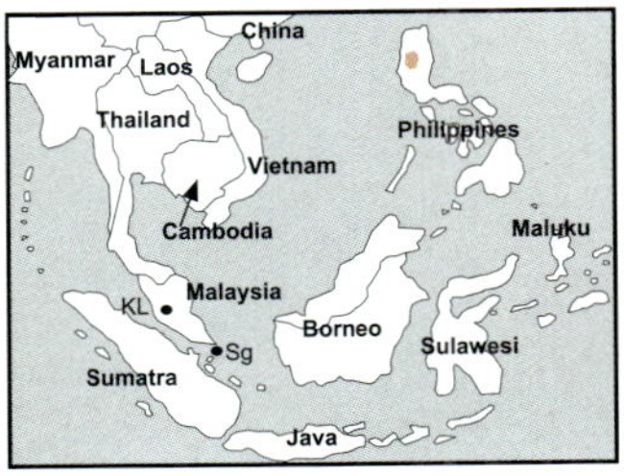


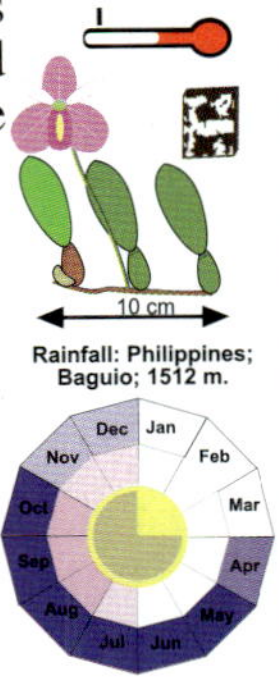

# C

## Calanthe

Some members of this genus have been cultivated for over 1000 years in Japan and China, yet today few of these showy, easily-grown plants are seen in collections outside Asia. The genus extends from Africa to Asia and the Pacific Isles (there is a single species in the American tropics) with the majority of estimated 100 species being terrestrial plants with large or medium-sized pseudobulbs. Species from regions with a long dry (or cold) season are often deciduous. *Calanthe* species all have the lip firmly fused to the column, and a spur projecting backwards from the lip. There are usually a group of warty keels on the lip underneath the column.

### *Calanthe ceciliae*

Rchb.f.

This attractive terrestrial species is found growing in forested areas from sea-level to almost 2000 metres altitude; in some places it is common on the forest floor and in crevices on rocky outcrops. A white-flowered colour form is known from East Java.

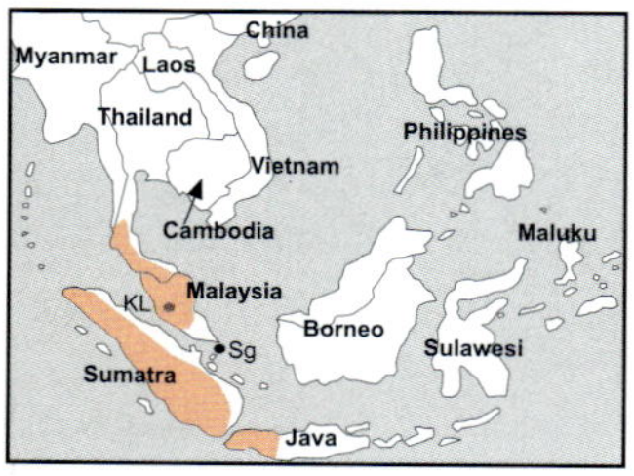


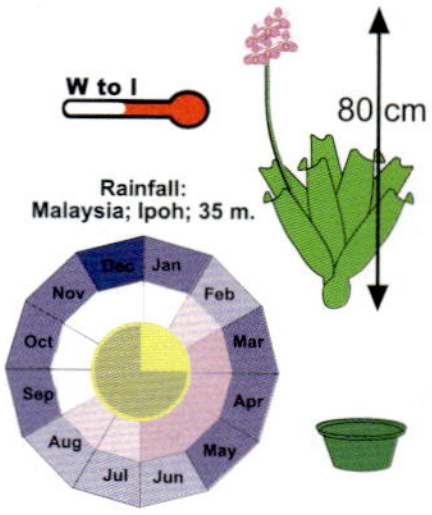

### *Calanthe rosea*

(Lindl.) Benth.

This terrestrial species, which is limited to the Tenasserim Range of Thailand and Myanmar, is probably the most beautiful *Calanthe* in S. E. Asia. It is deciduous, losing its leaves and then flowering during the long dry season. The inflorescence is covered in white hairs, and the flower is also externally white-haired. The trumpet-shaped lip is unusual in this genus; the basal margins overlap above the column.

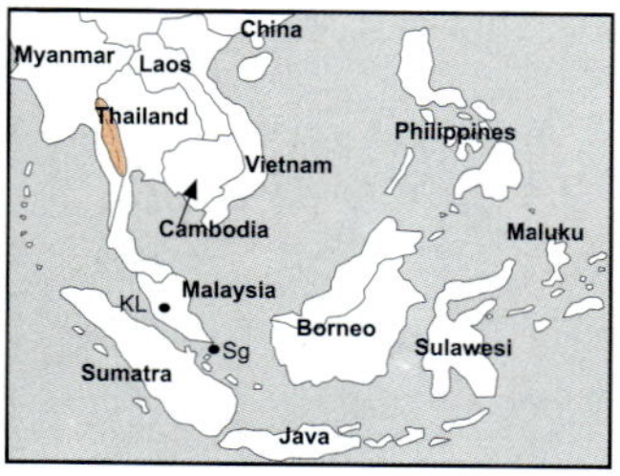


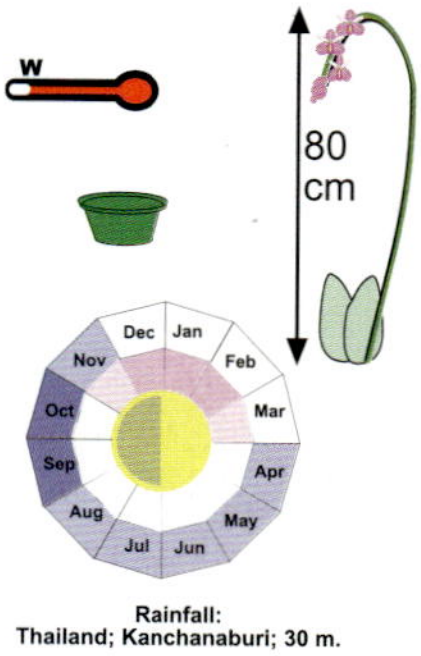

# C

## *Calanthe rubens* Ridl.

This attractive deciduous terrestrial species is found in limestone regions with a seasonal climate. The 10-12 flowers open 2-3 at a time and are well separated on the long arching inflorescence. A white colour form of this species has been recorded. This species is known as *C. elmeri* Ames in the Philippines.

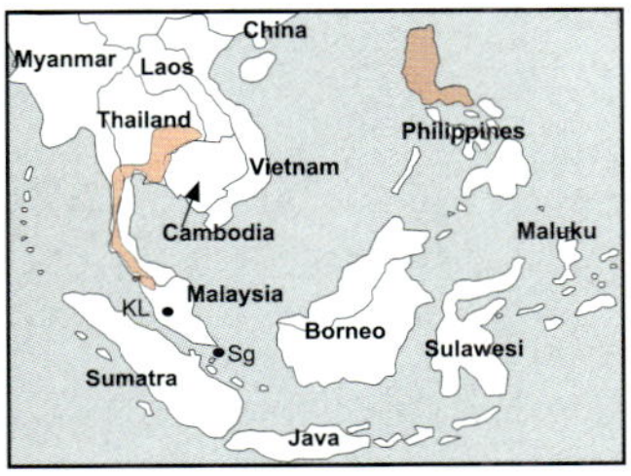


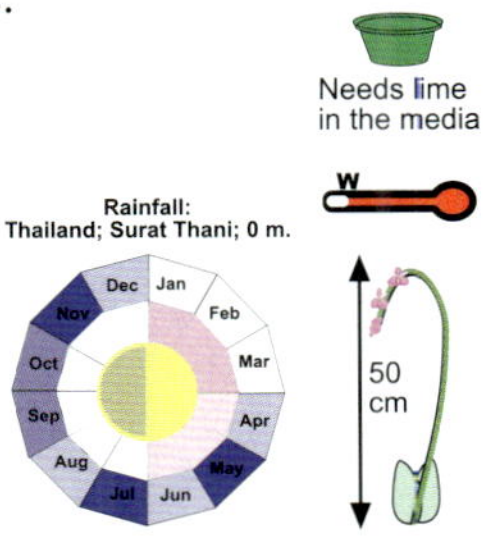


## *Calanthe sylvatica* (Thou.) Lindl.

This attractive terrestrial species is distributed from Africa through most of Asia, although nowhere is it common. It has a long history of cultivation, often under the name *C. masuca*, with the pink and "blue" flowered forms of the plant being particularly sought after. Both flowers and plants are highly variable in size and shape.

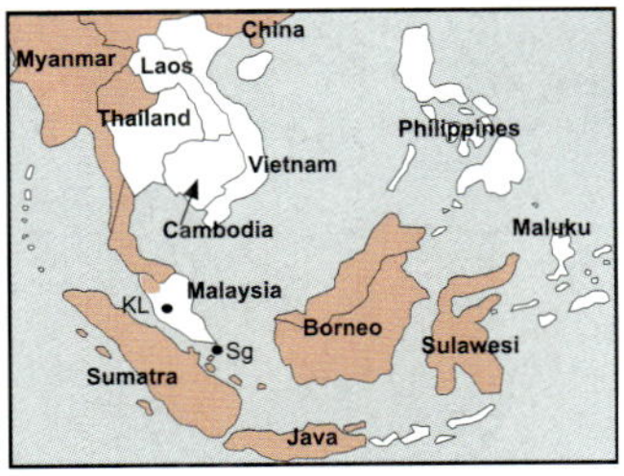


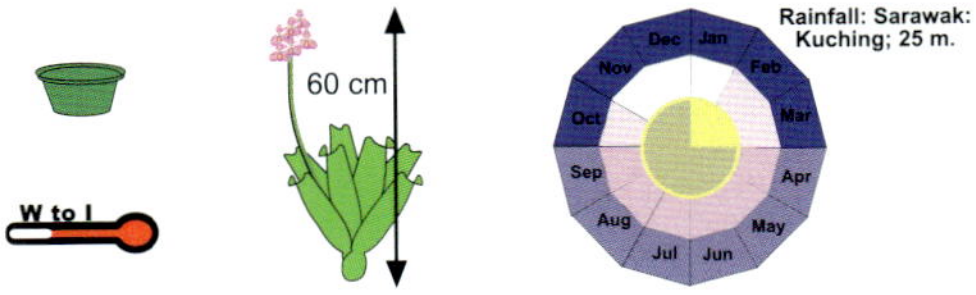


## *Calanthe triplicata* (Willem.) Ames

This attractive terrestrial species is often seen in gardens and collections found growing throughout the area. It occurs from sea-level to at least 1500 metres altitude but is most common at around 500-900 metres in rich moist habitats. It is a highly variable species and has a long list of synonyms. The flowers turn blue if crushed or damaged.

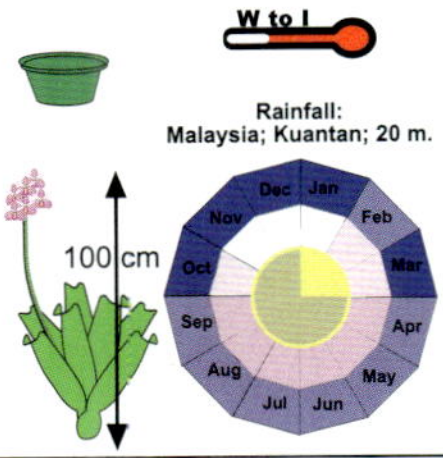


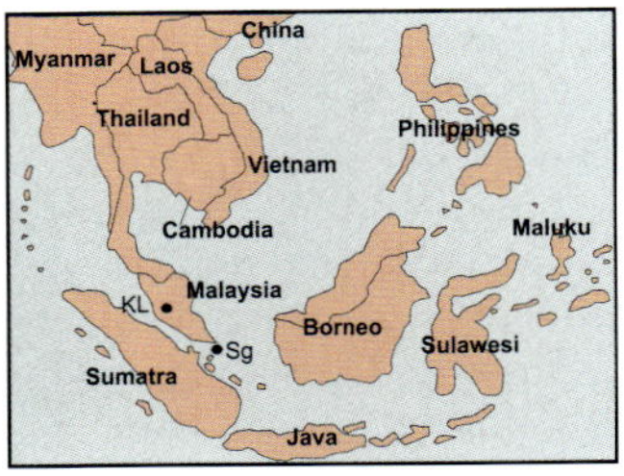

# C

## ***Calanthe vestita*** Lindl.

Although widespread through the area, this terrestrial species is confined to limestone regions in climates with a dry season. It has large silvery pseudobulbs which drop their leaves and initiate flowering at the start of the dry season. The white-haired arching inflorescence can get up to 70 cm long and carries up to 12 well-spaced flowers. Some plants have a bright red callus, while on others the lip is flushed rosy pink. *C. vestita* is very easy to grow and is widely cultivated.

Rainfall:
Thailand; Phuket; 30 m.

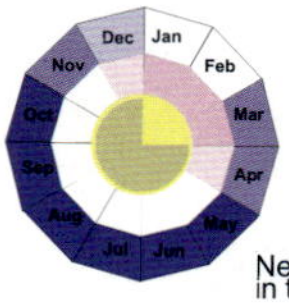


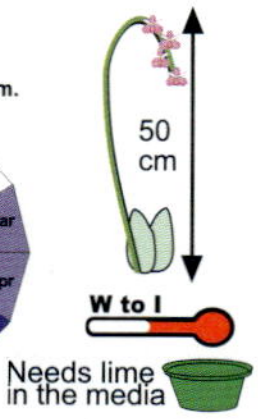


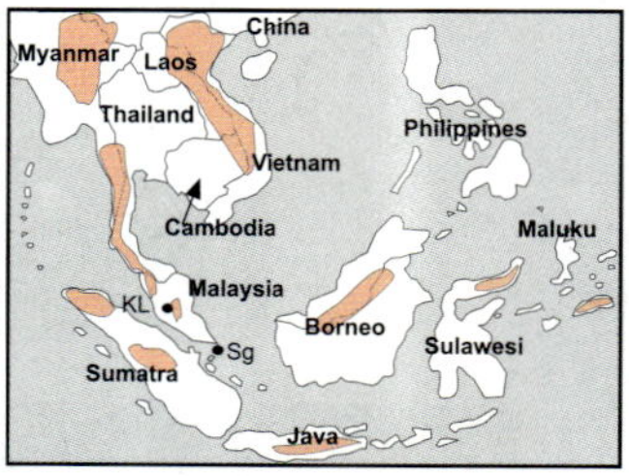

## ***Calanthe zollingeri*** Rchb.f.

This terrestrial species is quite common growing in moist humus alongside rivers in parts of Borneo up to about 700 metres. The flowers, which are initially white with a yellow spot on the lip, turn orange as they age.

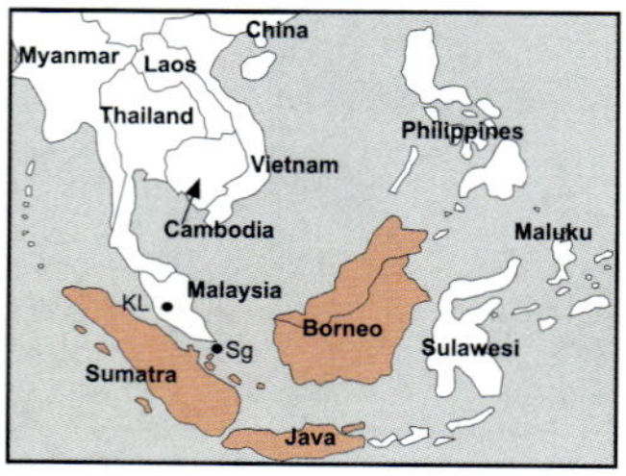


Data: Sabah:
Danum Valley; 20 m.

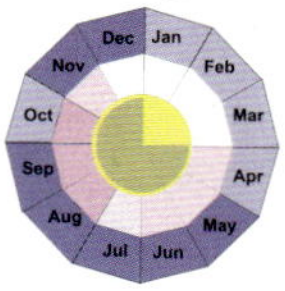


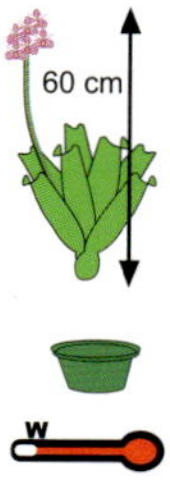

## Ceratostylis

Most of the 100 or so members of this Asia-Pacific genus look like clumps of rushes growing on tree trunks, branches or mossy rocks; they have thin tubular stems and leaves, and bunches of small, dull tubular flowers with swollen lips. ... as shown in the photo on the right. They are not cultivated; in fact, most people would not recognise them as orchids if they encountered them in the wild. However, a few species have broader, flatter leaves, and one of these, *C. retisquama*, has large red flowers and is often encountered in orchid collections around the world.

Unidentified *Ceratostylis* species.

# C

## *Ceratostylis retisquama* Rchb.f.

This species is usually known by the highly appropriate name *Ceratostylis rubra*. The plant, which is found in the hills and mountains up to at least 1500 metres altitude, has a hanging rhizome that is covered in reddish-brown sheaths. A long solitary leaf is formed on the end of a very short shoot. The brilliantly-coloured flowers are large for such a small plant.

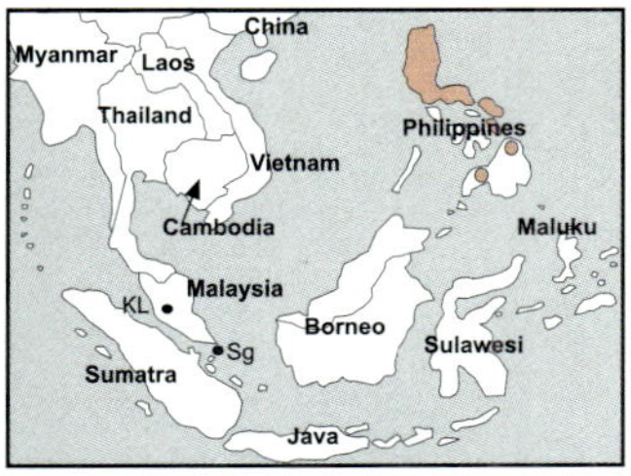

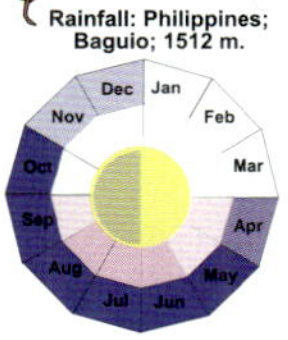


## Chelonistele

This genus of 11 small mountain species are closely related to *Coelogyne* and produce inflorescences on young shoots, often before the leaves develop. All species have pseudobulbs with one or two leaves. Only one species has a wide distribution; the other 10 are confined to the mountain regions of central and northern Borneo.

## *Chelonistele sulphurea* (Bl.) Pfitz.

This species has 2 recognised varieties. Var. *crassifolia* is twin-leafed terrestrial form, confined to Sabah between 2100-2700 metres altitude. Var. *sulphurea* (illustrated) has only one leaf and is more common and widespread, growing as an epiphyte at altitudes of 600-2300 metres; the flower colour varies considerably, from white to yellow, green, buff, brown, salmon pink or mauve, with a paler lip that has bold yellow to gold markings. Some plants are sweetly scented.

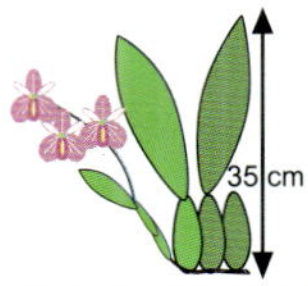


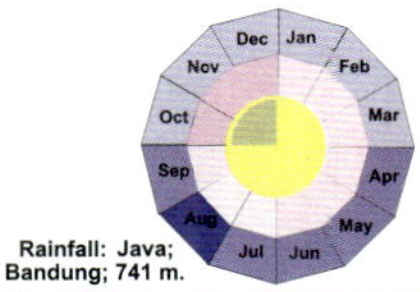


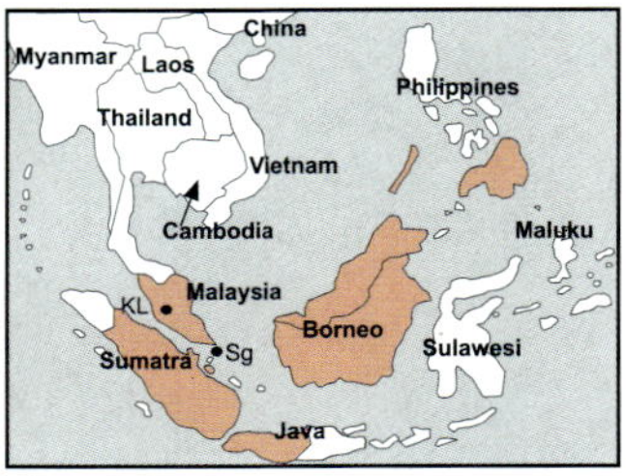

# C

## Chroniochilus

This genus of 4 small epiphytic species is distributed from Thailand to Borneo and Java, with 3 species being found in Malaysia. Although they are quite common in some areas, they are easily overlooked due to their small size and preference for growing on twigs. The inflorescences are shorter than the leaves and are often concealed by them; only a few flowers open at a time and they only last for a few days.

### *Chroniochilus virescens* (Ridl.) Holttum

This common lowland species can be found up to 1000 metres altitude. The short inflorescence is hidden under the leaves and forms single flowers intermittently for several months. The flowers are strongly vanilla-scented but only last for one day.

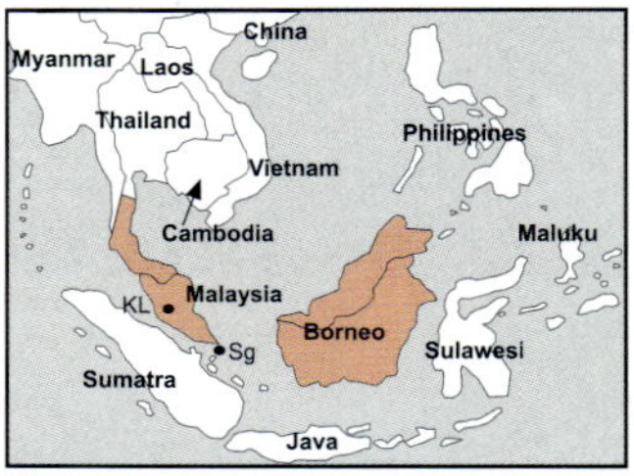


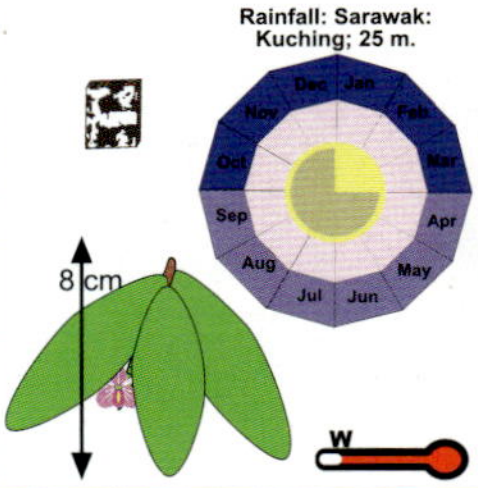

## Cleisostoma

This genus of 80 to 100 epiphytic species is widespread throughout Asia from India to the Pacific Isles. *Cleisostoma* species are often seen in private orchid collections in S. E. Asia, and some species are available in nurseries, particularly those in Thailand. Although the individual flowers of most species are rather small, the inflorescence extends considerably so the plants are in flower for several months. Many species in this genus have terete leaves.

### *Cleisostoma complicatum* (Seidenf.) Garay.

This tall pendulous species is uncommon and restricted to the Thai south-west and northern Malaysia. The stems grow up to 2 metres long and branch frequently, with long terete leaves at about 4 cm intervals. The flowers are highly distinctive, not only because of their unusual coloration, but because the architecture of the lip and column is probably the most complex of all S. E. Asian orchids.

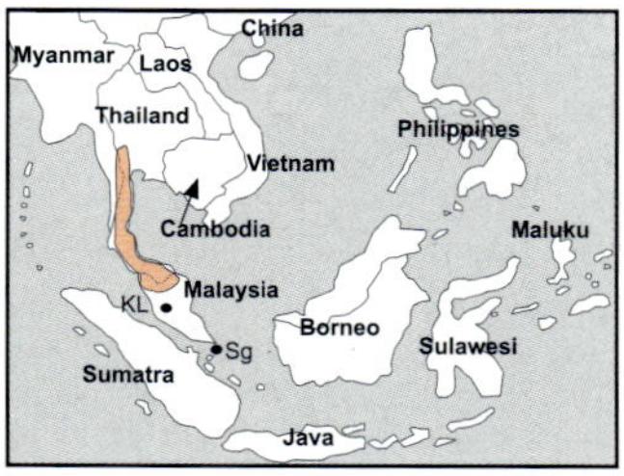


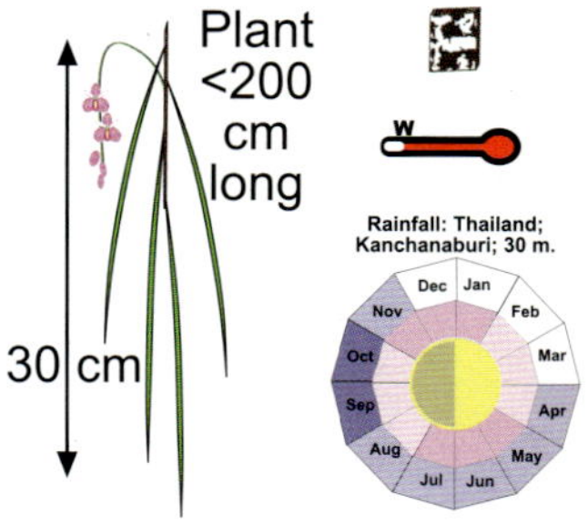

# C

## *Cleisostoma discolor* Lindl.

This common lowland species can be found from sea-level to about 1000 metres altitude. It has an thin pendulous branched inflorescence that will extend until it is about 30 cm long, producing flowers for six to seven months non-stop, although only a few are open at any time. The flowers are about 1 cm across and can be distinguished from other *Cleisostoma* species by their large sidelobes, which turn inwards and almost meet each other.

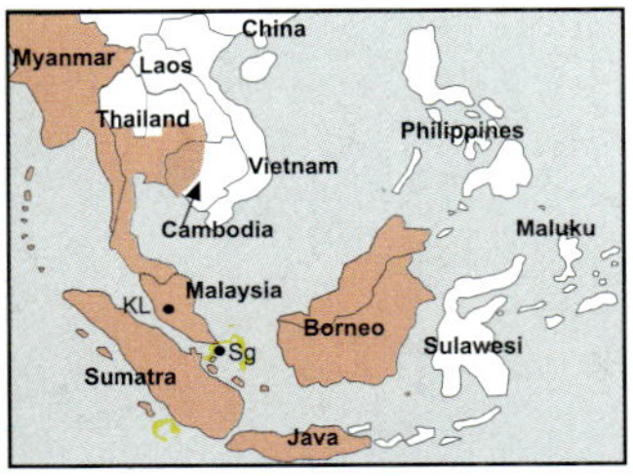


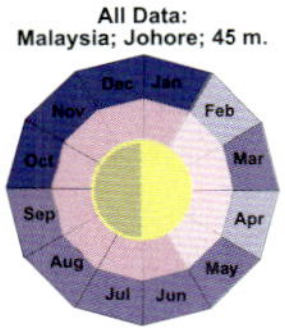


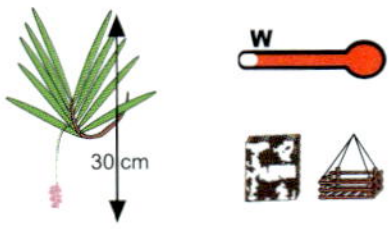


## *Cleisostoma nieuwenhuisii* (J.J.Sm.) Garay

This rare lowland species grows on forested hillsides in Borneo and is easily distinguished by the unusual colour of the flowers. It has a long slender inflorescence that blooms continuously for several months, although only a few flowers are open at a time.

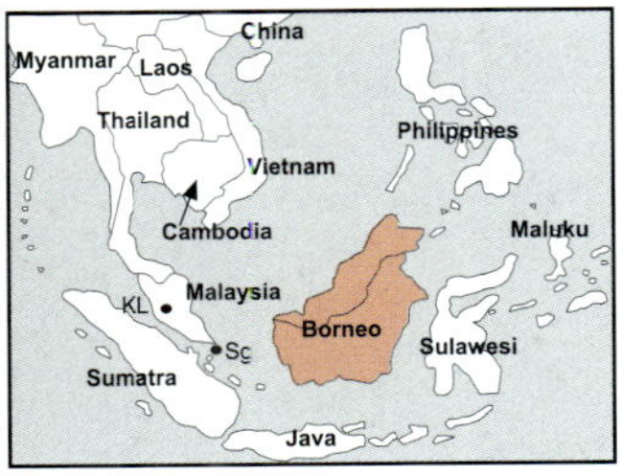


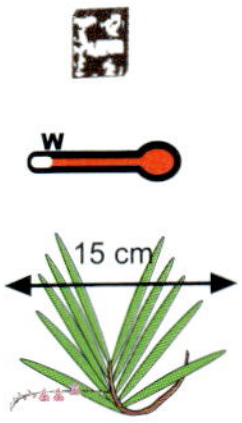


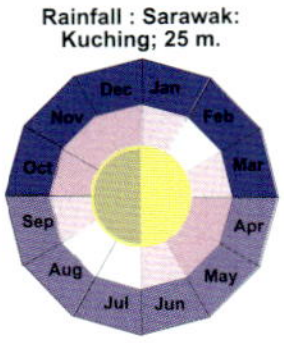


## *Cleisostoma simondii* (Gagnep.) Seidenf.

Closely related to *C. complicatum*, this erect terete-leaved species is distributed from Sikkim to Hong Kong. The attractive flowers are small but quite long-lasting.

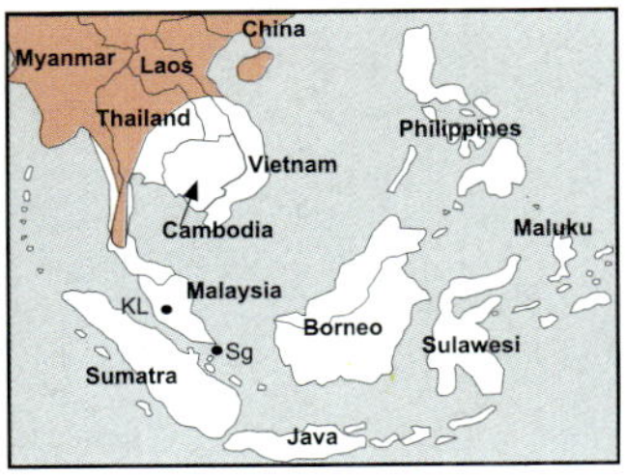


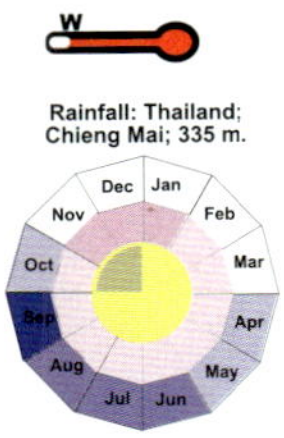


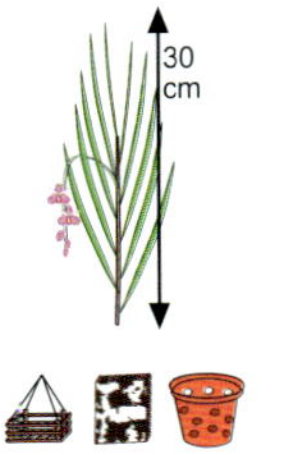

# C

## Coelogyne

A genus of about 300 species, distributed from tropical Asia to the Pacific Islands. If ever a genus deserved to be discovered by modern orchid enthusiasts, this is it. In the first decades of the 20th century *Coelogyne* species were deservedly popular, but the advent of hybrids changed that. Some of the cooler-growing Himalayan species such as *Coelogyne cristata* remain sought-after, but most S. E. Asian *Coelogyne* species are hardly known outside the region, probably because their flowers are relatively short-lived. To offset this drawback, many species flower several times a year, and their flowers are usually large, ornately decorated and beautifully coloured.

### ***Coelogyne asperata*** Lindl.

This large showy species occurs from Sumatra and Malaysia to New Guinea and the Solomon Islands. In some areas it is very common, growing as either an epiphyte or a lithophyte, often forming clumps several metres across. The arching inflorescence is produced on a young shoot as the leaves emerge, and bears up to 40 flowers that are about 6-7 cm across and all open at the same time.

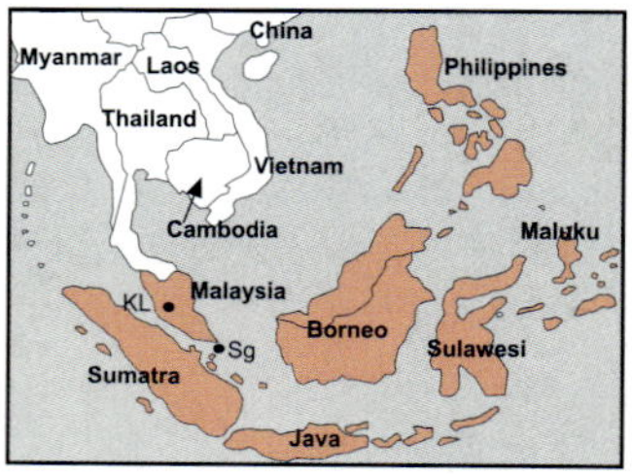


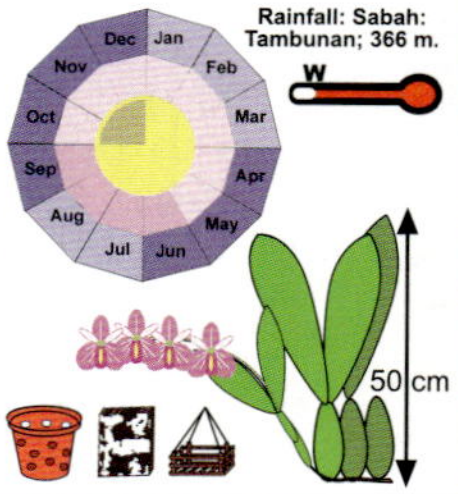

### ***Coelogyne bicamerata*** J.J.Sm.

This epiphytic species is found in the mountains of southern and central Sulawesi where it is common in mossy forests above 1000 metres altitude. It produces a stiffly erect inflorescence from a young shoot when the leaves are starting to emerge; there are up to 30 greenish-white flowers that are each 2-3 cm high and have a white-tipped green lip with 3 keels. The flowers seem to attract insects; every specimen I have seen in the wild has been partly-eaten by beetles and grubs. This does not seem to hinder pollination, as seed pods are commonly seen on mature plants.

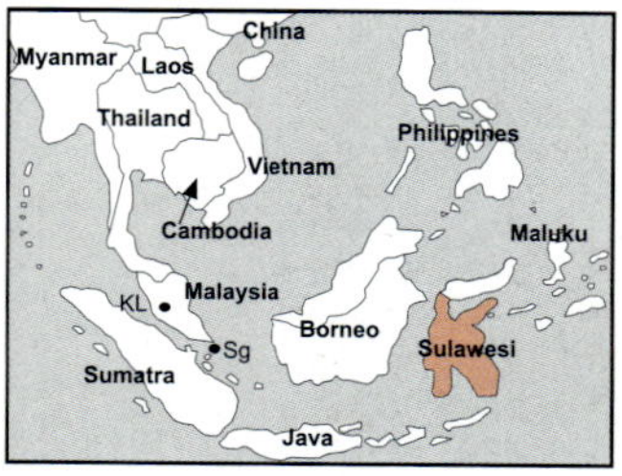


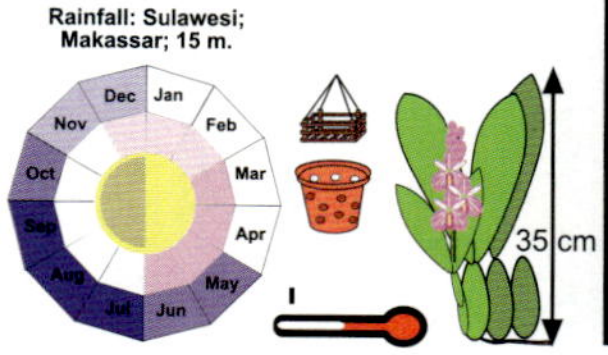

# C

## *Coelogyne celebensis* J.J.Sm.

This large showy species is common in parts of Sulawesi from 500-1200 metres, although it is occasionally found growing at sea level. It is known locally as "*anggrek kelapa*" or "coconut orchid" because of the shape of the very large leaves. The arching inflorescence bears 4-12 flowers that are about 6-7 cm across and open one at a time.

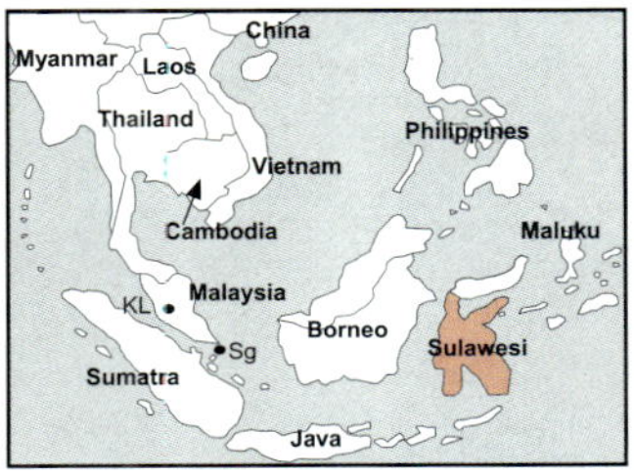


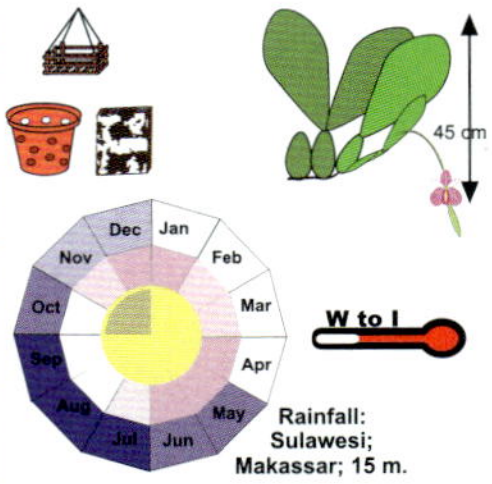


## *Coelogyne incrassata* (Bl.) Lindl.

This delicate species is one of a group that has a lip with thickened midlobe (epichile) margins. While most of this group are confined to Borneo, *C. incrassata* has a wider distribution, growing in forests from 400-1200 metres altitude.

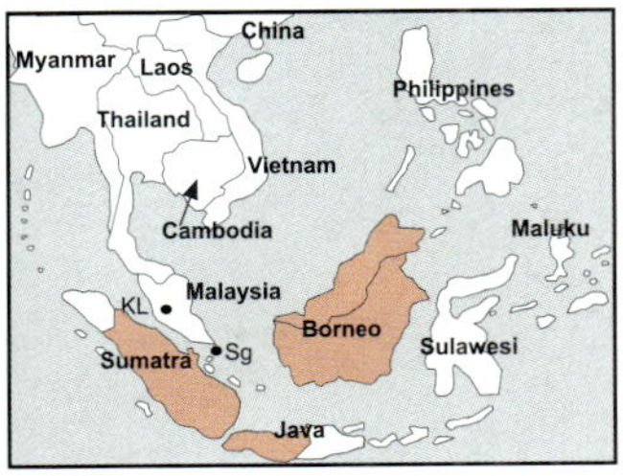

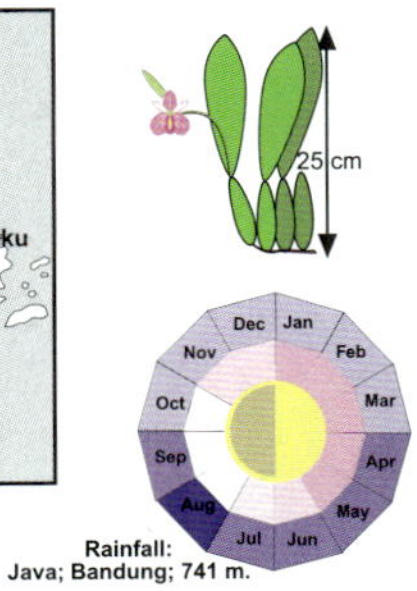

## *Coelogyne lawrenceana* Rolfe

*C. lawrenceana* has the largest flowers of any *Coelogyne*; they are usually 7-8 cm wide. Although only one flower is formed at a time, they are relatively long-lasting and the inflorescence keeps on producing more flowers, up to a total of about 6.

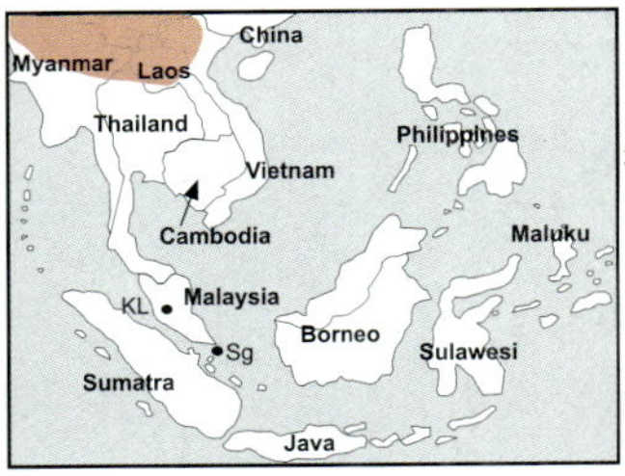


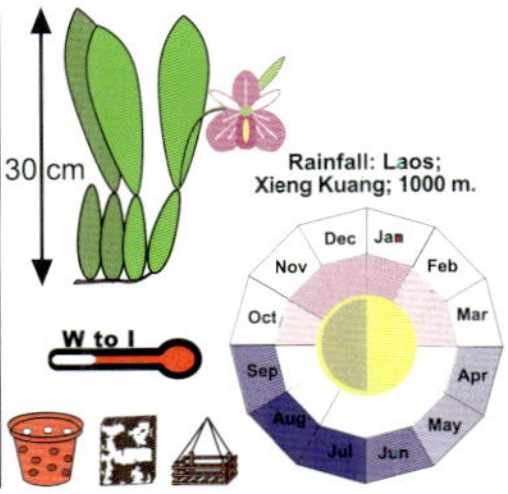

*Coelogyne rochussenii* De Vr.

# C

## *Coelogyne motleyi* Rolfe ex Wood, Clayton & Chan

Although first collected in 1858, this slender 2-leaved species remained unnamed until 1998. The inflorescence produces 2-3 flowers, but only one is open at a time. *C. motleyi* belongs to the group that have lips with thickened midlobe margins.

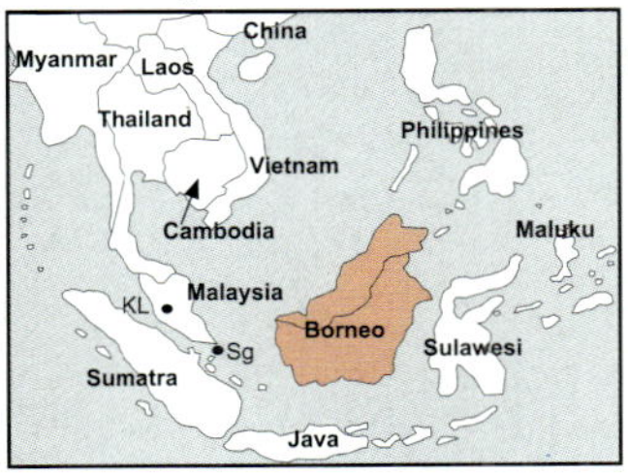


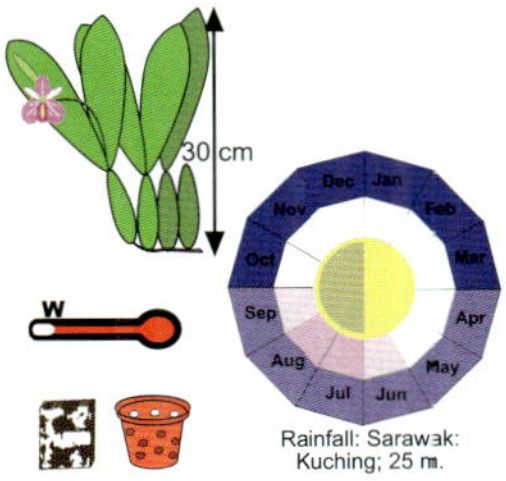

## *Coelogyne pandurata* Lindl.

This very large showy epiphytic species used to be common on trees alongside streams and rivers from Sumatra to Borneo, but is seldom seen in the wild today. It produces a 30-40 cm long horizontal inflorescence with up to 12 dramatically-coloured flowers, each of which is about 7 cm across. *C. pandurata* has a lip with a black central patch and prominent white warts; characters shared with several other *Coelogyne* species from Borneo.

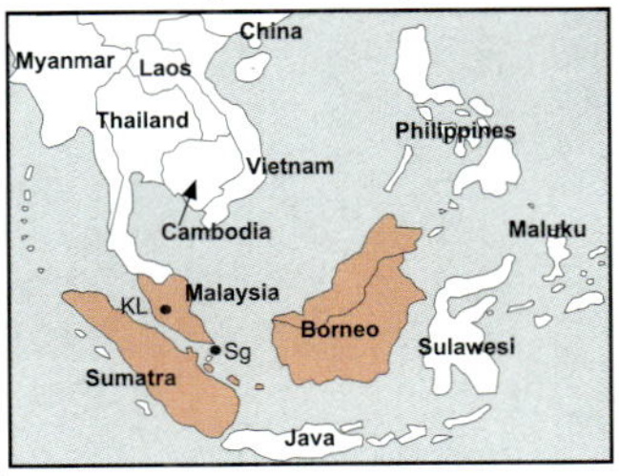


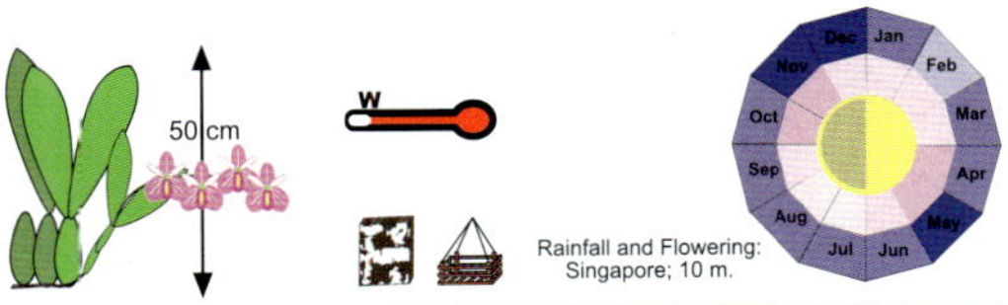


## *Coelogyne rochussenii* De Vr.

This species is common across much of S. E. Asia from sea-level to 1500 metres altitude, preferring to grow on tree trunks and branches next to water. The long, pendulous inflorescences are formed on non-leafy shoots that emerge from the rhizome. The flowers are lightly lemon-scented. In cultivation *C. rochussenii* can flower at almost any time of year, but occasionally a well-grown plant will put on a spectacular display; the plant illustrated on the facing page had over 2500 flowers on 63 flower spikes !

# C

## *Coelogyne speciosa* (Bl.) Lindl.

This showy species is common in the forested mountainous regions of Sumatra, Java and Flores. It forms an inflorescence from the new shoot when the leaves are about half developed. Each inflorescence can produce up to eight flowers which open one at a time and are usually about 6 cm across. There are 3 distinct subspecies which vary in colour, size and midlobe margins. The illustrated plant is a rare *alba* form.

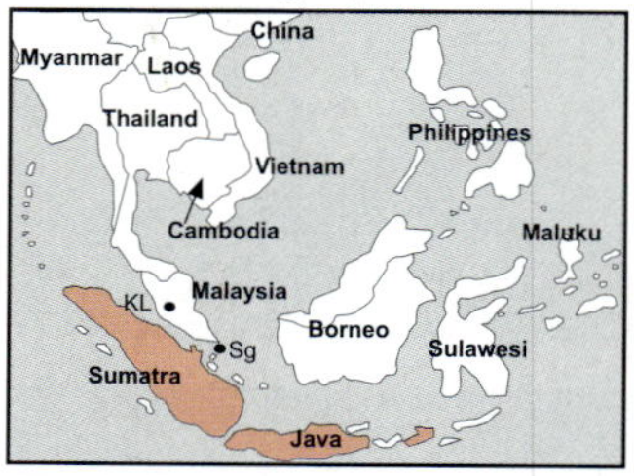


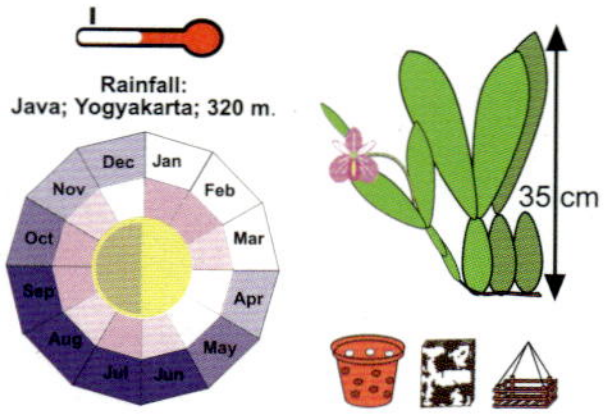


## *Coelogyne swaniana* Rolfe

This species grows as an epiphyte from 650-1600 metres altitude and can also be found growing on rocks or on the ground, especially on roadside cuttings. The pendulous inflorescence is formed on a non-leafy shoot at the base of the newest mature pseudobulb. It reaches up to 60 cm long and carries up to 17 flowers, all of which are open at once. The 4 cm wide flowers are white to cream with beautiful markings on the lip.

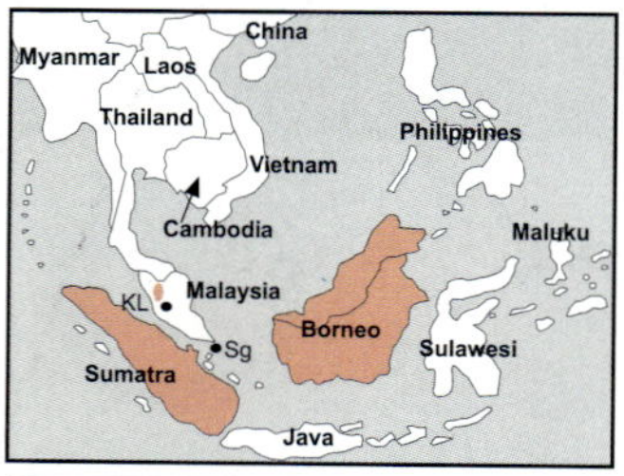


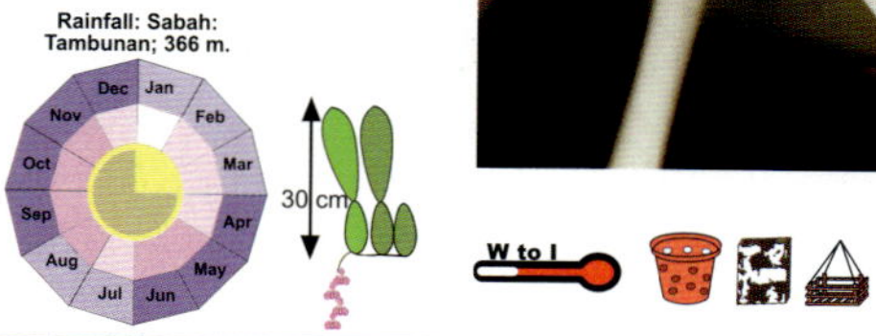

## *Coelogyne tiomanensis* Henderson

This rare species is only found above 700 metres altitude on the mountains of one small Malaysian island, where it grows on trees and rocks. The short stiff inflorescence emerges when the new shoot is more than half-formed, and usually carries 2-3 flowers that open one at a time. The 6-7 cm wide flowers have a lip with a dark chocolate-brown centre.

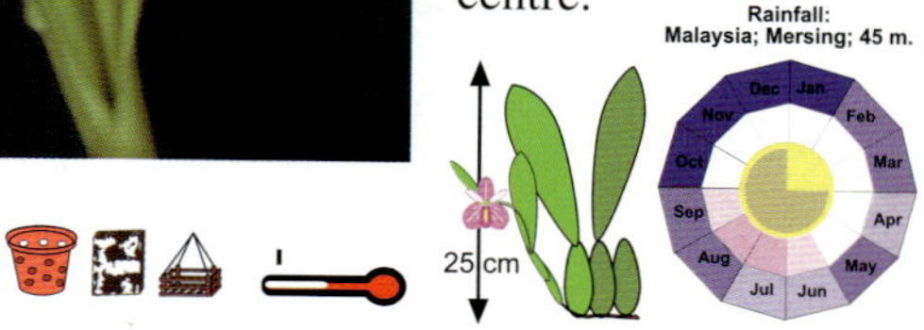


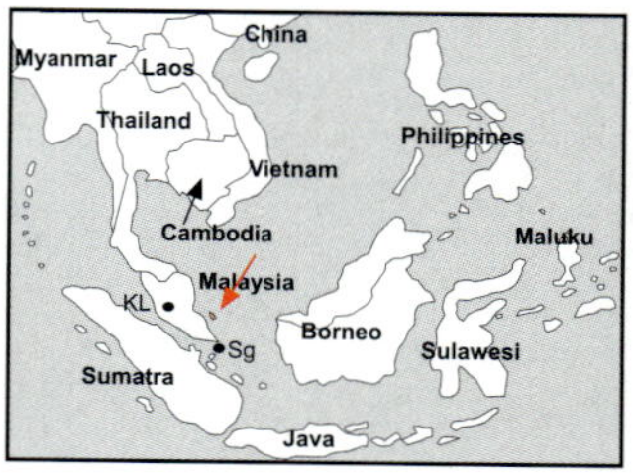

# C

## *Coelogyne tommii* Gravend. & O'Byrne

This rare species was first noticed growing in a collection in Singapore and was described in 1995. The origins of the plant are lost, but it is thought to have come from Java. The species flowers from the new shoot before the leaves develop, and carries 4-5 flowers most of which are open at the same time. The lightly-scented 4 cm wide flowers last for 7-10 days.

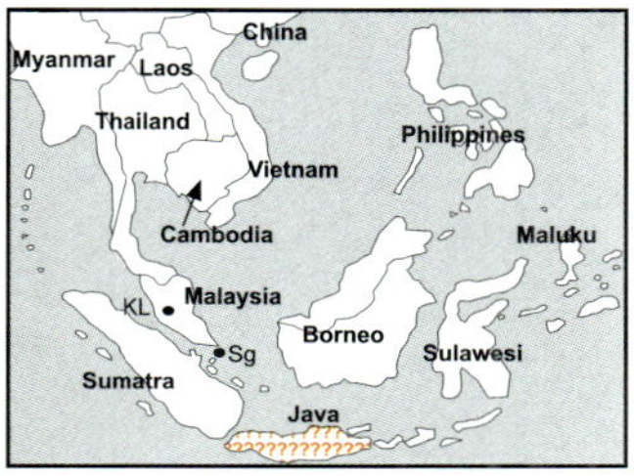


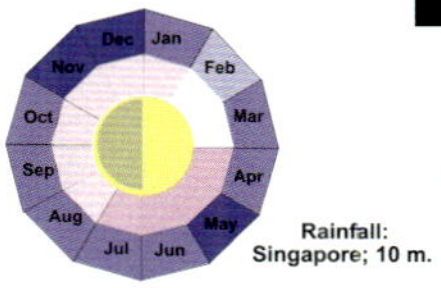

## *Coelogyne trinervis* Lindl.

This well-known species is distributed across much of S. E. Asia, growing on trees and rocks from 100-1500 metres altitude. It has large egg-shaped leathery pseudobulbs, and bears up to 10 flowers on a rather short inflorescence that is formed on the new shoot before the leaves develop. All the flowers are open at the same time.

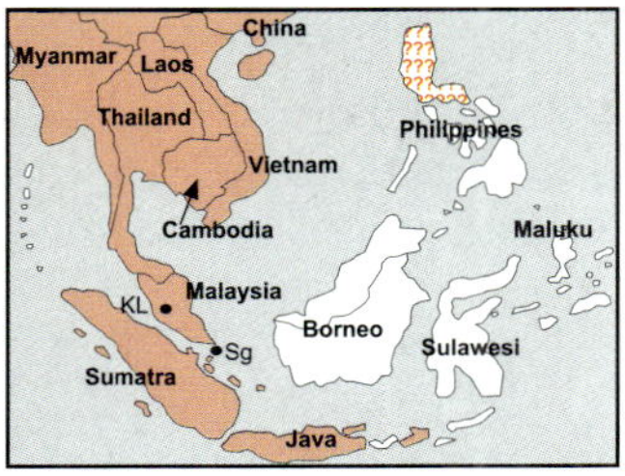


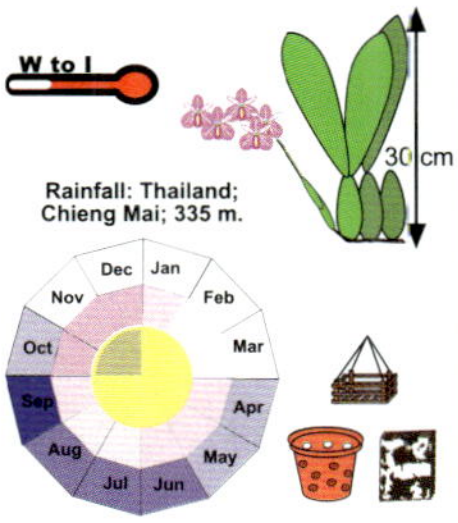

## *Coelogyne verrucosa* S.E.C. Sierra

This species was only described in 2001 because it has been confused with *C. peltastes* for over a century. It has flattened pseudobulbs that are an adaptation to its tree-climbing habit. It has up to 10 flowers on a long inflorescence; all are open at the same time. The flowers are 7 cm across and have a large patch of cream-coloured warts on the lip.

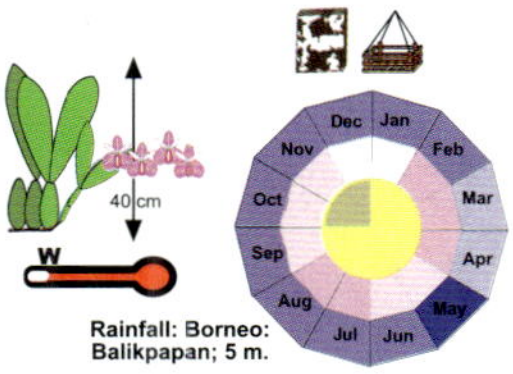


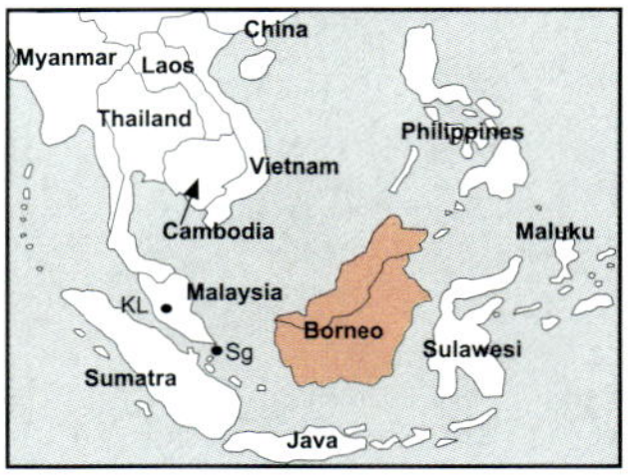

# C

## Cordiglottis

This genus of 7 small twig epiphytes is distributed from Thailand and Sumatra to Borneo, with 5 of the species found in Malaysia and 4 in Borneo. All have short stems and terete or laterally-compressed leaves, and slender inflorescences that produce a succession of single flowers which are very short-lived. Consequently these plants are seldom seen in collections, although *C. filiformis* is quite common in the wild.

### *Cordiglottis filiformis* (Hk.f.) Garay

This very small species is quite common in some parts of lowland Malaysia. It has recently been discovered in Jambi Province of Sumatra. It grows unobtrusively on twigs and small branches of trees in forests, swamp forests and plantations. It has stems that are only 4-5 cm long and carry a few 3 cm terete leaves that turn purple-coloured if the plant is exposed to the sun. The inflorescence grows up to 10 cm long and flowers intermittently over a several-month period. The flowers are about 1.5 cm wide and last for only one day.

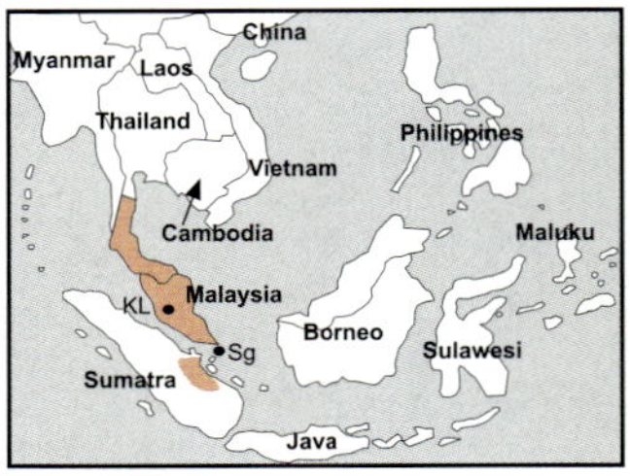


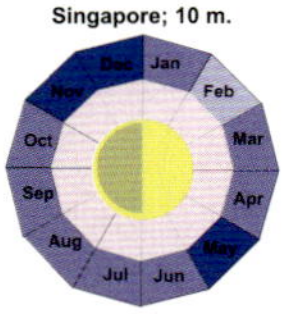

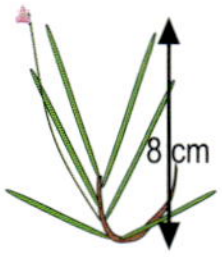

Sumatran plant, courtesy of Tan Heok Hui.

## Corymborkis

This pantropical genus of 5 tall terrestrial species has one species distributed from India throughout S. E. Asia to the Pacific Islands. The plants have a short underground rhizome that sends up stems at intervals; these have a number of large leaves along the length of the stem. The inflorescences emerge from the leaf petioles and have several branches. The large white and green flowers are displayed "upside down" with the lip uppermost; the sepals and petals are similar, and the lip is simple.

### *Corymborkis veratrifolia* (Reinw.) Bl.

Although the natural habitat of this widespread species is on the floor of the forest from sea-level to 1500 metres altitude, nowadays it is more likely to be found in rubber or fruit plantations, taking advantage of the rich well-watered soils found there. It also occurs in secondary forest. The 5 cm wide flowers produce a strong fragrance in the early evening.

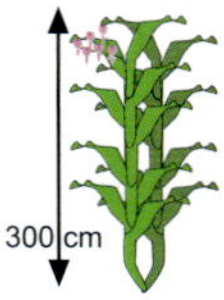


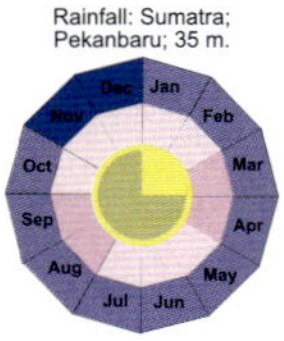


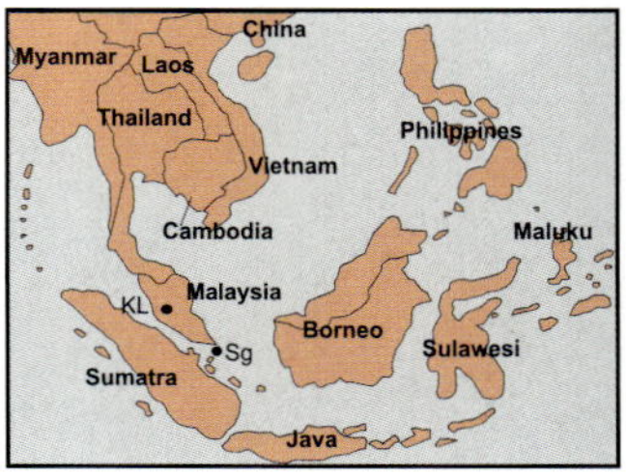

# C

## Cryptostylis

This genus of about 15 terrestrial species is distributed from Sri Lanka and India across S . E. Asia to Australia and the Pacific Islands, usually found in evergreen montane forest. All species have a shoot that emerges from a cluster of thick roots (there is no tuber or rhizome) and carries one to a few leaves on distinct stalks. The inflorescence arises from the base of the plant and carries several medium-sized flowers that have the lip uppermost. Despite the fact that the plants are easy to grow and the flowers are long-lived, *Cryptostylis* species are not commonly cultivated and are seldom seen in collections.

### *Cryptostylis arachnites* (Bl.) Hask.

This widespread species grows on the forest floor from 500-2200 metres altitude. It is quite common in acidic peaty soil in some area above 1000 metres altitude, where it likes to grow by paths or in moss-pads at the foot of trees. It has up to 4 long-stalked leaves and a single 40-60 cm tall inflorescence with up to 15 long-lasting flowers. The 3-4 cm tall flowers are dominated by a rather narrow velvety lip that points upwards and has red or purple spots near the tip.

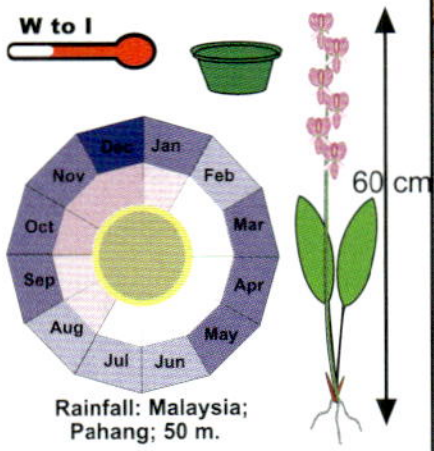


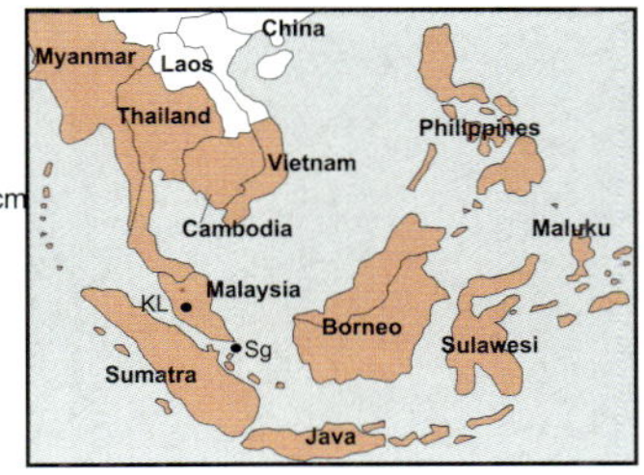

### *Cryptostylis conspicua* J.J.Sm.

This uncommon species has a limited distribution in Malaysia and Java but is more widespread in Sumatra. It is found only in montane forest from 1200-1800 metres altitude, where it prefers to grow on moss pads at the foot of trees. It has 2-4 long-stalked dark green leaves and a single 40-60 cm tall inflorescence with up to 10 long-lasting flowers. The 3-4 cm tall flowers are dominated by a velvety lip that can be green to yellow or cream and is flushed red to maroon near the margins, and lacks spots but has a red-brown central streak.

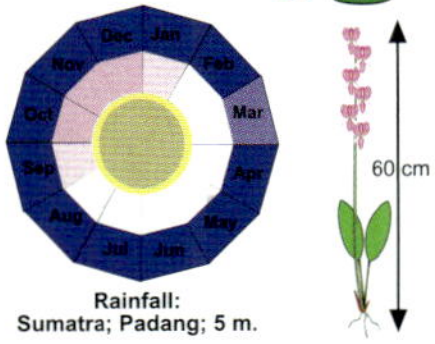


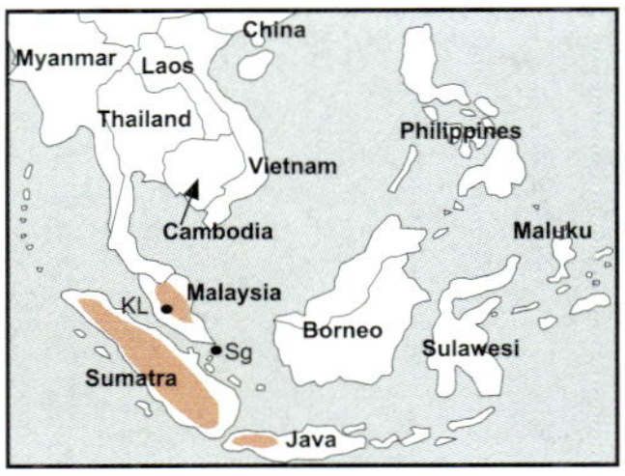

# C

## Cymbidium

This genus of about 50 species is distributed from the Himalayas to Australia. Many *Cymbidium* species are well-known to orchid enthusiasts, and the genes of several of these species (particularly the cooler-growing Himalayan and Chinese species) have given rise to the huge array of modern *Cymbidium* hybrids. In S. E. Asia the majority of *Cymbidium* species are large tough epiphytic plants that are not widely grown outside the region because their pendulous inflorescences of rather drab flowers do not compare well with the more showy erect-flowered hybrids. There are, however, some exceptions.

### *Cymbidium aliciae* Quis.

This rare terrestrial species is easily the most beautiful of the S. E. Asian *Cymbidium* species, with an erect inflorescence that holds the delicately-coloured lightly-scented flowers just clear of the leaves. Although usually listed in books as growing in the Philippines from 500-1500 metres altitude, *C. aliciae* also occurs in the hilly country of Central Sulawesi. Some botanists consider *C. aliciae* to be synonymous with the widespread Asian mainland species *C. cyperifolium*.

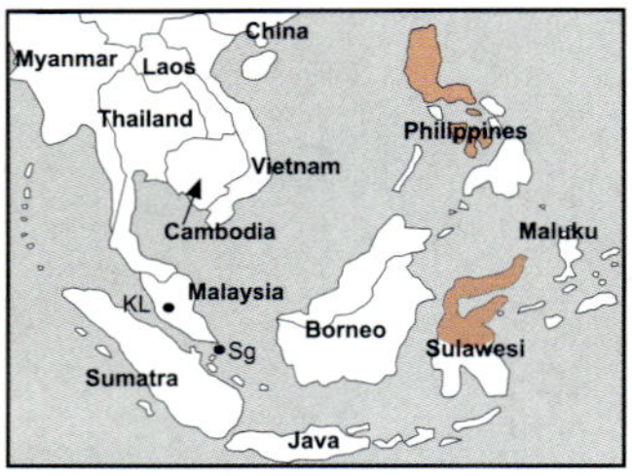


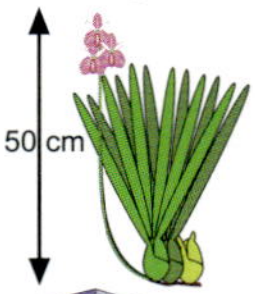

### *Cymbidium bicolor* Lindl. *ssp. pubescens* (Lindl.) Du Puy & Cribb

This medium-small epiphytic species has a wide distribution and is quite common in some areas between 500-1200 metres altitude. It grows on forest trees but also colonises trees in plantations and parkland. It has rather tough leathery leaves and a pendulous inflorescence with up to 12 flowers. The lip of this subspecies has a covering of short hairs. The normal form of *C. bicolor* is only found in southern India and Sri Lanka.

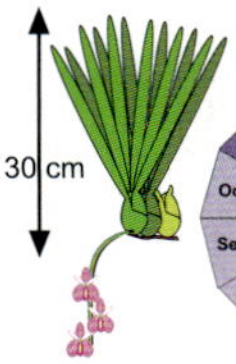


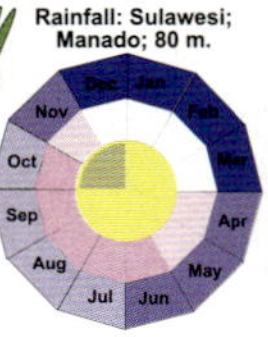

# C

## *Cymbidium chloranthum* Lindl.

This epiphytic species is one of the showiest of Malaysian *Cymbidiums* as it produces several erect inflorescences at a time. The flowers gradually turn bright pink as they age; the white lip-margins change first, then the rest of the flower over the next few days.

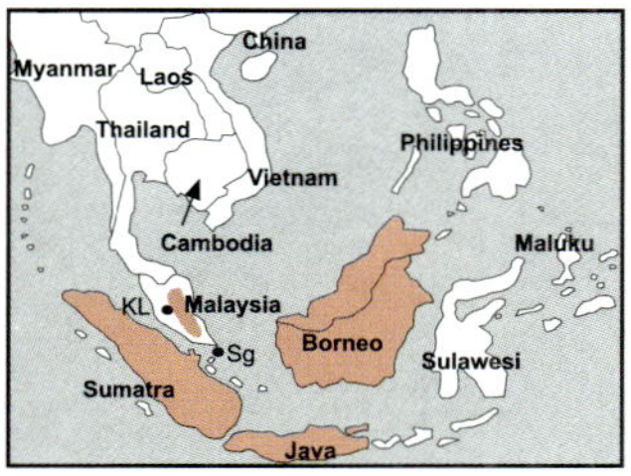


Rainfall: Malaysia; Pahang; 50 m.

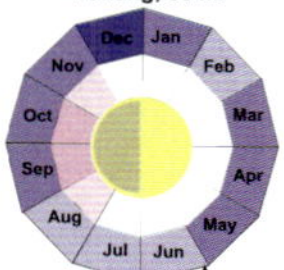


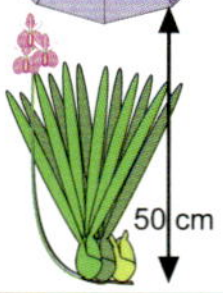


A well-drained mix containing partly-rotted wood chips.

## *Cymbidium ensifolium* (Lindl.) Sw.

*C. ensifolium* subspecies *haematodes* is a terrestrial species found through much of the area from 500-1700 metres altitude. It has small pseudobulbs and an erect inflorescence with up to 8 flowers. The illustrated specimen is a rare *concolor* form from a private collection; wild plants usually have red markings on the tepals, lip and column.

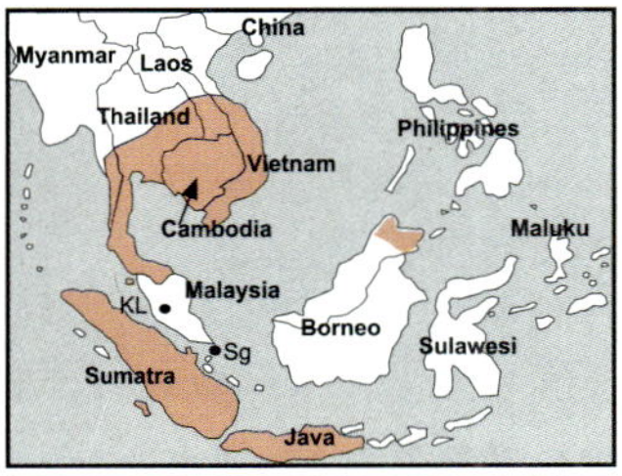

Rainfall: Thailand; Kanchanaburi; 30 m.

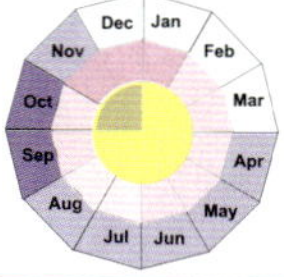

## *Cymbidium finlaysonianum* Lindl.

This epiphytic species is abundant in many coastal and lowland habitats and can be found up to 1200 metres altitude. It forms huge masses with hard leathery leaves and pendulous inflorescences that may be up to 90 cm long. The illustrated flower is unusual in that it has red markings rather than purple.

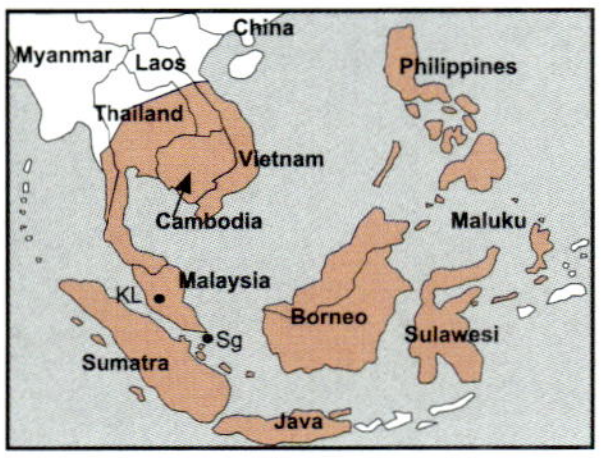

Rainfall: Malaysia; Ipoh; 35 m.

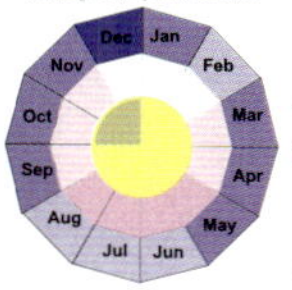


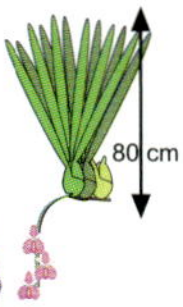

# C

## ***Cymbidium lancifolium*** Hk.f.

This small terrestrial species grows in humus on the forest floor at altitudes of 300-1800 metres. It has slender pseudobulbs with 2-4 lanceolate leaves and an inflorescence with up to 8 flowers.

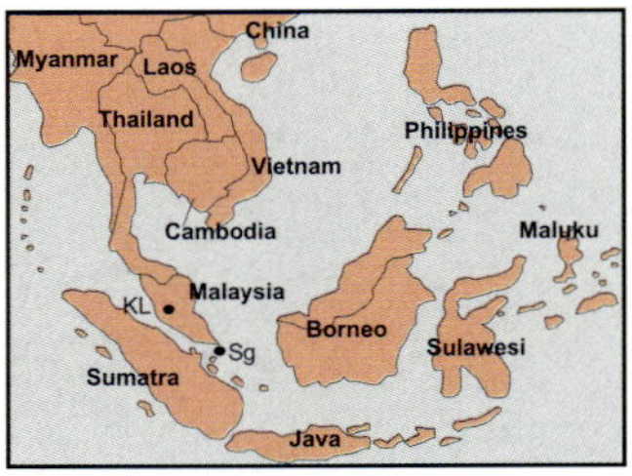


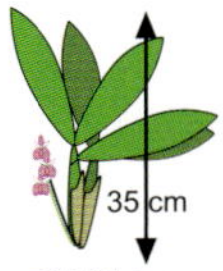


Rainfall:Java; Maduin; 110 m.

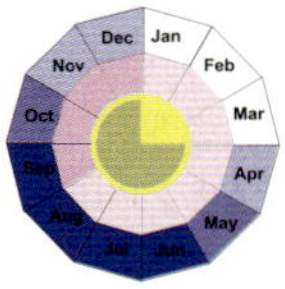


A well-drained mix containing partly-rotted wood chips.

W to I

## ***Cymbidium rectum*** Ridl.

This uncommon epiphytic species is found from sea-level to 500 metres altitude in rainforest, swamp forest and heath forest. It was originally found in Malaysia, but has recently been discovered in Sabah and Sumatra. It has stiff leathery leaves and an erect inflorescence with up to 10 slightly-fragrant flowers that are about 4 cm high. Some Sumatran plants are autogamous (self-pollinating).

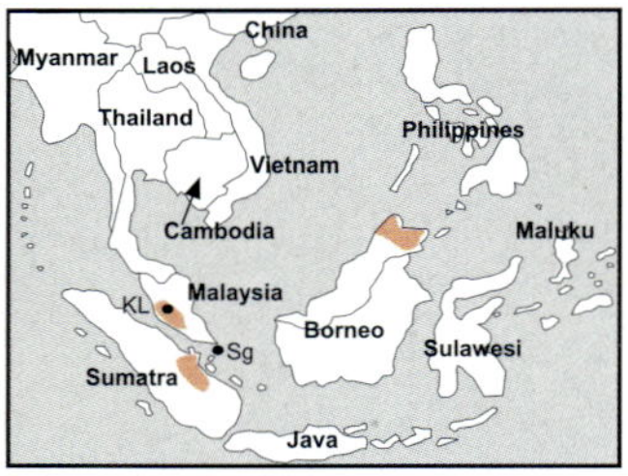


Rainfall: Sumatra; Pekanbaru; 35 m.

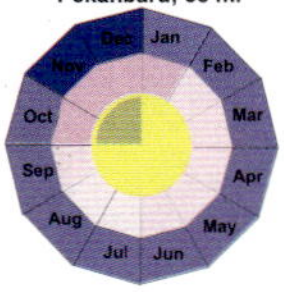


W

A well-drained mix containing partly-rotted wood chips.

## ***Cymbidium sinensis*** (Jackson in Andr.) Willd.

This tall highly-scented terrestrial species is found in woodlands from 250-2300 metres altitude. It has been cultivated in China and Japan for centuries. A large number of forms exist, differing principally in their colour. Wild plants are usually maroon to brown with darker stripes and a yellow to brown lip with red spots. Thai plants are usually pale yellow with red veins. An *alba* form from Taiwan is illustrated.

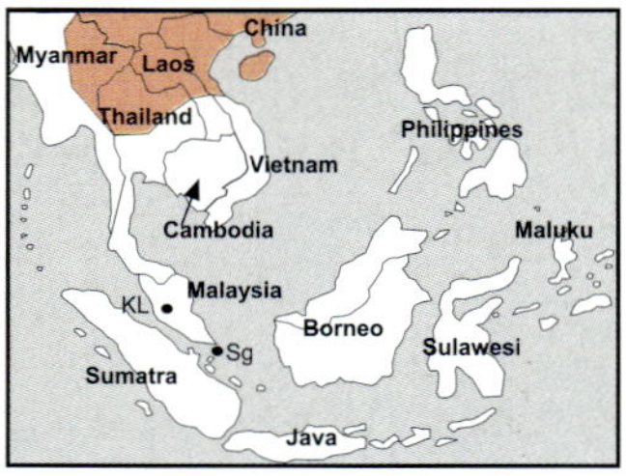


W to I

Rainfall: China; Yunnan; 1300 m.

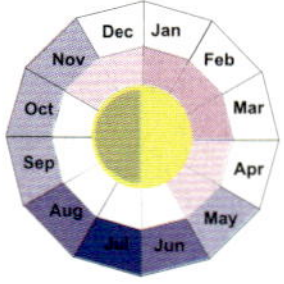


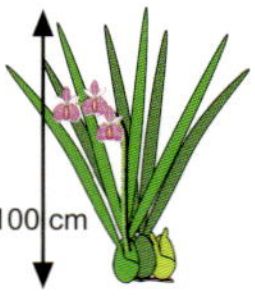

# D

## Dendrobium

*Dendrobium*, with an estimated 1400 species, is the second largest genus in S.E. Asia and is probably the most diverse in terms of the different plant habits and flower shapes that it contains. It is distributed throughout S. E. Asia from India to Japan, New Zealand and the Pacific Islands. It is the most important genus in the region in both economic terms and in the number of species in cultivation. In order to make the genus easier to handle, it is often subdivided into "Sections", of which sections *Spatulata* and *Phalaenanthe* have been particularly significant in the development of hybrids for the orchid cut-flower industry. These two sections (and their hybrids) are long-standing favourites with hobby growers, but in recent years species from other sections have become more popular, so some of the S.E. Asian species from sections *Bolbidium*, *Breviflores*, *Callista*, *Calcarifera*, *Dendrobium*, *Distichophyllum* and *Formosae* have become increasingly well-known. Since it is impossible to show the full extent of *Dendrobium* in a book of this nature, I have chosen a mixture of many of the well-known showy species and a good number of rare or poorly-known species.

### ***Dendrobium acaciifolium*** J.J.Sm. Section: *Rhopalanthe*

This easily-recognisable species is known from only a few collections. It grows on trees in forests close to the coast, from sea-level to about 300 metres altitude. The stems have a slight swelling above the base, and the leaves are flattened sideways. The bare upper part of the stem has many inflorescences which each produce 1-2 flowers at a time. The 13 mm long flowers last several days.

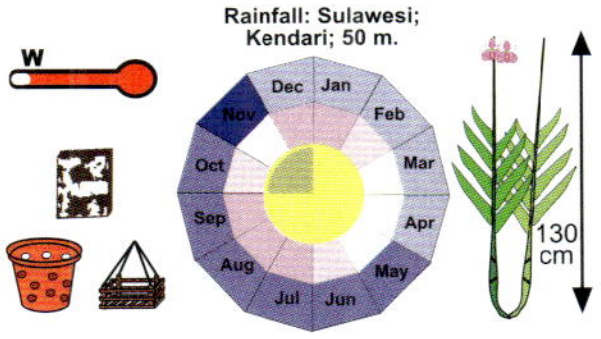


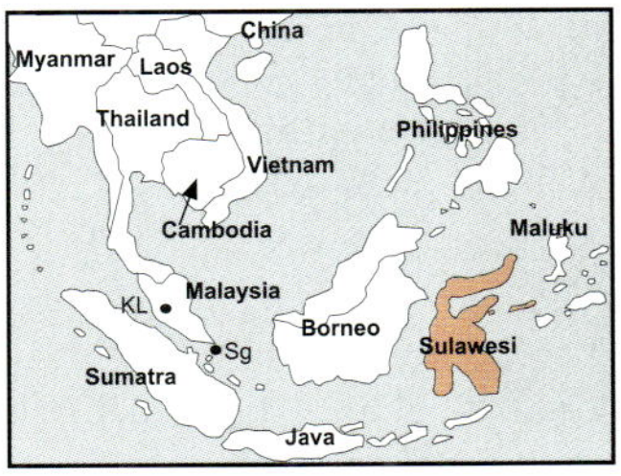

### ***Dendrobium acerosum*** Lindl. Section: *Strongyle*

This common species can be found in most types of lowland forest and also grows on trees in parks, golf courses and cultivated land. It forms tufts of stems with short curved terete sharp-pointed leaves. The bare upper part of the stem has many inflorescences which each produce 1- 4 flowers at a time. The 12 mm long flowers last a couple of days.

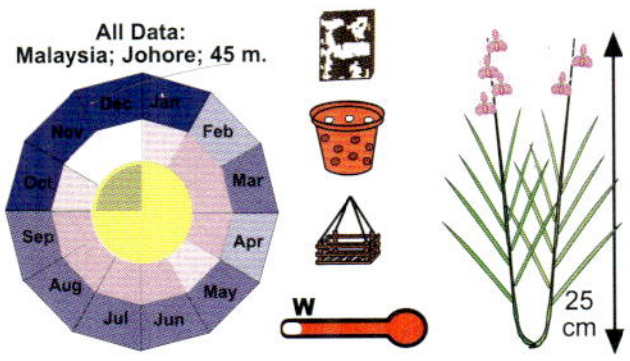


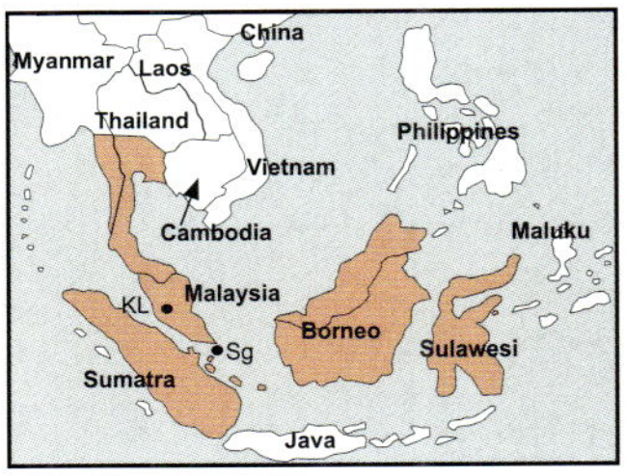

# D

## *Dendrobium affine* (Decne.) Steudel    Section: *Phalaenanthe*

This well-known species, often grown under the name *D. dicuphum*, has been extensively used in hybridization programs. The stems may grow to 100 cm long but are usually much shorter. Each stem can produce a few inflorescences, each with up to 20 flowers that last 4-6 weeks.

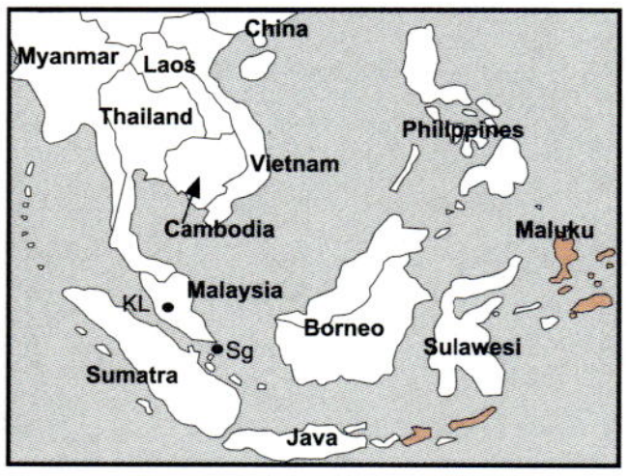


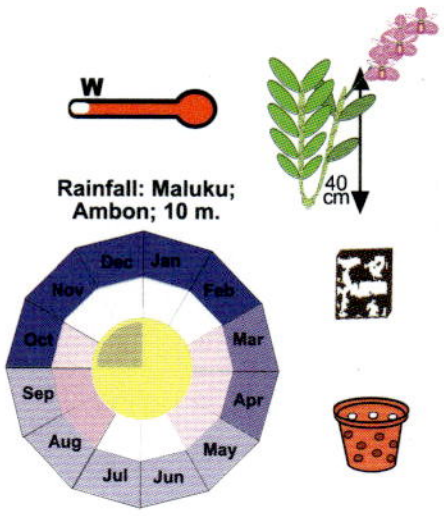


## *Dendrobium anosmum* Lindl.    Section: *Dendrobium*

This well-known species has deciduous stems that may grow to 300 cm long but are usually much shorter. Many inflorescences are formed on the leafless stems, each with 1-2 large flowers that smell of raspberry jam. White-flowered forms are common in cultivation.

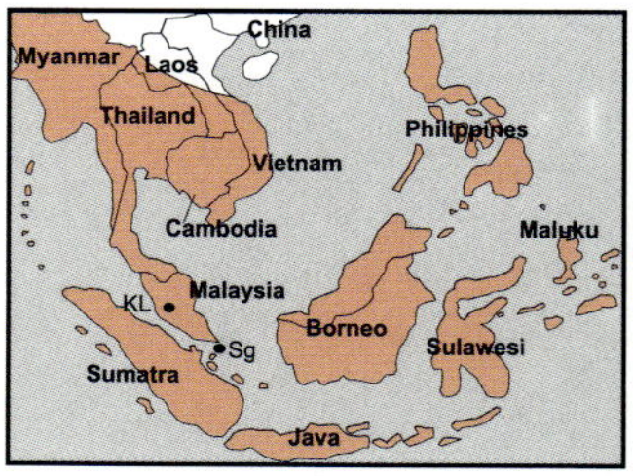


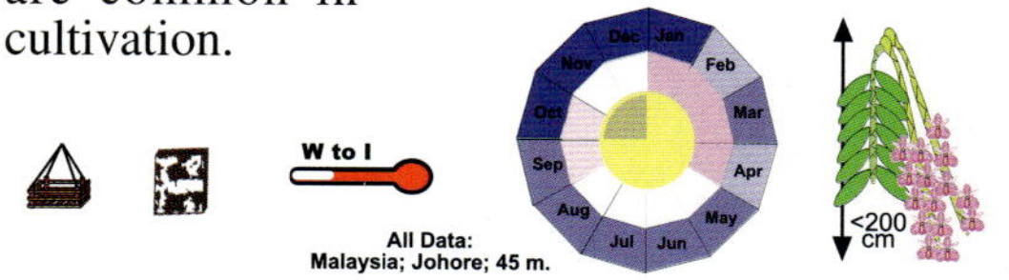


## *Dendrobium auriculatum* Ames & Quis.    Section: *Calcarifera*

This uncommon species is starting to appear in orchid collections in Europe and the USA. It grows on trees and rocks in mossy montane forests from 800-1200 metres altitude. It forms tufts of slender unbranched rather grasslike stems that are almost encircled by the bases of the narrow leaves. A succession of single-flowered inflorescences are formed on the leafless upper part of the stem. The 6 - 7 cm wide flowers last for 10-14 days.

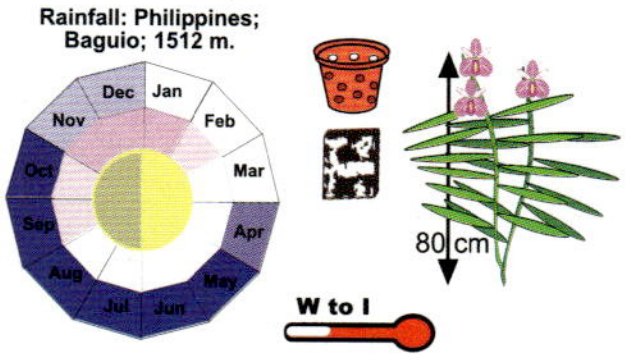


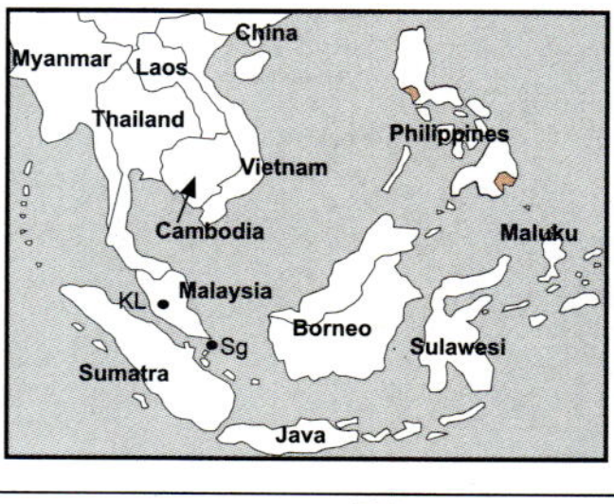

# D

## *Dendrobium babiense* J.J.Sm. Section: *Aporum*

This species has been reported growing on trees in heath-forest at 900-1000 metres altitude, but also grows on tree branches overhanging water in the lowlands. The lower part of the stem carries short laterally-flattened leaves; the long leafless upper part has many inflorescences which produce a succession of tiny flowers.

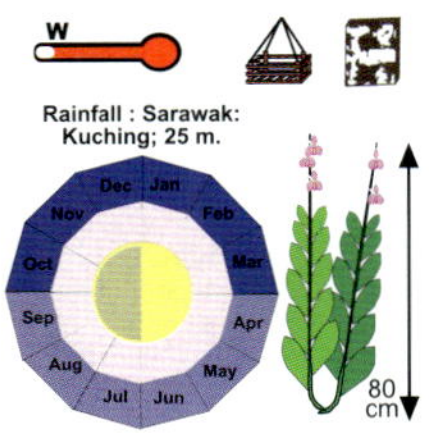


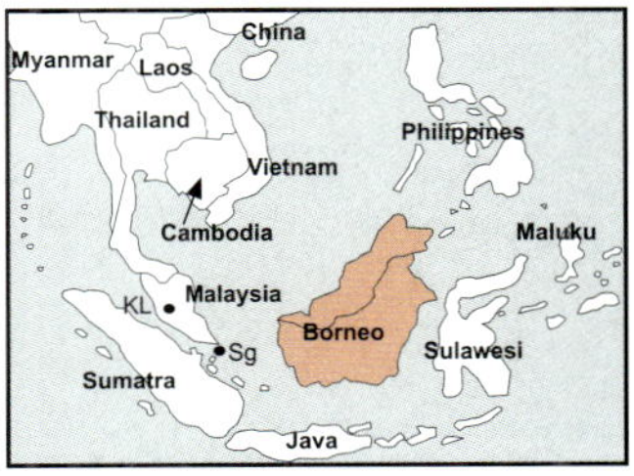

## *Dendrobium bicaudatum* Reinw. ex Lindl. Section: *Spatulata*

This is one of the smaller and less well-known members of this popular section. It grows on the upper branches and forks of trees in the lowland rainforest, and is fairly common in some areas. The 40 cm long stems are slightly swollen in the lower leafless third, then taper gradually to the apex. A few 10-20 cm long inflorescences are formed at a time, each with up to 10 flowers that last several weeks. The 3-4 cm tall flowers are rather variable in colour; in some specimens the petals and sepals have more green or purple than the illustrated plant.

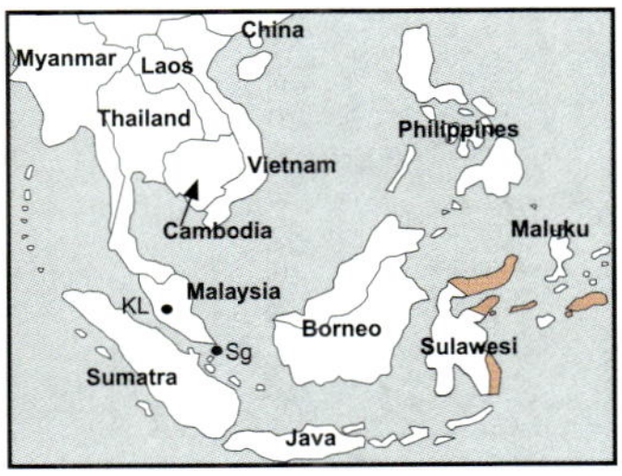


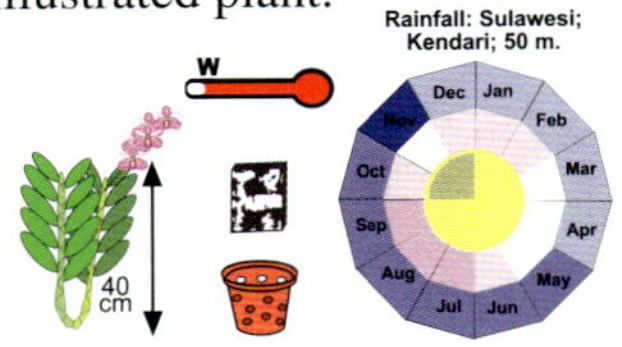


## *Dendrobium brinchangense* Holttum Section: *Pedilonum*

This rare species appears to be confined to one mountain in the Malaysian central range, where it grows on mossy trees and shrubs from 1800 - 2200 metres altitude. It forms tufts of rather slender stems that are leafy for most of their length; the sheaths (and often the undersides of the leaves) are purple. One to three flowers are produced on each very short inflorescence; the flowers are 15 mm long and last for 7-10 days. Some flowers are tinged purple on the outside.

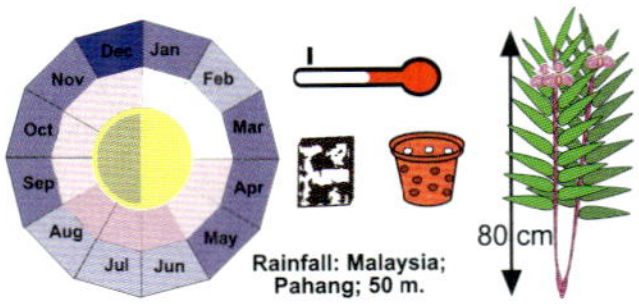


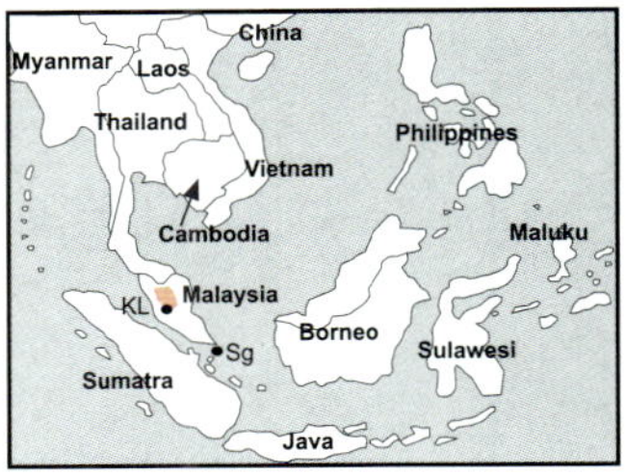

# D

## *Dendrobium bullenianum* Rchb.f. Section: *Pedilonum*

This dramatic species has a scattered distribution, growing in regions that have a distinct dry season. It forms stout stems that lose their leaves after 2 years, after which they start forming flowers. Many short inflorescences are formed, each crowded with 2 cm long striped yellow to orange flowers that last 10-14 days.

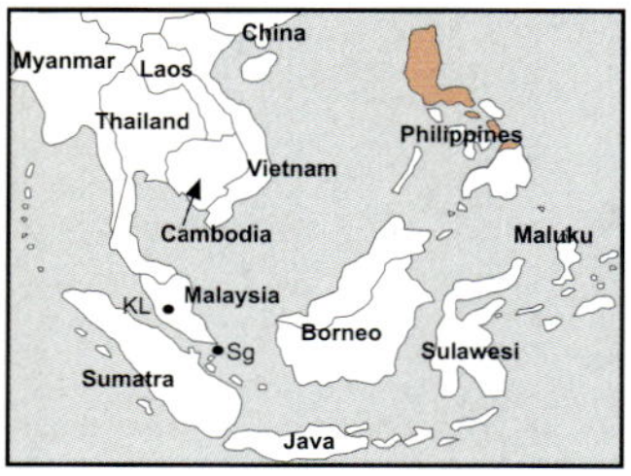


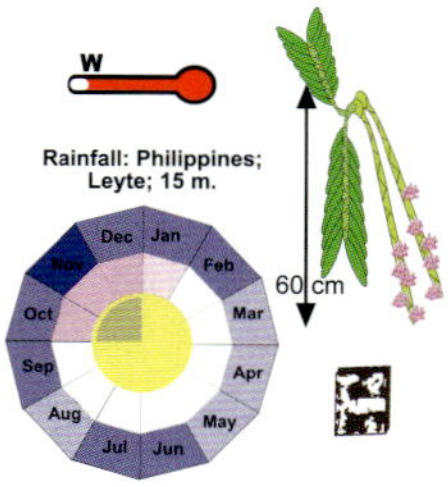


## *Dendrobium busuangense* Ames Section: *Spatulata*

This rare species was considered "lost" until very recently, when it was re-discovered on Palawan Island, growing on a ridgetop tree at about 300 metres altitude. The inflorescences are up to 70 cm long with 15-25 large flowers that last 6-8 weeks. The species show some variation in colour; some plants have pale lavender sepals and lavender-tinged greenish petals.

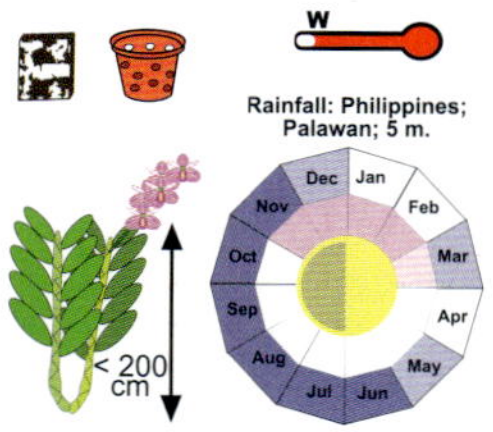


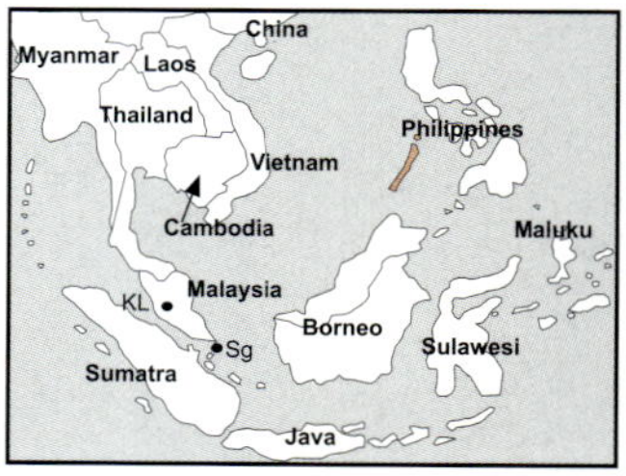

## *Dendrobium christyanum* Rchb.f. Section: *Formosae*

This diminutive species grows on rocks and trees in deciduous woodland between 1000-1300 metres altitude in regions with a long dry season. The small stems have a few large leaves that have small black hairs on their sheaths and undersides. The fragrant single flowers are 5 cm across and last for 2-3 weeks.

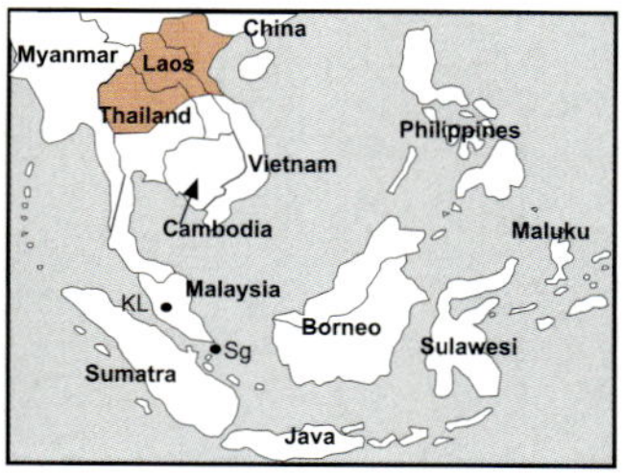


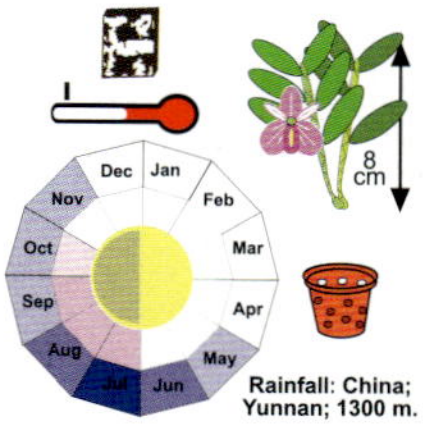

# D

## *Dendrobium chrysotoxum* Lindl. Section: *Callista*

This species is a favourite garden plant in the villages of Northern Thailand and Laos and provides a splash of bright colour at the end of the long dry season. It grows in deciduous forests from 400-1600 metres altitude. The arching inflorescences carry several 5 cm wide fragrant flowers that last 2 weeks.

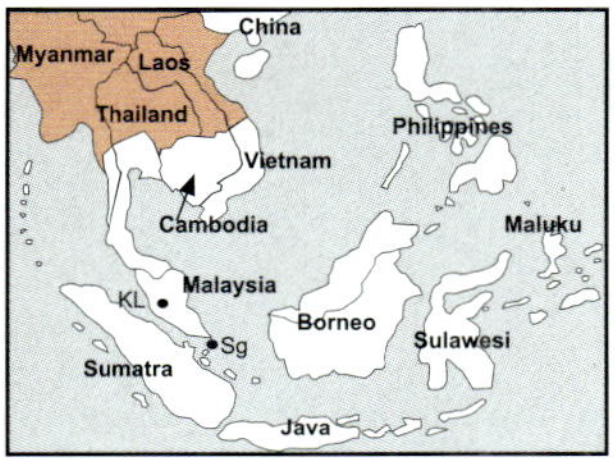


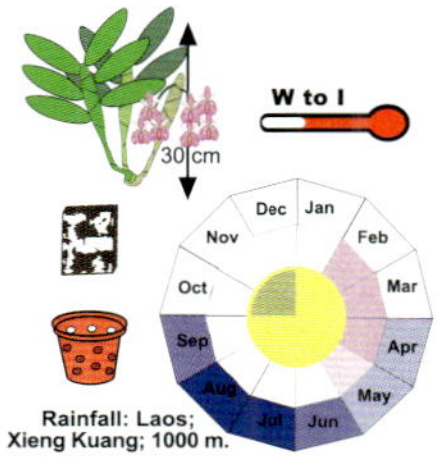

## *Dendrobium cinerium* J.J.Sm. Section: *Calcarifera*

This little-known species has been reported from stunted ridge-top forest and montane forest at 900-1500 metres altitude, but has recently been found growing on trees in heath-forest at 350 metres altitude. It has long grooved stems with oblong leaves that are maroon underneath and grey-green above; the leaves fall after 1 year. The 1-2 flowered inflorescences are formed on the leafless stems; the petals on the 25 mm wide fragrant flowers slowly recurve as the flower ages.

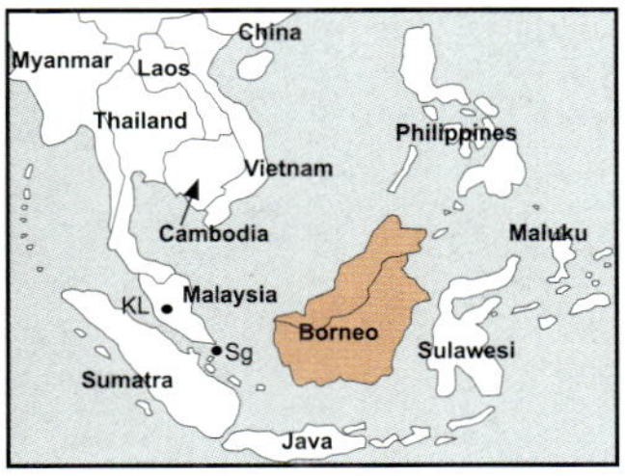


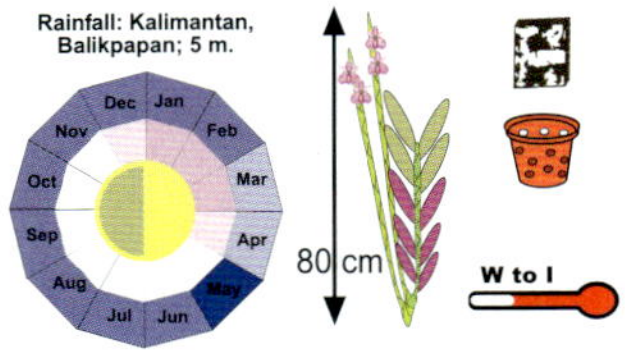


## *Dendrobium consanguineum* J.J.Sm. Section: *Grastidium*

This free-flowering species is related to the New Guinea species *D. insigne*, but lacks bristles on the side of the lip. It forms erect tufts of slender stems that are leafy to the apex and retain their leaves for several years. Throughout the year a steady succession of twin-flowered inflorescences emerge from thorn-like bracts on the stem. The 3 cm flowers last for 1 day.

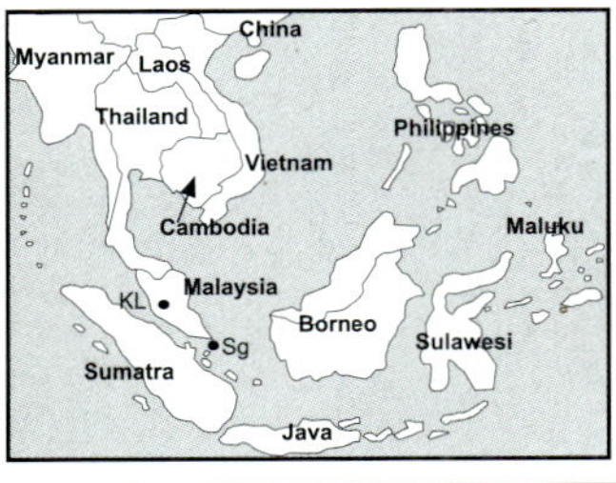


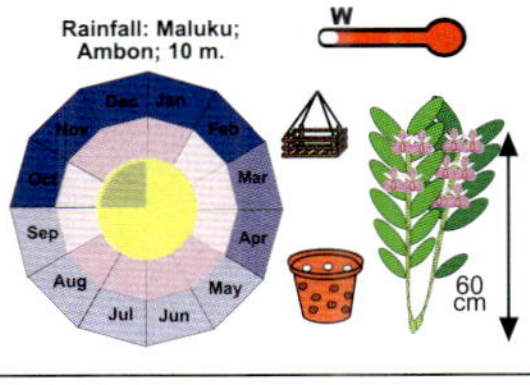

# D

## *Dendrobium corrugatilobum* J.J.Sm. Section: *Conostalix*

Almost nothing is known about this extremely rare species, which has recently been 'rediscovered' after being 'lost' for almost a century. Unlike other species in the section, it has an inflorescence with several flowers, 1-3 of which are open at a time. The distinctive 3 cm wide flowers are held upside-down and have a thick mealy layer on the midlobe.

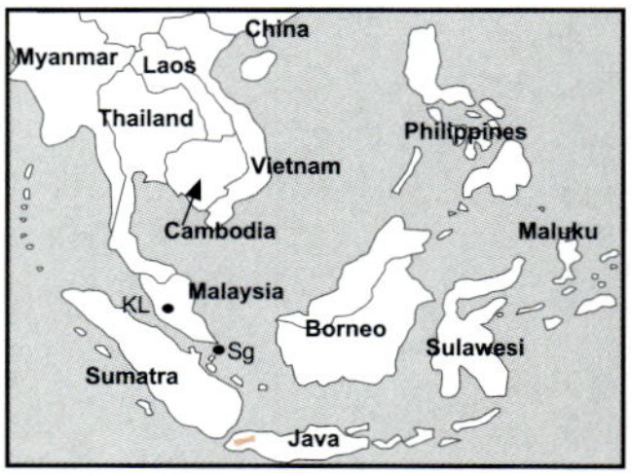


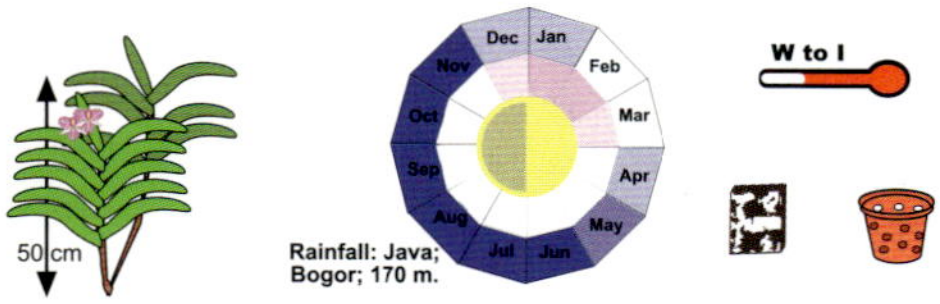


## *Dendrobium crocatum* Hook.f. Section: *Calcarifera*

This easily-identified species is found on trees near water in the lowland forests. It forms clumps of long slender stems which are usually purple to maroon. The deciduous leaves are often of a similar colour. Several 5 cm long inflorescences are formed at a time; they carry up to 8 bright orange flowers that are 4-5 cm long and last 7-10 days.

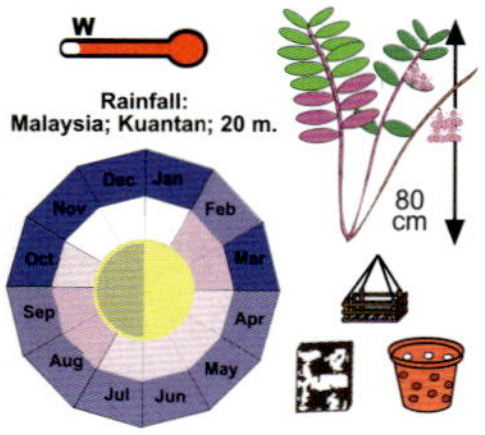


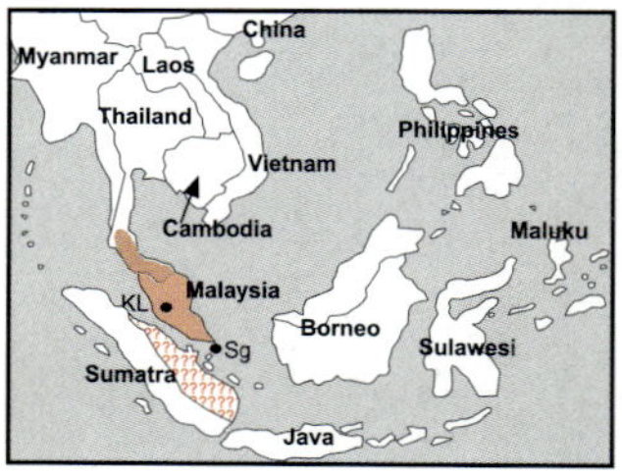


## *Dendrobium cruentum* Rchb.f. Section: *Formosae*

This well-known species is endangered in the wild and is listed on CITES Appendix 1. It is common in cultivation, where it is prized for its free-flowering qualities. It has frequently been used in hybridization programs. It has a tidy growth habit; both the stems and leaves are covered in small black hairs. The short inflorescences carry 1-3 fragrant flowers that are 4-5 cm wide and last for 4-5 weeks.

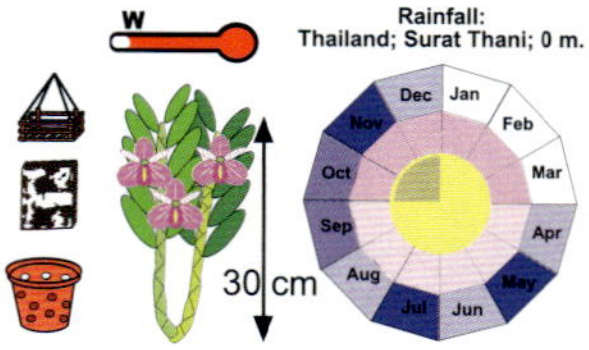


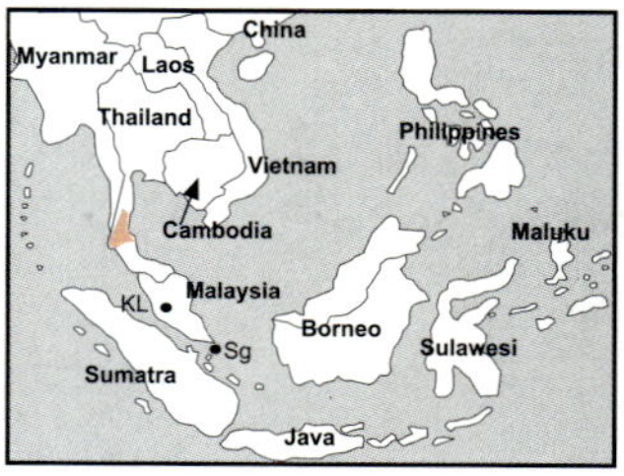

# D

## *Dendrobium crumenatum* Sw.

Section: *Rhopalanthe*

The "Pigeon Orchid" is probably the most common orchid species in S. E. Asia. In some areas it is found on every large tree, forming untidy curtains that hang from forks and branches. Flowering occurs 9 days after a heavy rainstorm causes a sudden drop in temperature; every plant in the area produces quantities of fragrant flowers that last for less than a day.

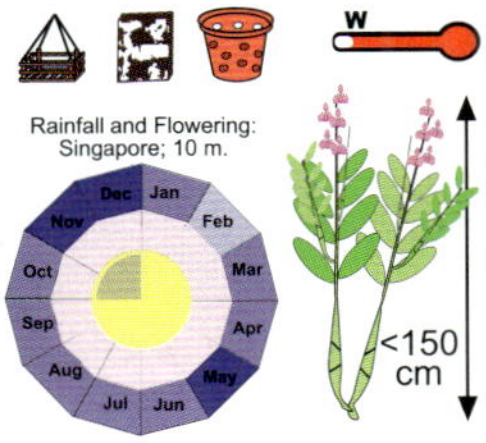

## *Dendrobium dearei* Rchb.f. Section: *Formosae*

This lowland species has a long history in cultivation, where it produces thin-textured but long-lasting flowers through most of the year. Plants from the Philippines (left side of photo) can be 90 cm tall but are usually much shorter. Plants from Borneo are much larger (up to 110 cm tall) with longer leaves and inflorescences and have minor differences in the lip shape (right side of photo).

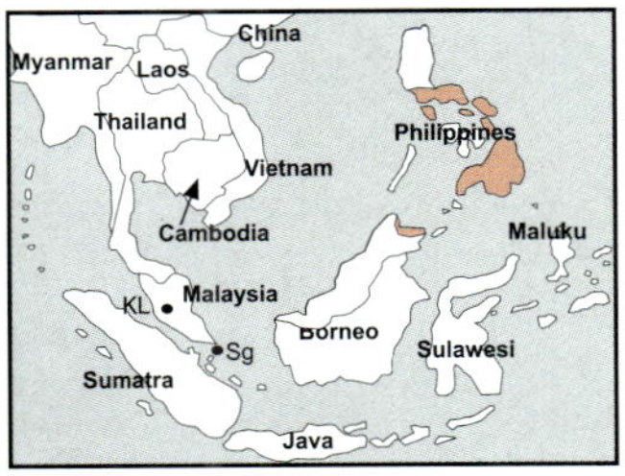


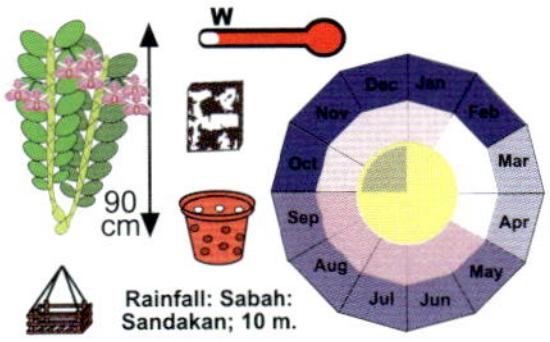


## *Dendrobium draconis* Rchb.f. Section: *Formosae*

This beautiful species is widespread in forests from Vietnam to the Himalayas between 300-1300 metres altitude. The leaf sheaths and young leaves are covered in small black hairs. It usually produces several inflorescences at a time, each with 2-5 flowers that are 5 cm wide and last for 2-3 weeks.

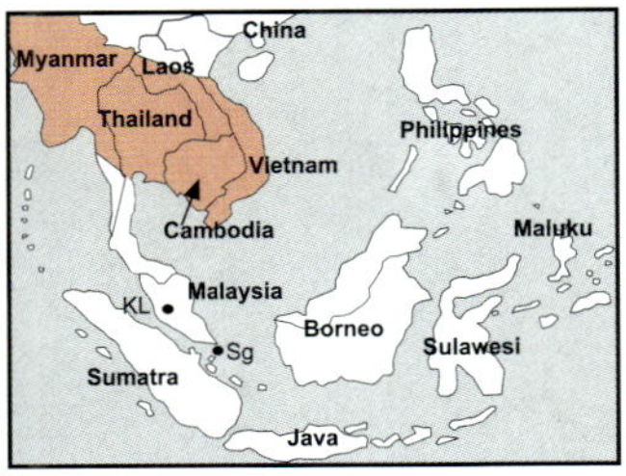


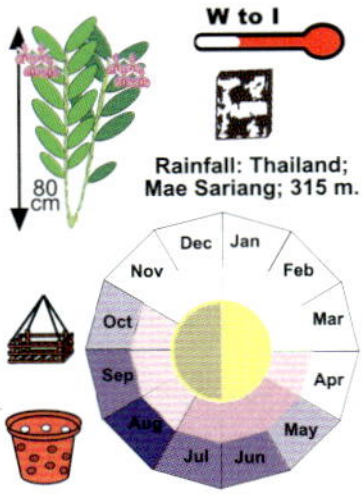

# D

## *Dendrobium ellipsophyllum* Tang & Wang

Section: *Distichophyllum*

This species, which is often confused with *D. uniflorum*, grows on tree branches in forests and deciduous woodlands from 300-1000 metres altitude. It is a tall-growing species with rather short broad leaves along most of the stem. In cultivation it grows and flowers continuously throughout the year and is seldom without a display of 2 cm long flowers that each last several weeks.

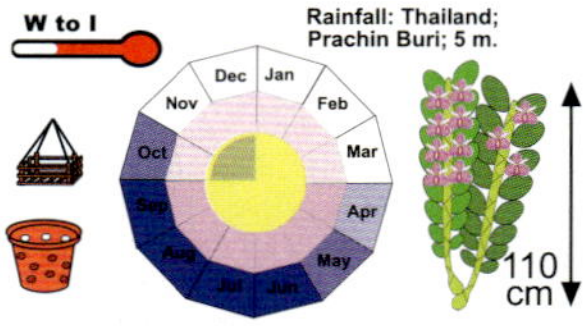


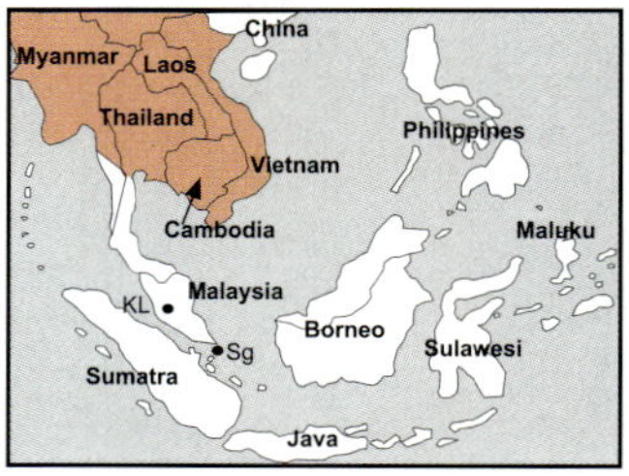

## *Dendrobium endertii* J.J.Sm. Section: *Calcarifera*

This uncommon species can be found from 400-1100 metres altitude, growing on trees near water in rainforest and montane forest. It is a medium-sized species with thin leaves that drop after about 1 year. The rather short descending inflorescences have up to 10 flowers that are 4 cm long and last 7-10 days.

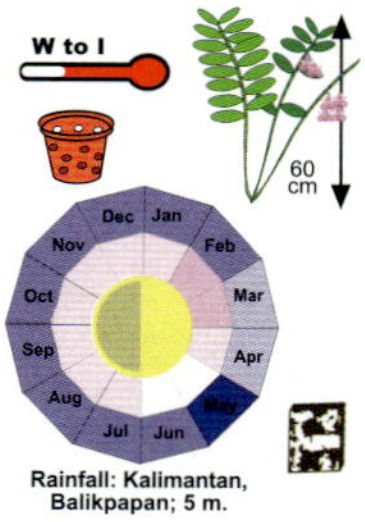


## *Dendrobium farmeri* Paxton Section: *Callista*

This free-flowering species can be found from sea-level to about 1000 metres altitude, growing in deciduous woodland or rainforest, usually in situations that remain shady throughout the year. The 15-20 cm long inflorescence has up to 40 flowers that are 3 cm wide and last for a week. The flower colour varies from white to pink, always with a yellow to orange lip.

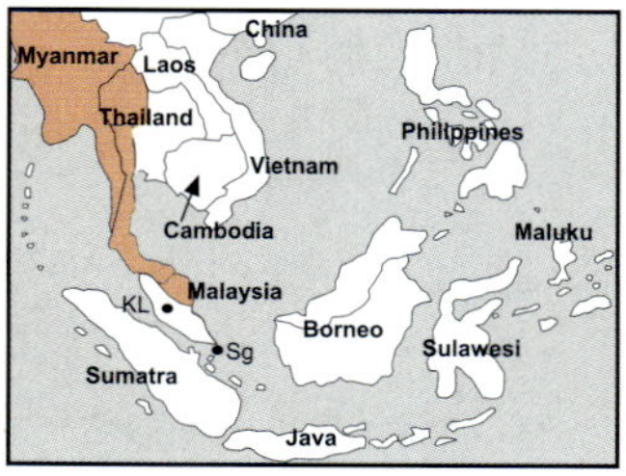


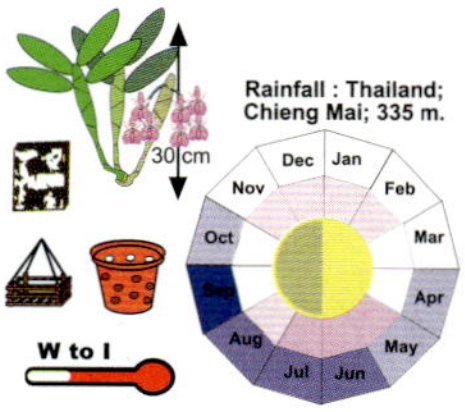

# D

## *Dendrobium friedericksianum* Rchb.f. Section: *Dendrobium*

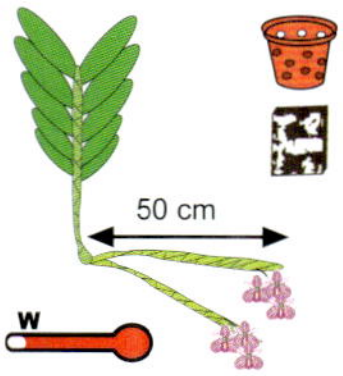


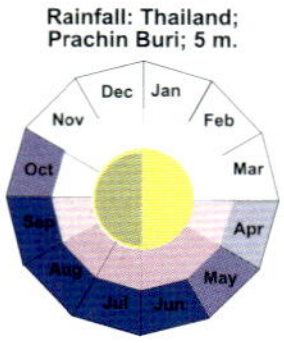


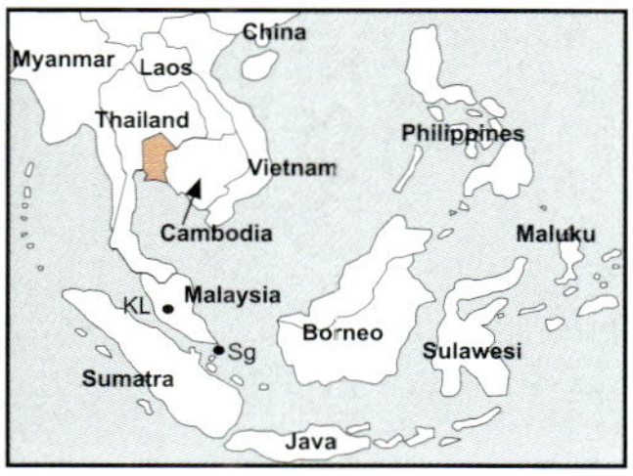


This large-flowered species grows on trees in semi-deciduous lowland forests in regions with a distinct dry-season. The tall stems are initially leafy and erect, but droop and drop their leaves after 1 year. The inflorescences are formed on bare stems and have up to 5 large flowers that last about a month. Specimens with 2 dark patches on the lip are often called "var. *oculatum*".

## *Dendrobium glomeratum* Rolfe
Section: *Calyptrochilus*

Most books place the species in section *Pedilonum*, but the upturned fringed lip apex is characteristic of *Calyptrochilus*. Little is known about this uncommon species, which is frequently confused with a series of New Guinea plants. This dramatically-coloured form occurs in Maluku; it was originally described as *D. crepidiferum* and is occasionally grown under that name. The flowers are 3 cm wide and last for 7-10 days.

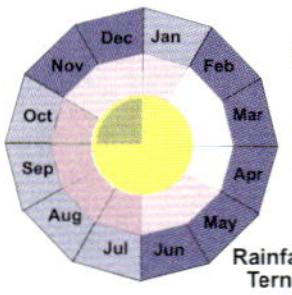


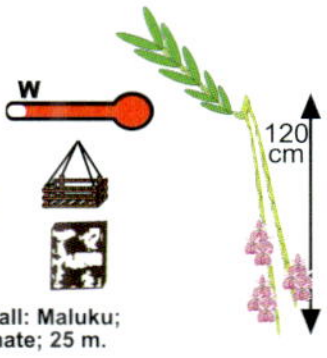


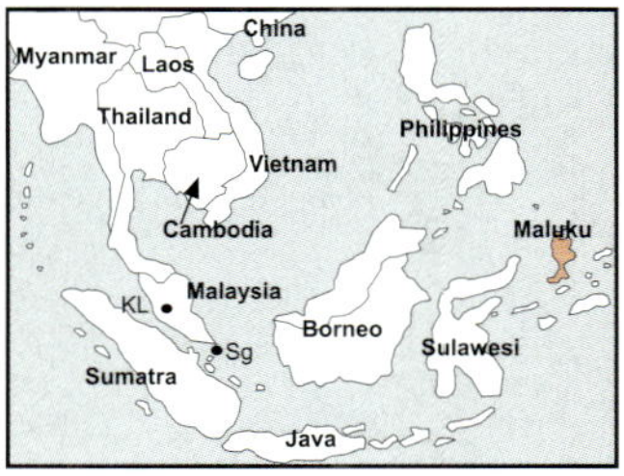

## *Dendrobium goldschmidtianum* Krzl.
Section: *Pedilonum*

This species is found growing from sea-level to about 300 metres altitude on forest trees in Batan Island in the Philippines and is also found in Taiwan. It is quite common in cultivation, often under the old name *D. miyakei*. The stems lose their leaves after 1 year, then produce clusters of 15-20 mm wide flowers that last for about 1 week.

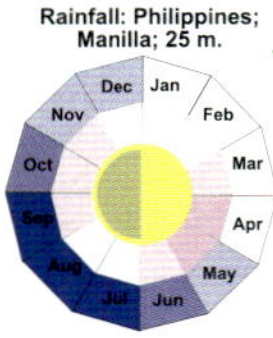

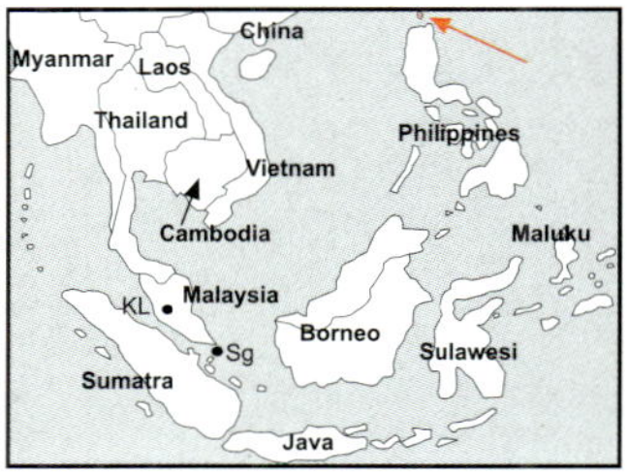

# D

## *Dendrobium grande* Hook.f.
Section: *Aporum*

This species is found growing on the upper branches of tall forest trees from sea-level to about 1000 metres altitude, and is quite common in some areas. It has completely flattened leaves that cover the wiry stems. The inflorescences push through the leaf-bases and produce a succession of 12 mm long waxy flowers that last up to 2 weeks. As the name suggests, this species has very large flowers for this section.

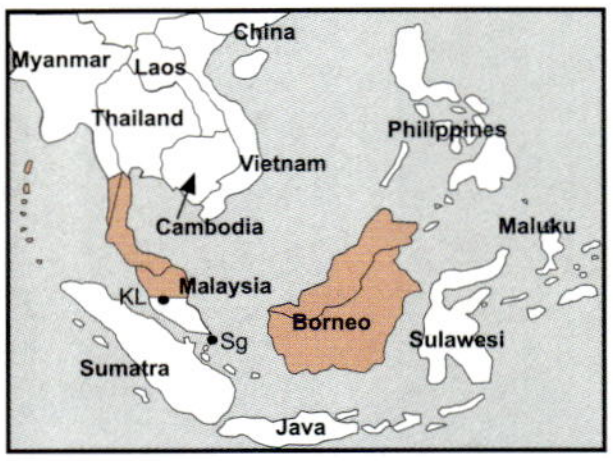


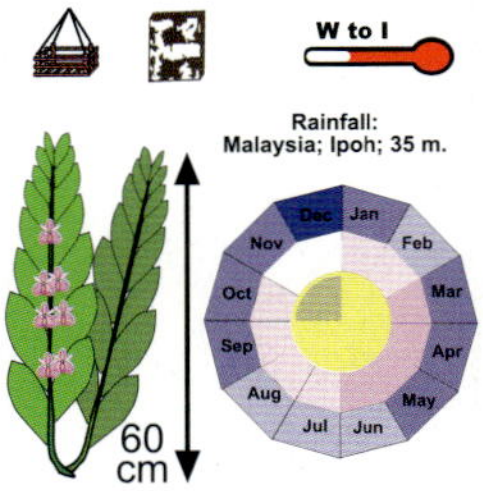


## *Dendrobium hallieri* J.J.Sm.
Section: *Formosae*

Little is known about this dramatic species, which was 'rediscovered' recently after being 'lost' for 85 years. It grows in Kalimantan, but nothing is known about the locality or environment. It has zig-zag stems that are 5-7 mm in diameter. The sheaths and young leaves are covered in short black hairs.The 2-4 cm tall inflorescence carries up to 4 stiff waxy 4 cm high flowers that last 10-14 days.

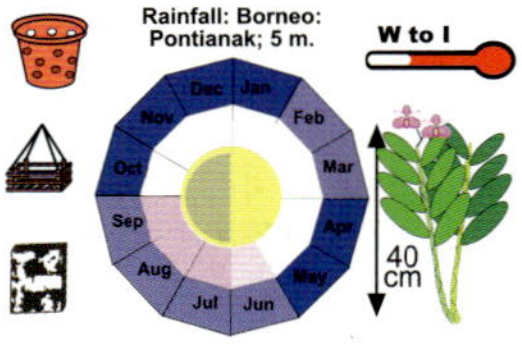

## *Dendrobium hepaticum* J.J.Sm. Section: *Distichophyllum*

Little is known about this rare species, which is related to *D. uniflorum*. It has finger-thick grooved stems that grow continuously for several years and start flowering when about 20 cm tall. Single-flowered inflorescences are produced throughout much of the year. Each flower is about 2 cm wide and lasts for several weeks, gradually changing colour from cream to liver-brown as it ages.

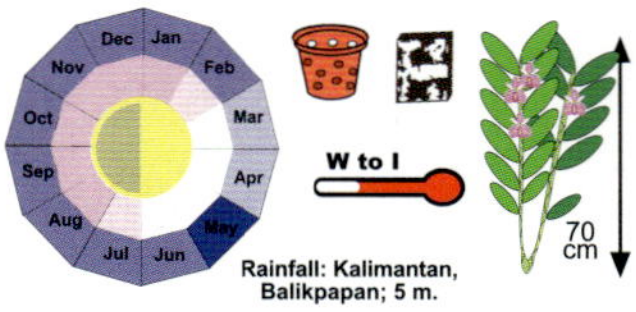


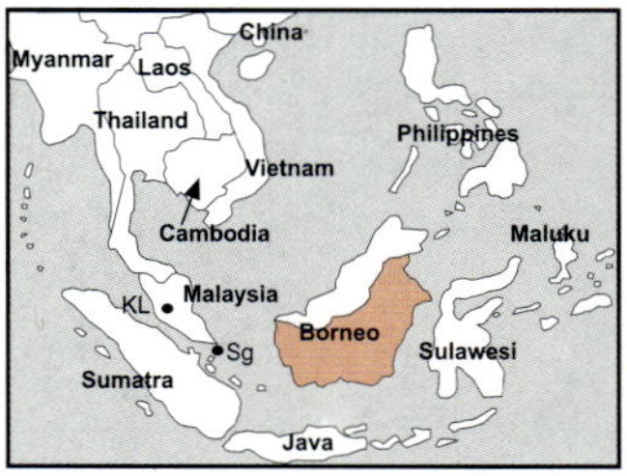

# D

## *Dendrobium hercoglossum* Rchb.f. Section: *Breviflores*

This small species is usually found growing on forest trees in regions with a distinct dry season. In cultivation it grows rapidly and flowers frequently. It has club-shaped stems with thin leaves that drop after 1-2 years. Many short inflorescences are formed at a time, each with up to 5 delicate flowers that last about 7 days.

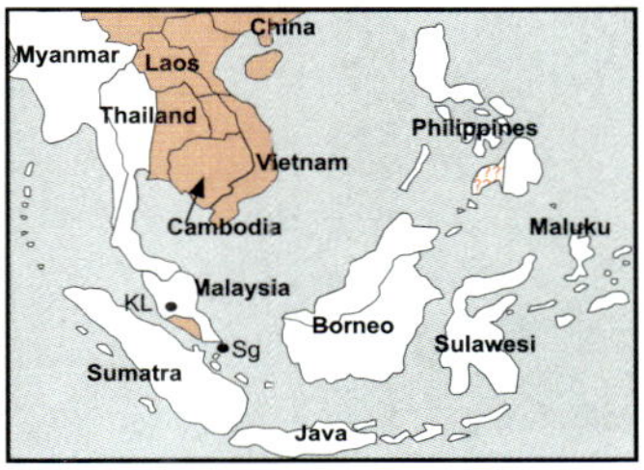


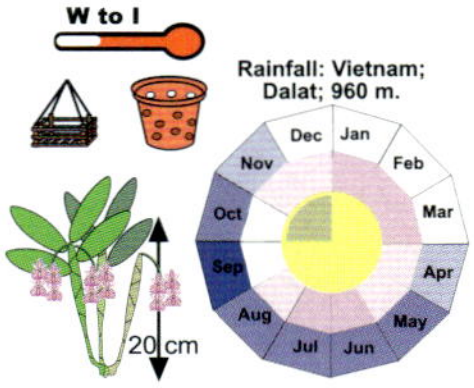


## *Dendrobium hymenanthum* Rchb.f. Section: *Bolbidium*

This miniature species is widely grown under the name *D. quadrangulare* Par. & Rchb.f. It is frequently confused with *D. pachyphyllum*, which usually has pink-striped flowers. It has a short 4-angled stem with 1-2 tough leathery leaves at the tip. A succession of single flowers are formed at sporadic intervals. The illustrated plant is unusual in having red lip-margins.

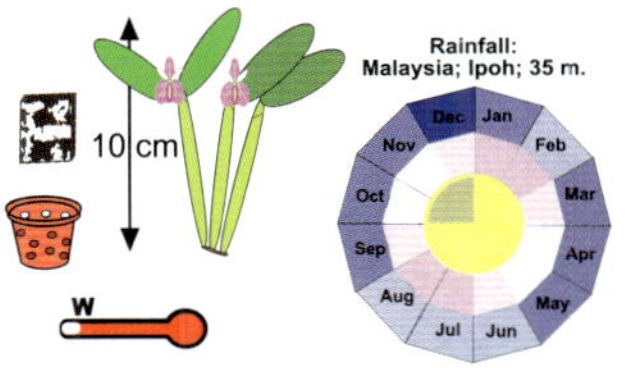


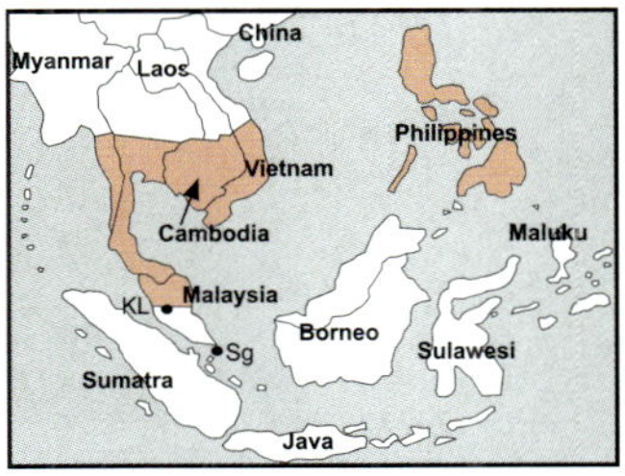

## *Dendrobium indragiriense* Schltr. Section: *Grastidium*

This species grows in light shade or full sun on trees from sea-level to about 300 metres altitude. It has slender grasslike stems and leaves. A drop in temperature, caused by heavy rain, triggers the production of the twin-flowered inflorescences. Each small flower lasts for one day.

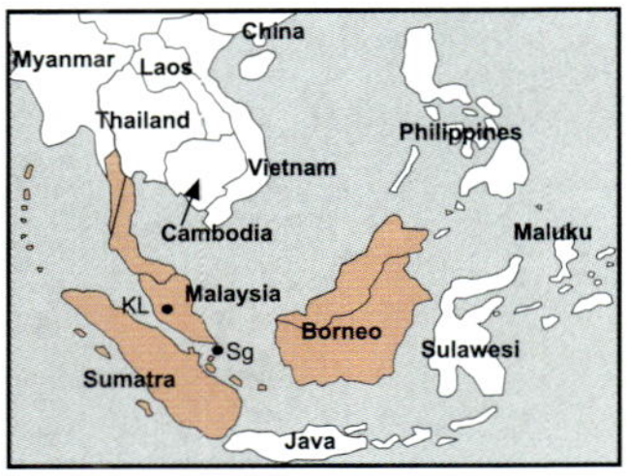


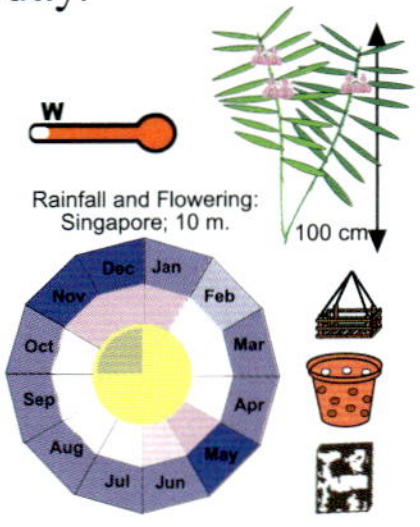

# D

## *Dendrobium ionopus* Rchb.f.
Section: *Calcarifera*

This species grows in montane forests at around 700-900 metres altitude. It is known to growers as *D. epidendropsis* and has long been confused with *D. guerreroi*. It has slender stems with deciduous leaves that are purple when young. The inflorescence has 10-15 flowers that are 5 cm long and last for 7-10 days.

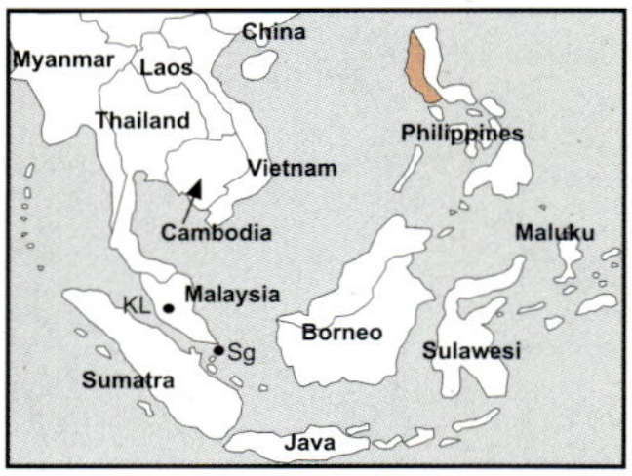


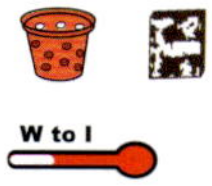


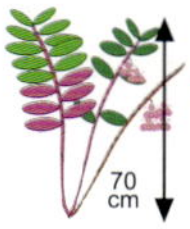


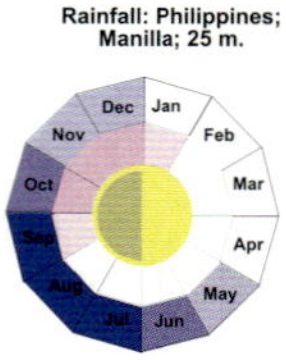


## *Dendrobium kenepaiense* J.J.Sm.
Section: *Distichophyllum*

This rare lowland species, which is related to *D. revolutum*, has slender rather 4-angled stems that grow continuously for many years, eventually reaching a length of over 1 metre. Once a stem starts producing flowers (at about 30 cm long) it flowers continuously and seldom has less than 3-4 flowers open at any time. The pea-shaped single flowers are 25 mm long and last for many weeks.

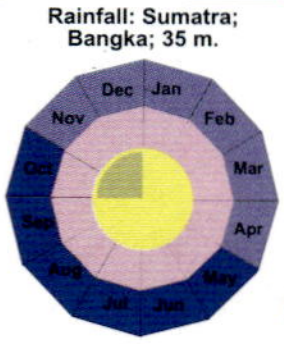


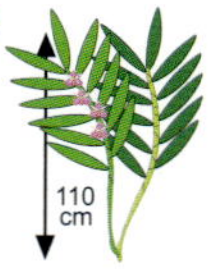

## *Dendrobium kruiense* J.J.Sm. Section: *Calcarifera*

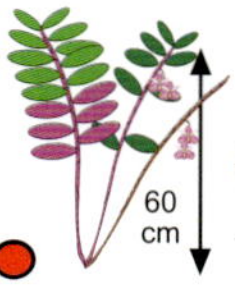


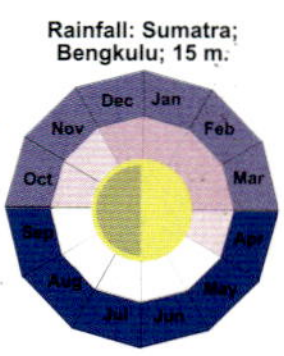


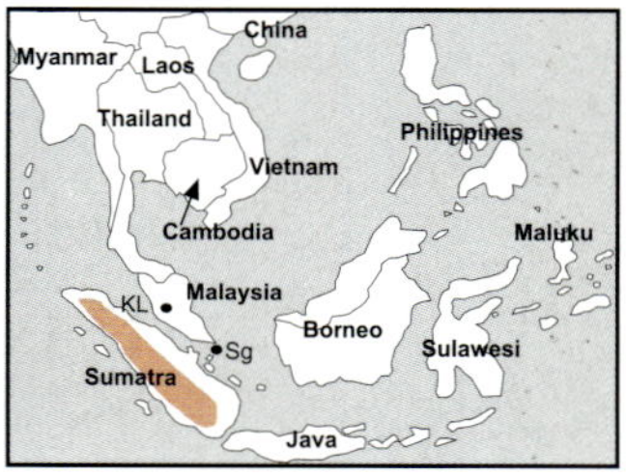


This species is quite common in the mountains of Sumatra, where it grows on trees in montane and secondary forests. It has long grooved stems that are about 4-5 mm diameter. The narrow deciduous leaves are purple on the underside when young. The short descending inflorescence carries 1 or 2 flowers that are 2.5 cm wide and last about 7 days. The white flowers are suffused a very delicate lavender.

# D

## *Dendrobium lamellatum* Bl. Section: *Calcarifera (Platycaulon)*

This common species grows on forest trees from sea-level to 1000 metres altitude. It is frequently confused with *D. platygastrium* and *D. platycaulon*. It is an extremely variable species, and probably deserves to be split into several new taxa. It has 3-15 cm long stems shaped like a flattened club, with 2 rows of thin deciduous leaves. It is a very free-flowering species, producing large numbers of short inflorescences which bear up to six flowers that are 2 cm long and last about 7 days.

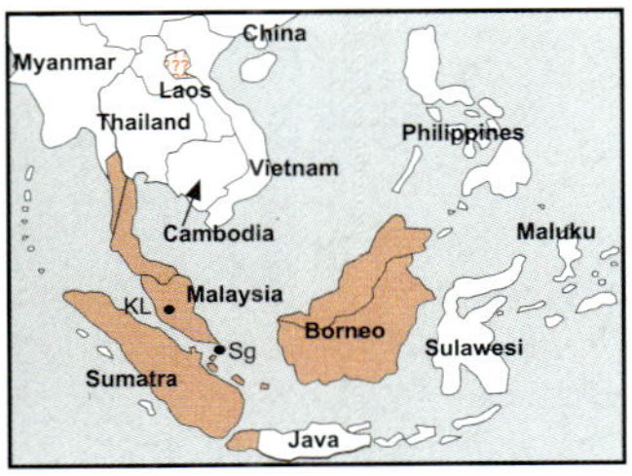


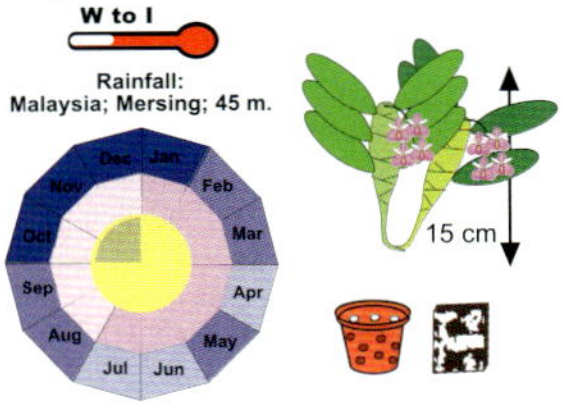


## *Dendrobium lancifolium* A.Rich. Section: *Calcarifera*

This unusual species is common in parts of the Sulawesi Highlands, where it colonises moist grasslands and road-cuttings. It also grows as an epiphyte on tree branches. It has rather brittle branching stems and short inflorescences that produce a succession of 3 cm wide flowers throughout the year.

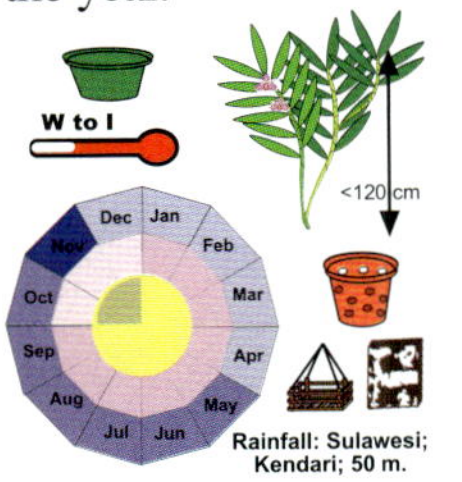


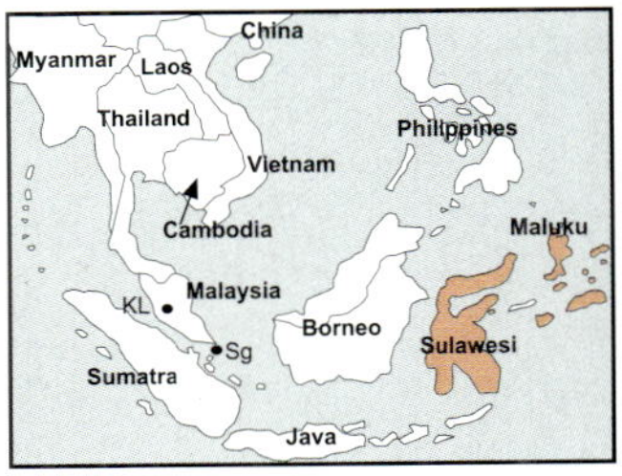

## *Dendrobium lanyaiae* Seidenfaden Section: *Dendrobium*

This small species, which is sometimes sold as *D. unicum* or *D. arachnites*, can be distinguished from these closely-related species by having much larger flowers and a narrow lip that has a tall central keel. It has very short stems with up to 6 deciduous leaves, and flowers freely after the leaves have fallen. The bright flame-red flowers are about 6 cm wide and last for up to 10 weeks.

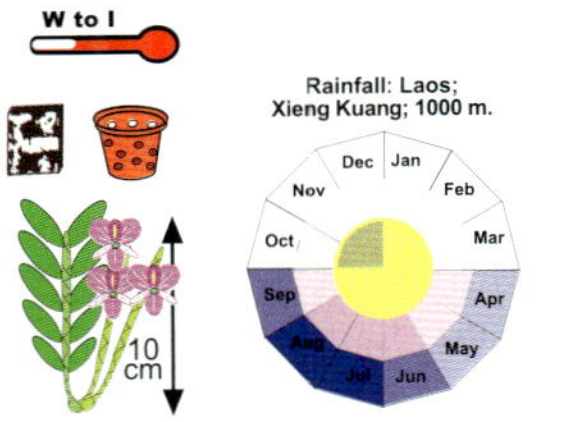


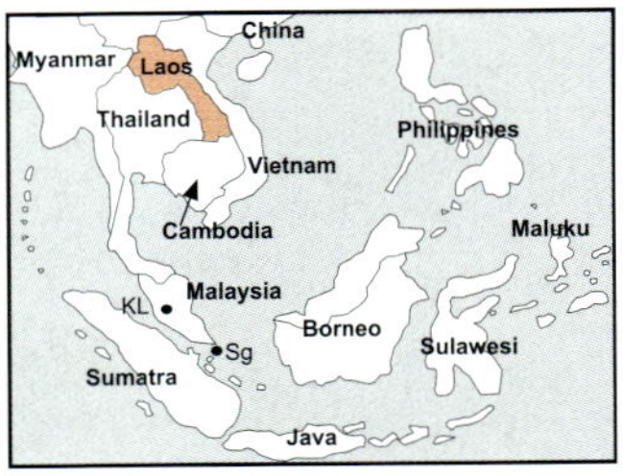

# D

## *Dendrobium leonis* (Lindl.) Rchb.f. Section: *Aporum*

This common lowland species grows on trees in rainforest and cultivated land. It has a short stem obscured by laterally-flattened leaves. The inflorescence at the stem apex produces a succession of single honey-scented flowers.

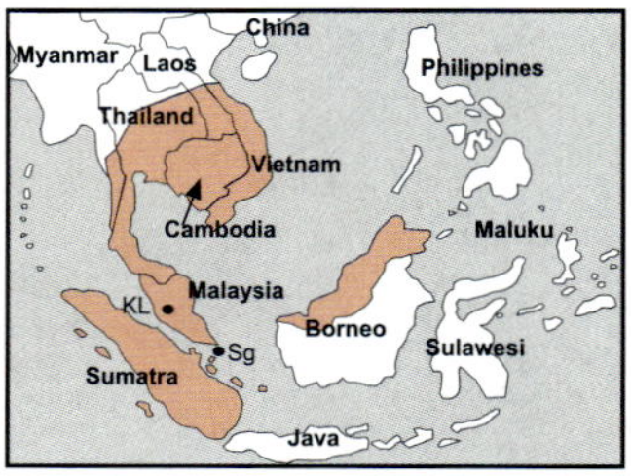


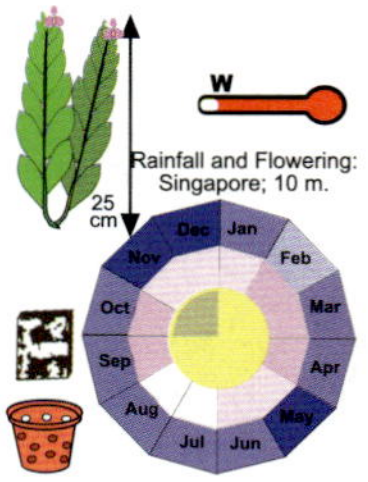

## *Dendrobium leporinum* J.J.Sm. Section: *Spatulata*

This showy large-flowered species is less well-known and more compact than most others in the section. It grows in the lowlands of Halmaheira and Irian Jaya, but otherwise little is known about its habitat. The cane-like stems are 40-80 cm tall and are swollen below the first leaf. Flowering occurs throughout the year; the inflorescences are up to 20 cm long and have up to 6 flowers that are 6 cm tall and last several weeks.

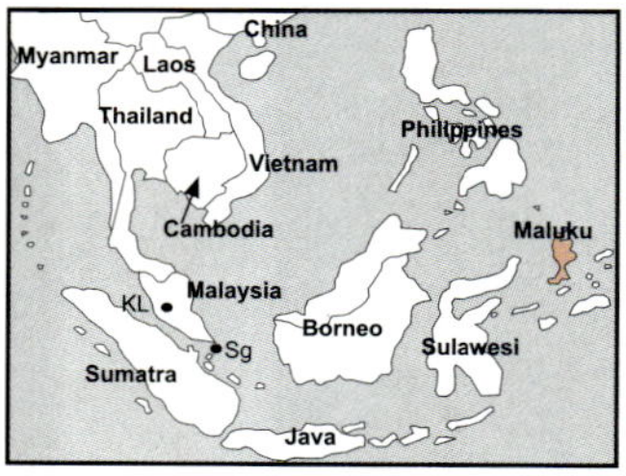


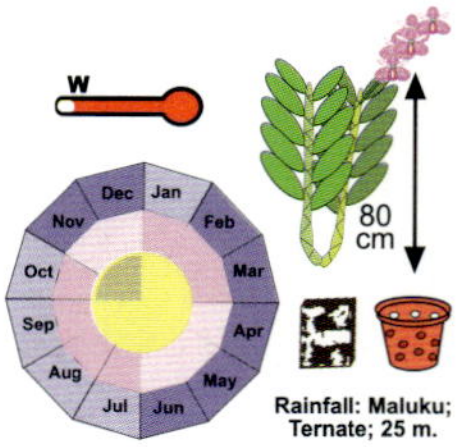


## *Dendrobium lindleyi* Steud. Section: *Callista*

This popular species is widely grown under the old name *D. aggregatum*. It is found on trees in deciduous forests from 300-1000 metres altitude, usually in areas where the dry season brings higher light levels. It forms clusters of short stout pseudobulbs, each with a single leathery leaf. The plants produce many pendulous inflorescences, forming a curtain of 3 cm wide fragrant golden-orange flowers that last for a week.

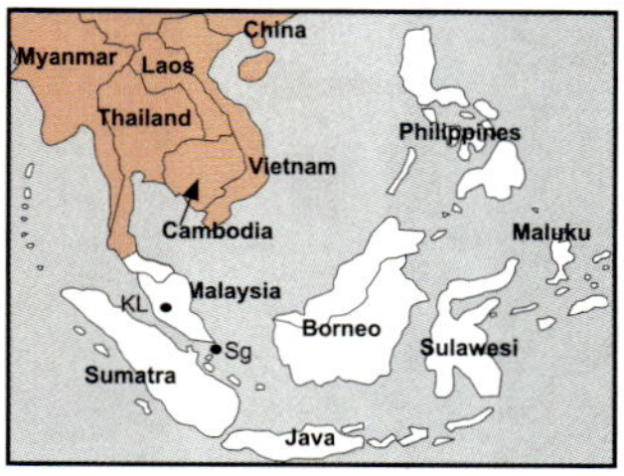


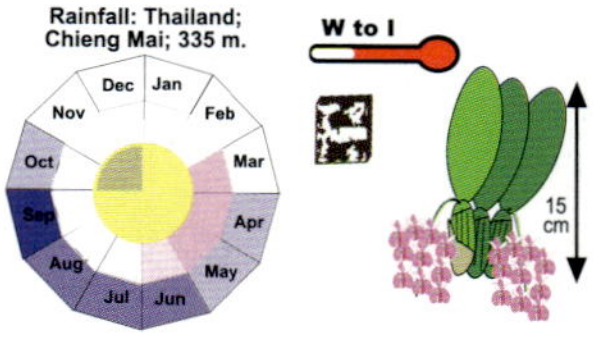

# D

## *Dendrobium linguella* Rchb.f. Section: *Breviflores*

This pendulous species is quite common in regions with a year-round rainfall. It grows on tree-branches from sea-level to over 1000 metres altitude. It has long slender stems that drop their leaves within 1 year, after which they produce short inflorescences of 2-6 flowers that are 2 cm wide and last for a week. The colour varies from white to pink or pale violet, with a pale yellow lip.

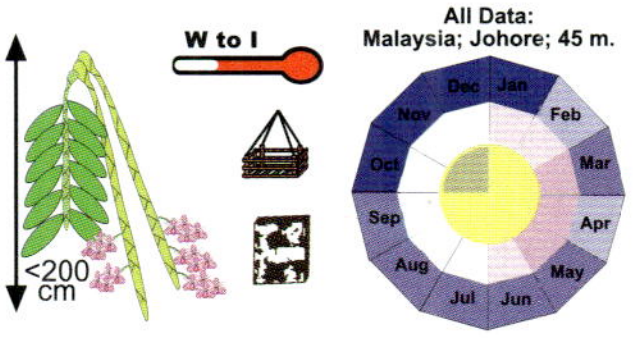


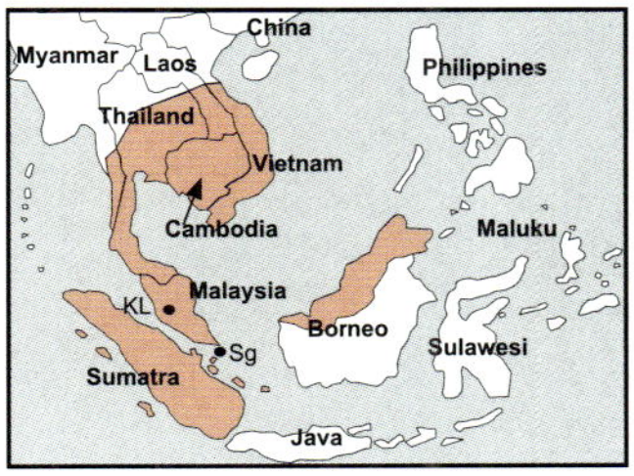

## *Dendrobium macrophyllum* A. Rich. **var. *ternatense*** Section: *Latourea*

This is a miniature form of the widespread species *D. macrophyllum*. Some authors consider it to be a good species called *D. ternatense* J.J.Sm. It grows on trees in cultivated land, parkland and other open areas from 500-800 metres altitude. It has short club-shaped stems with 2 tough leathery leaves. It flowers freely during the drier part of the year, producing short inflorescences of 1-6 flowers. The 3 cm wide flowers are covered in soft bristles and last 2-3 weeks.

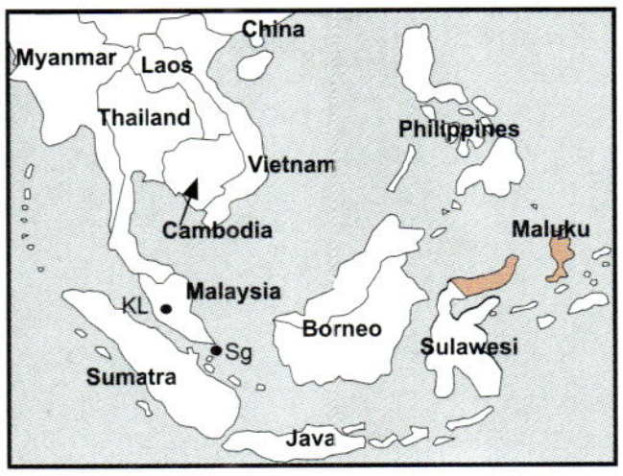


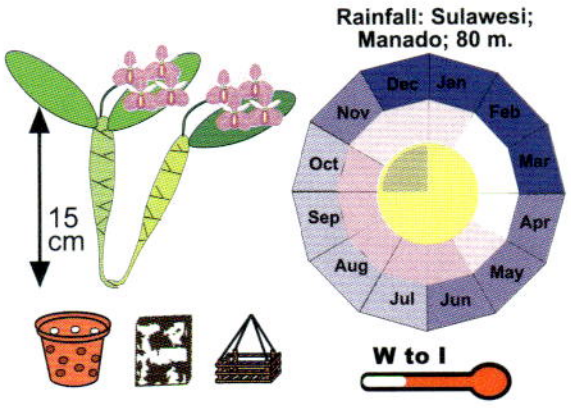


## *Dendrobium macrostigma* J.J.Sm. Section: *Grastidium*

This uncommon lowland species looks more like a *Vanda* than a *Dendrobium*. It has a woody cane-like stem with strap-shaped leaves. Several 2-flowered inflorescences are produced at a time, emerging in the leaf axils. The flowers are 2-3 cm across, strongly fragrant, but only last for 1 day.

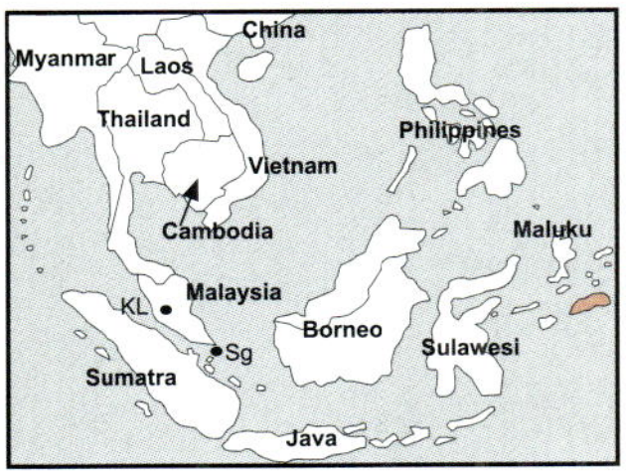


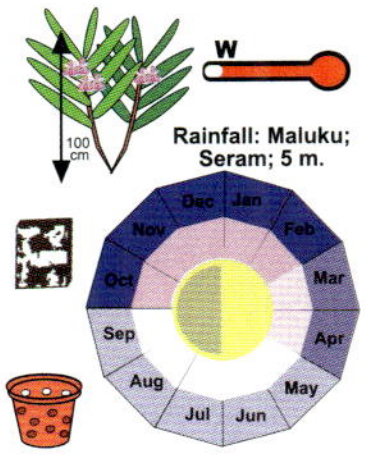

*Photo courtesy of S.Y. Alsagoff*

# D

## *Dendrobium malvicolor* Ridl. Section: *Pedilonum*

This delightful little species is common in montane and ridge-top forests above 1300 metres altitude in Sumatra, growing in locations ranging from full shade to fully exposed. It has narrow club-shaped stems with thin deciduous leaves. In Java it is rarer and much smaller, the stems usually being only 3-4 cm long. The short inflorescences bear up to 10 flowers in a tight cluster. The flowers are 3 cm long and last 2-3 weeks.

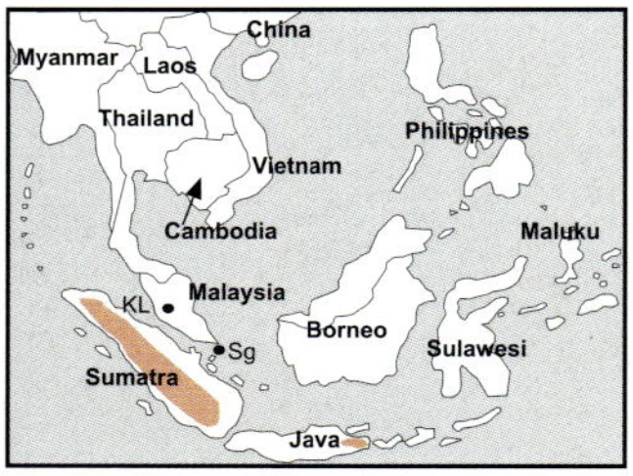


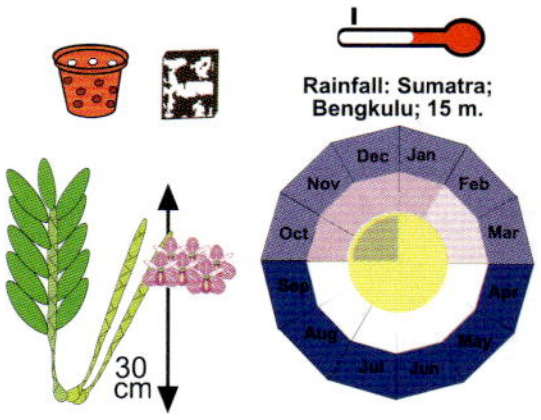


## *Dendrobium moquetteanum* J.J.Sm. Section: *Distichophyllum*

This uncommon species occupies a specialist niche in the rainforest, forming encircling masses on the exposed upper trunks of emergent trees, always well above the forest canopy layer. Its 4-angled bright yellow stems make it easy to identify. The stems grow and flower non-stop for several years. It is very hard to maintain in cultivation.

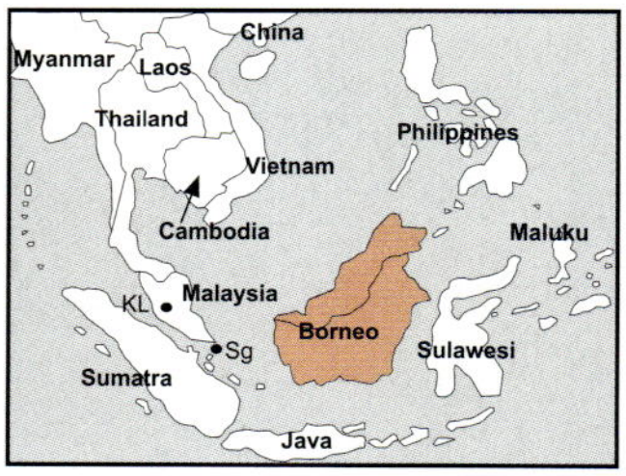


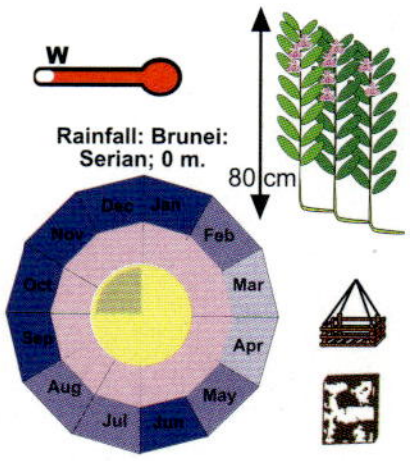


## *Dendrobium odoratum* Schltr.

Section: *Rhopalanthe*

This attractive species replaces *D. crumenatum* in North Sulawesi, where it is common from 100-800 metres altitude on trees in cultivated land, parkland and other open areas. It has robust erect wiry stems that are distinctly swollen just above the base. It produces large numbers of strongly scented flowers 10 days after a heavy rainstorm. The flowers are 3-4 cm wide and last for 2 days.

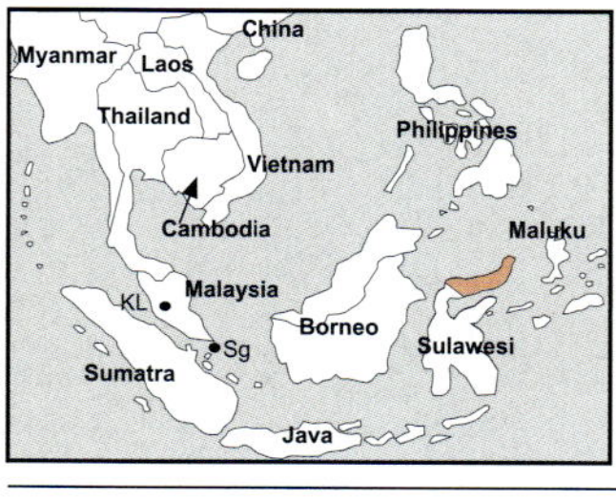


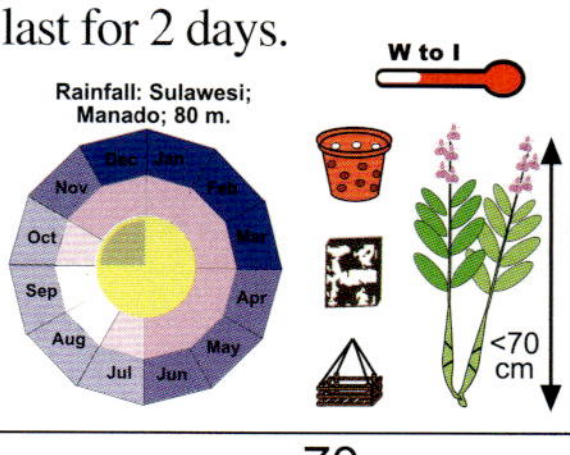

# D

## *Dendrobium oligophyllum* Gagnep. Section: *Distichophyllum*

This miniature species has recently become popular in cultivation (where it is often confused with *D. trinervium*) since it blooms through most of the year. It has stout zig-zagging stems that remain leafy for several years. The inflorescence has 1-4 flowers that are 2 cm wide and last for several weeks.

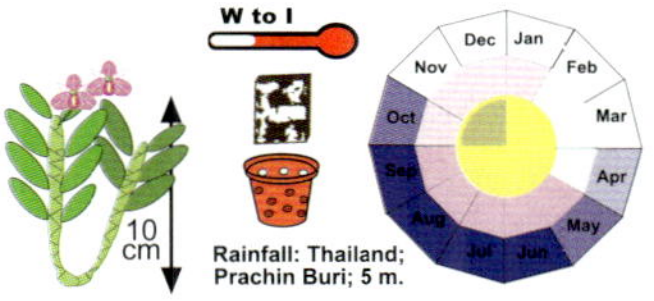


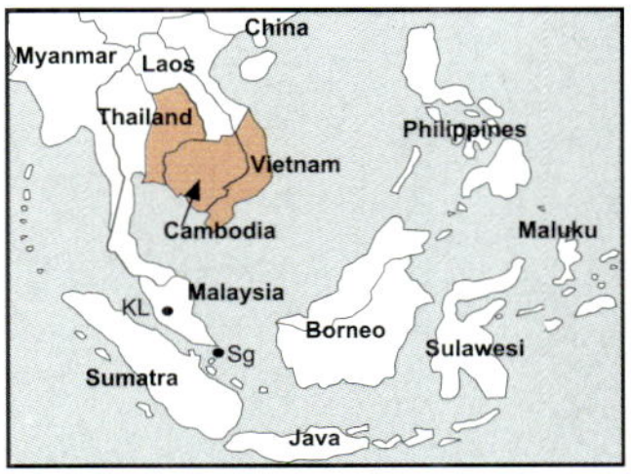

## *Dendrobium pachyphyllum* (Kuntz) Bakh.f. Section: *Bolbidium*

This common and widespread miniature species can be found from sea-level to at least 1500 metres altitude, growing on trees in well-lit positions such as roadsides, plantations, or village gardens. It has a creeping rhizome with close-set stems that form a dense mat. The club-shaped stems are usually 2- 4 cm long (longer if the plant is growing in shade) with 1-2 thick leathery leaves at the tip. The inflorescences are at the stem apex and produce a succession of single flowers that are about 15 mm long and last 1-2 days. Flowering appears to be triggered by sudden rainstorms.

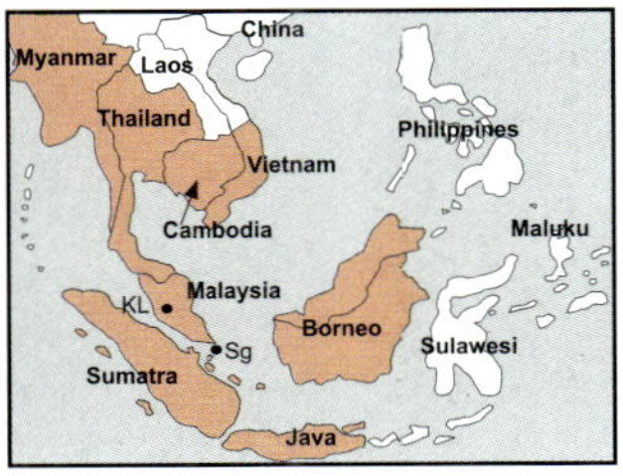


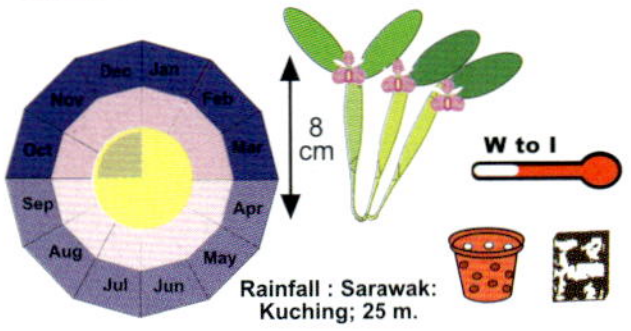


## *Dendrobium panduriferum* Hook.f.
Section: *Calcarifera*

This uncommon species can be found growing on forest trees at about 500-600 metres altitude. It has pencil-thick stems and large deciduous leaves. Inflorescences are formed on the leafless stems and have up to 12 flowers that are 3 cm long and last 7-10 days. This species can be distinguished from the similar *D. ionopus* and the Javanese *D. hymenophyllum* by the two ridge-like folds on the midlobe, and the small callus that sits between them.

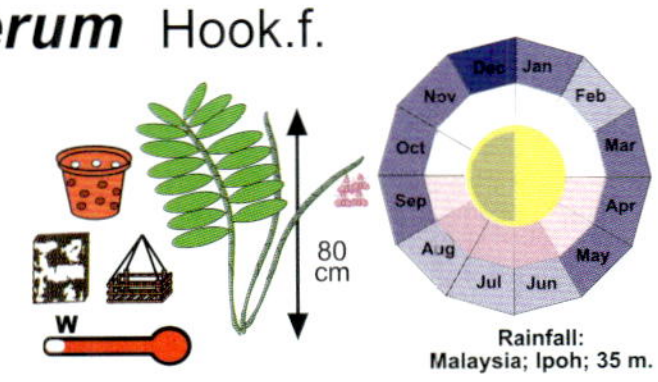


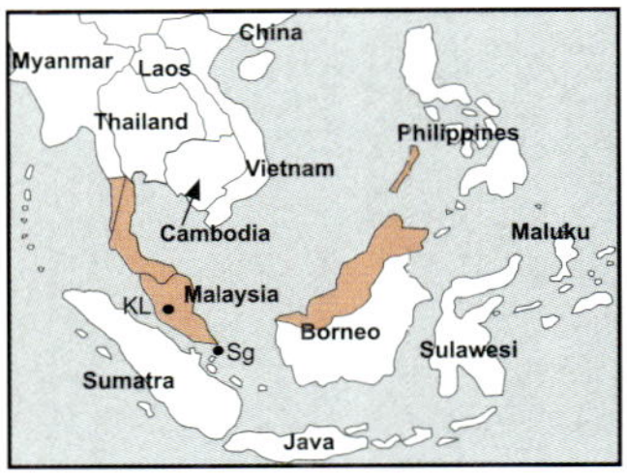

# D

## *Dendrobium philippinense* Ames. Section: *Rhopalanthe*

This lowland species can be found growing on trees in well-lit locations. It has long wiry stems with a few terete leaves. The stems are swollen just above the base and have a long leafless section above the leaves. Inflorescences on the upper stem produce a succession of flowers that are fragrant in the morning and last for 2-3 days.

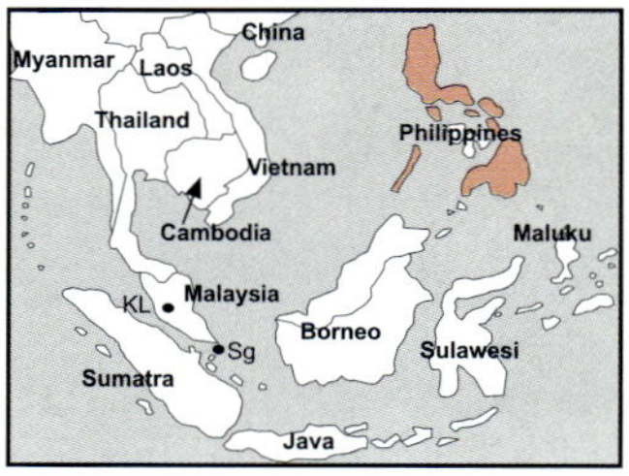


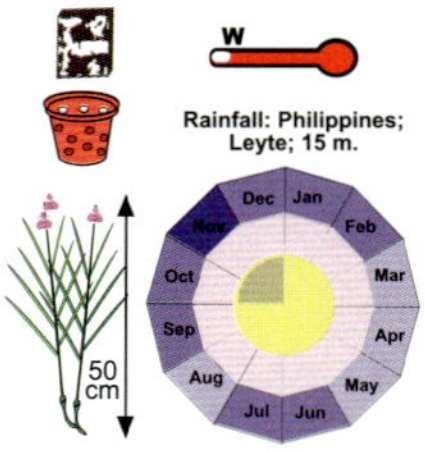

## *Dendrobium platycaulon* Rolfe Section: *Calcarifera (Platycaulon)*

This lowland species is often confused with the closely related *D. lamellatum*, but it is a larger plant and has a different-shaped lip. The club-shaped but flattened stems produce a number of short inflorescences from their edges. The inflorescences are up to 6 cm long and have up to 8 flowers that open to 25 mm wide and last 7-10 days.

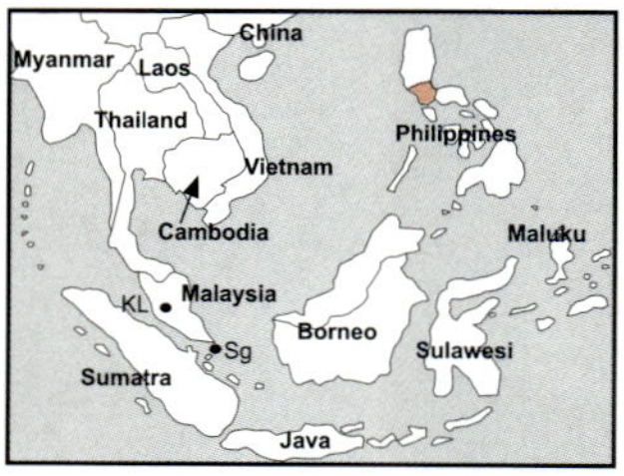


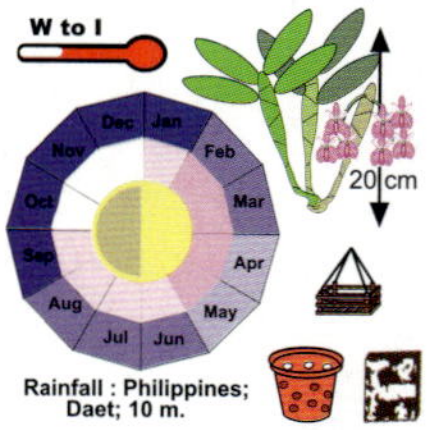


## *Dendrobium purpureum* Rxb.

Section: *Pedilonum*

This large species is seldom seen in collections. It grows on tree branches in dense forest from sea-level to 800 metres altitude. There are two subspecies, both of which are illustrated. The plant at the bottom is subspecies *purpureum*, (the normal form), while the top plant is subspecies *candidulum*. The two subspecies differ only in the lip shape and the position of the nectary. In both forms the flower colour can be purple to pink or pink and white or white, invariably with green tips.

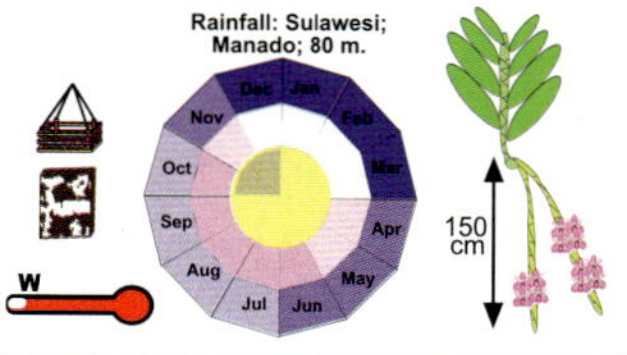


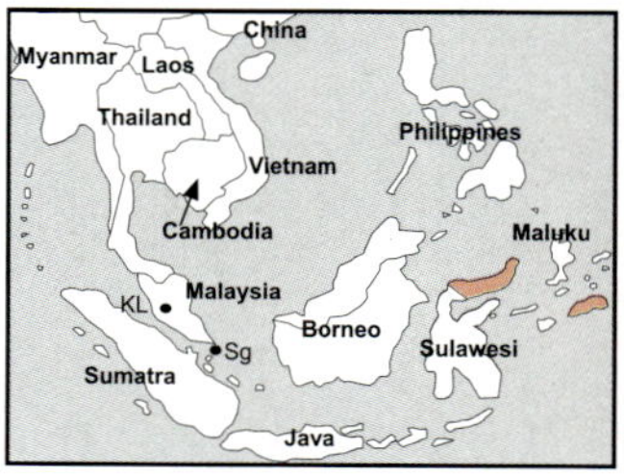

# D

## *Dendrobium roseatum* Ridl. Section: *Calcarifera*

This uncommon species is restricted to a few mountains in Malaysia, where it grows on trees next to streams from 1300-1600 metres altitude. The slender stems have large leaves that last 1-2 years. The short inflorescences have 35 mm long flowers that last about 7 days. Some specimens have flowers that are flushed pale pink.

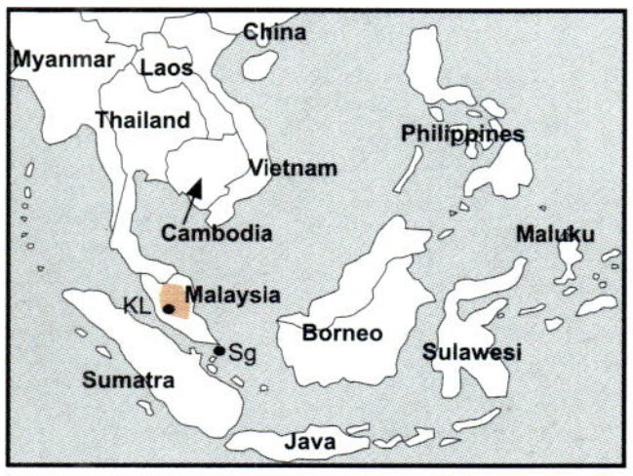


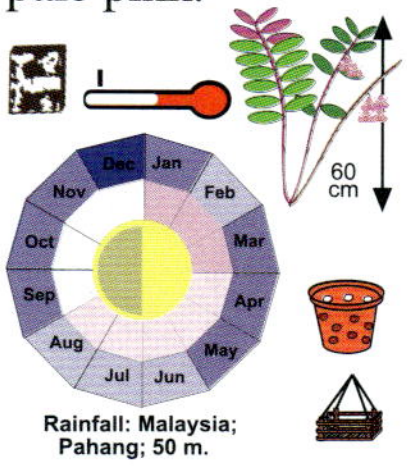


## *Dendrobium sanderae* Rolfe Section: *Formosae*

This beautiful large-flowered species grows on trees from 700-1600 metres altitude. The species has rather stout stems and leaves without any small black hairs. The species only flowers once a year, forming many inflorescences, each with up to 12 flowers that are 5-6 cm wide and last 10-14 days. There are 5 different varietal names used for plants in cultivation, of which 'var. *major*' is the most sought-after since the flowers are 9-10 cm wide.

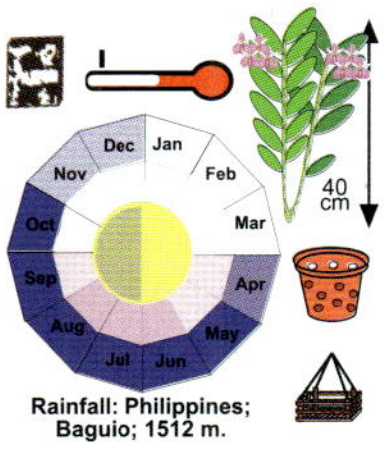


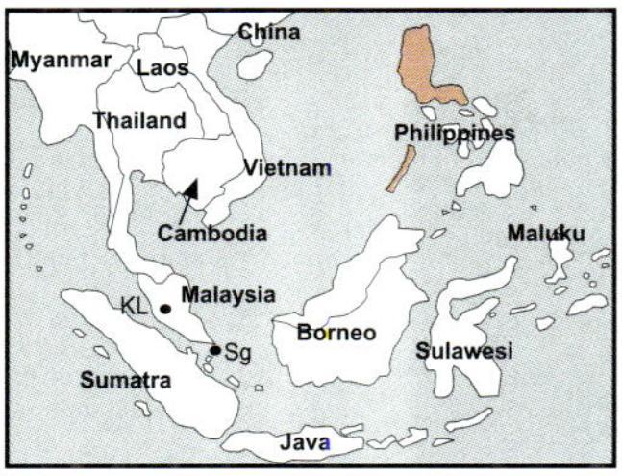

## *Dendrobium sanguinolentum* Lindl. Section: *Calcarifera*

This widespread species is fairly common in some lowland areas. It grows on forest trees from sea-level to about 900 metres altitude. The inflorescence has up to 6 flowers that are 2-3 cm wide. Some plants have a purple spot at the tip of the sepals, petals and lip.

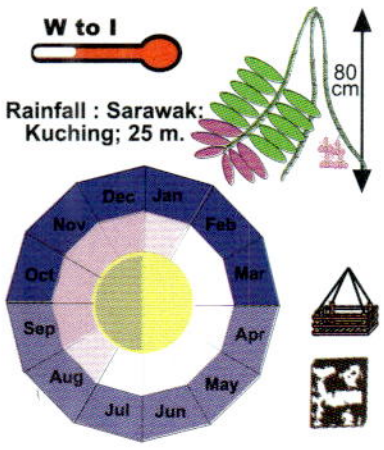


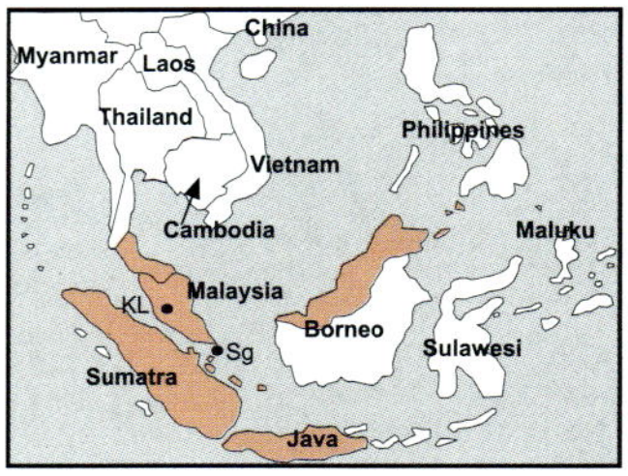

# D

## *Dendrobium scabrilingue* Lindl. Section: *Formosae*

This small species is found growing on trees in deciduous forests from 600-1300 metres altitude, usually in brightly-lit situations. It is occasionally seen in cultivation. The short robust stems have a few leaves, most of which are lost during the dry season. Both stems and leaves are covered in short black hairs. The inflorescences have 1-3 fragrant flowers which are about 2 cm wide and last for several weeks.

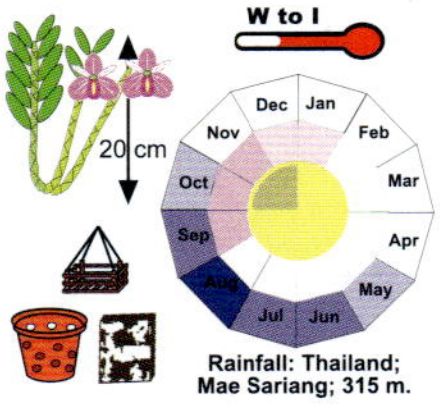


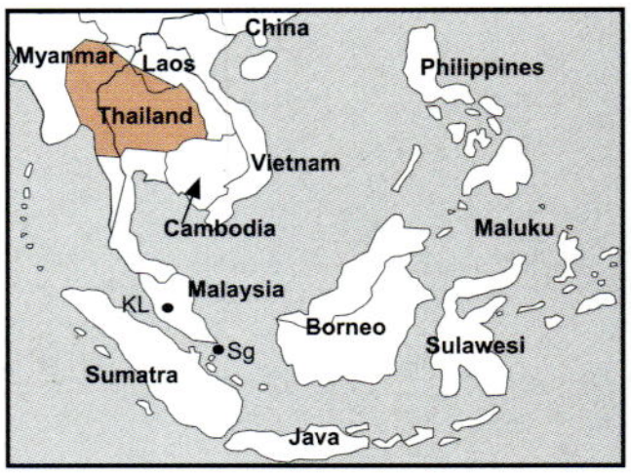


## *Dendrobium secundum* (Bl.) Lindl. Section: *Pedilonum*

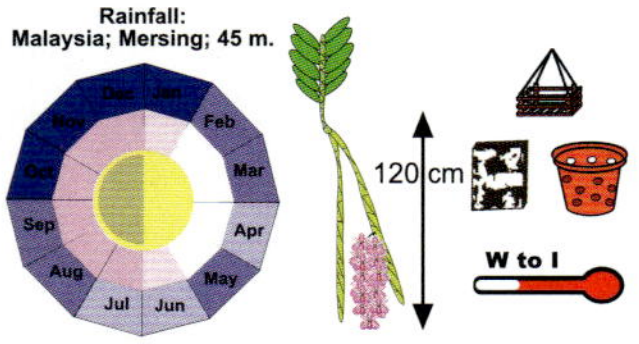


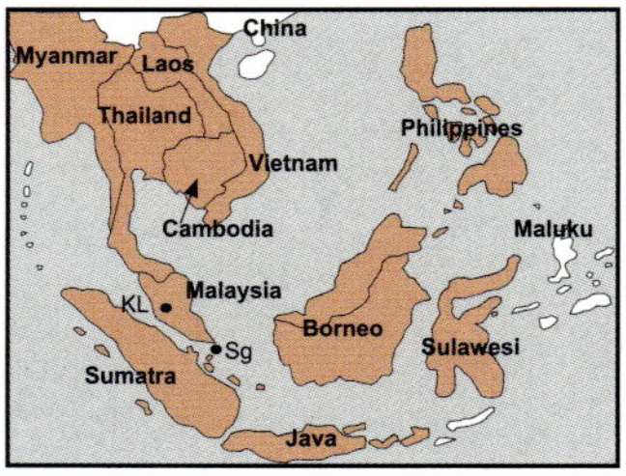


This distinctive species is well-known in cultivation. It is usually found in the lowlands, but has been recorded at 1200 metres altitude. It is quite common in areas where it has colonised rubber plantations. It has a long finger-thick stem that becomes pendulous with age. Some clones have short (30 cm) erect stems and can be grown in a pot. The small flowers, which are held on one side of the 10 cm long inflorescence, are usually purple or pink. White-flowered forms are common in cultivation.

## *Dendrobium serratilabium* L.O.Williams

Section: *Calcarifera*

This distinctive lowland species has narrowly club-shaped stems that can grow to 50 cm, although they are usually much shorter. The stems branch occasionally and droop with age. The spear-shaped leaves are deciduous, falling within 1 year. Inflorescences are produced for several months, each with up to 6 flowers that are 25 mm wide and last 7-10 days. Some plants have green-brown flowers, and most have red-brown stripes on the sepals and petals.

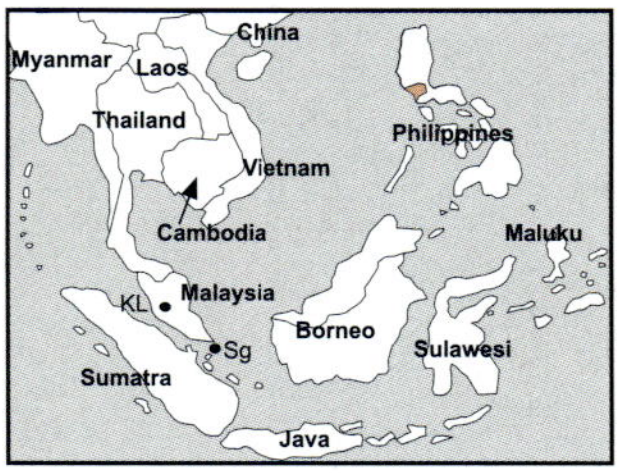


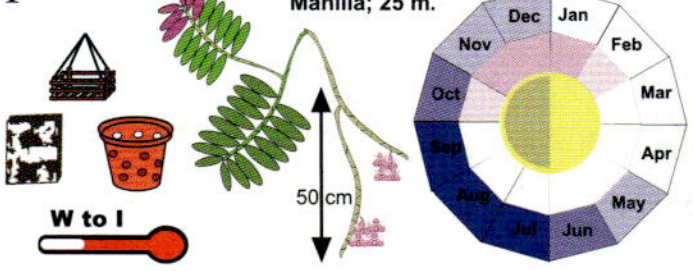

# D

## *Dendrobium singaporense* Hawkes & Heller
Section: *Strongyle*

This species is still quite common in parts of Malaysia and Indonesia, growing on trees in exposed situations. It has long terete leaves, and forms inflorescences on the leafless stem apex. The fragrant flowers are 25 mm long, do not open wide, and last for a few days.

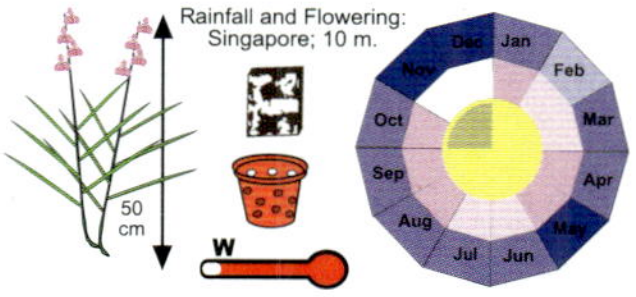


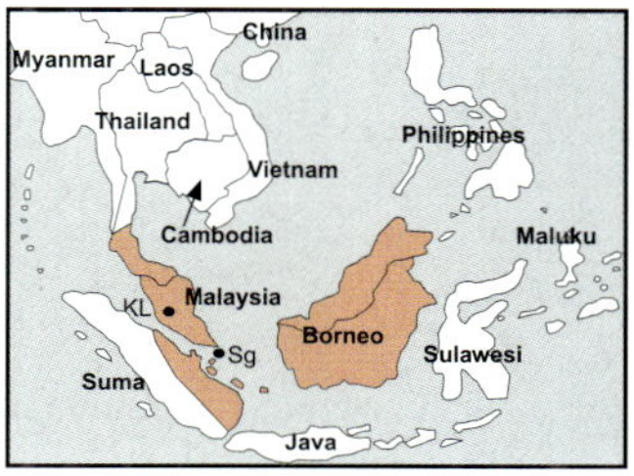

## *Dendrobium singkawangense* J.J.Sm. Section: *Formosae*

This free-flowering rare lowland species was recently "rediscovered" after being "lost" for 70 years. Nothing is known about it's natural habitat. The slender zig-zag stems and leaves are covered in tiny black hairs. Several inflorescences at a time are formed, each bearing up to 7 stiff waxy flowers that are 3 cm wide and last for up to 10 days.

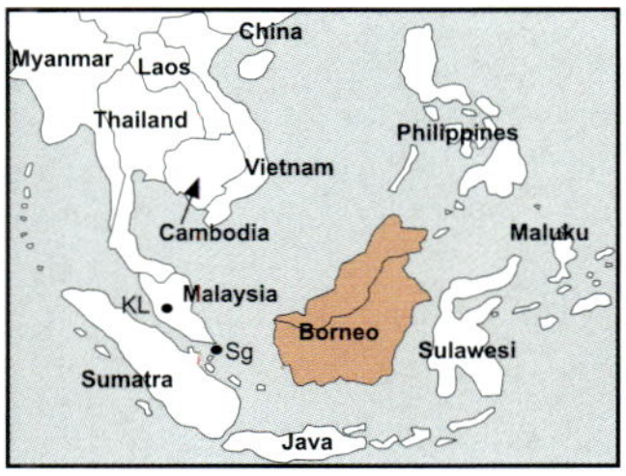


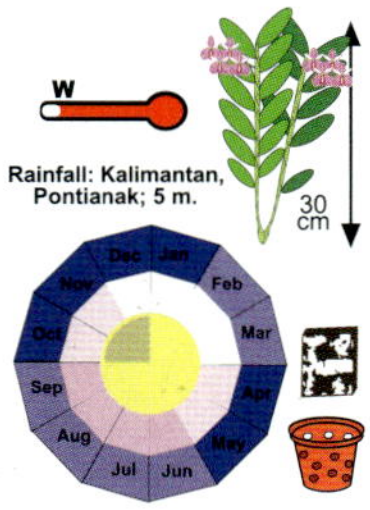


## *Dendrobium sororium* Schltr.
Section: *Grastidium*

This uncommon and unusual species has a scattered distribution across Sulawesi from 600-1300 metres altitude, where it grows on isolated trees. It is highly distinctive with heart-shaped blue-green to purple-green leaves. The stems grow continuously (but slowly) for several years, intermittently producing a number of twin-flowered inflorescences from a bract opposite a leaf. The flowers are 2 cm across and last for less than one day.

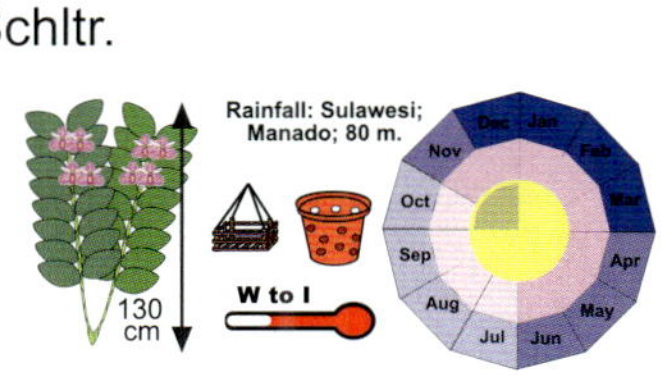


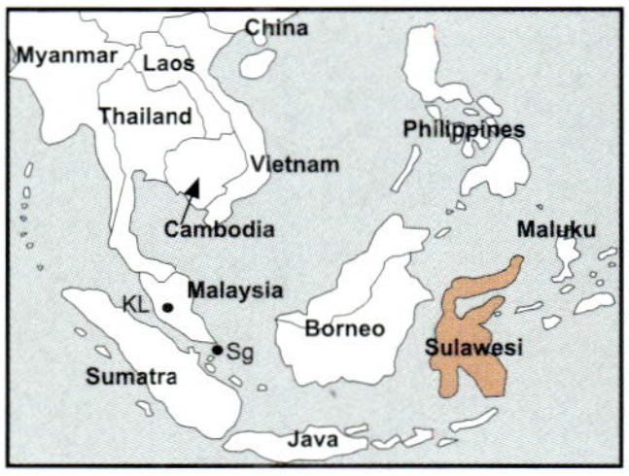

# D

## *Dendrobium spurium* (Bl.) J.J.Sm.

Section: *Euphlebium*

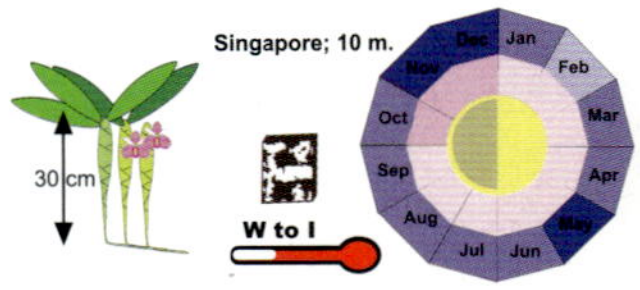


This widespread species is often seen growing on mangroves, but has been found in most types of forest up to 900 metres altitude. The erect club-shaped stems (usually about 15 cm long but sometimes much larger) produce a steady succession of single-flowered inflorescences. The 2- 3 cm flowers are open for less than 1 day.

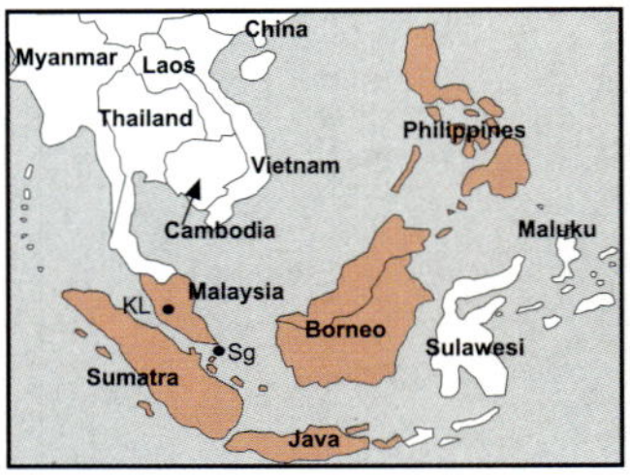

## *Dendrobium tobaense* J.J.Wood & Comber Section: *Formosae*

This very rare species was first described in 1993 and has only recently been seen in cultivation. It has a very restricted distribution as it is only found in one small area of Sumatra, growing from 700-1500 metres altitude in pine forest and montane forest. The slender zig-zagging stems and leaves are covered in soft black hairs. Inflorescences are produced several times a year and can bear up to 4 flowers which open at about 7 day intervals. The flowers, which are 5-6 cm wide and last about 10-14 days, have a remarkable lip structure.

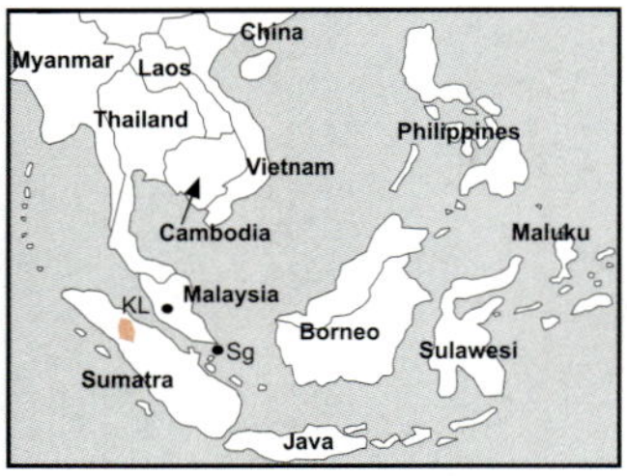


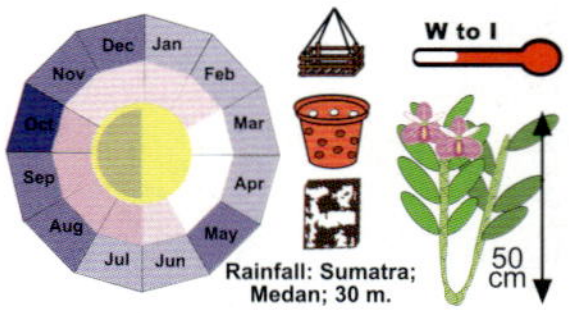


## *Dendrobium trinervium* Ridl. Section: *Distichophyllum*

This very rare miniature species is known from only two locations, growing on trees in rather open forest. Some plants have recently been seen in collections, labelled as *D. oligophyllum*. The species forms small clumps of stout zig-zagging stems that stay leafy for 1-2 years. The inflorescences have 1- 4 flowers that are 2 cm wide and last for several weeks.

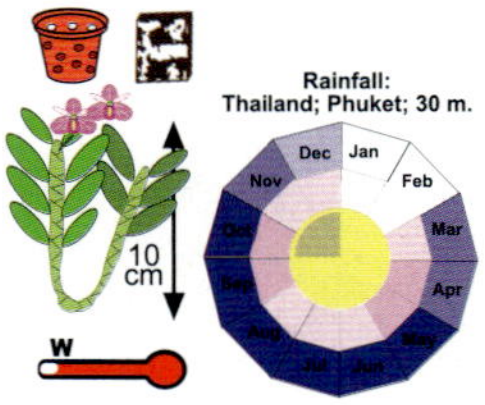


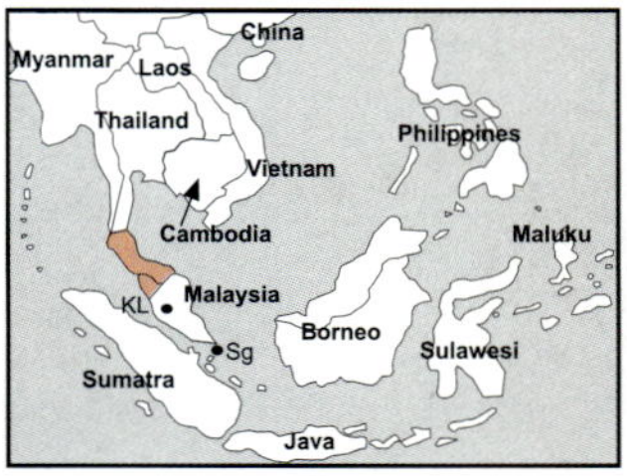

# D

## *Dendrobium unicum* Seidenf. Section: *Dendrobium*

This brilliantly-coloured deciduous species grows on trees, bushes or rocks from 400-1500 metres altitude. The short stems flower freely for several years after the leaves have fallen. The bright orange (or yellow) flowers are about 4 cm wide and last for 2-3 weeks.

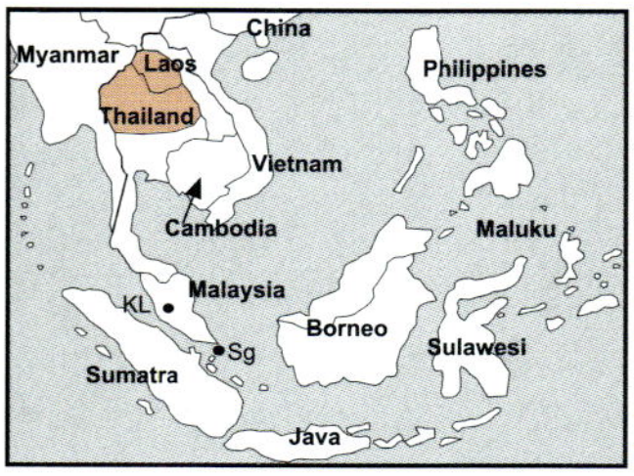


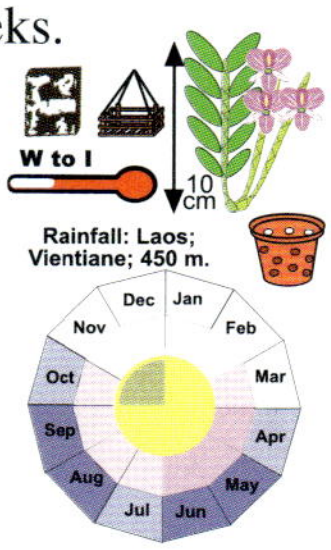


## *Dendrobium uniflorum* Griff.
Section: *Distichophyllum*

This highly variable species is found from sea-level to 1800 metres altitude. It is uncommon or rare in most parts of its range, although in the Philippines it can be locally common. In the wild the plant is seldom more than 40 cm tall, but some clones grow much taller in cultivation. The stems grow and flower continuously for several years. Flower colour and size are variable; flowers are usually 2-3 cm wide, but some specimens have flowers twice this size.

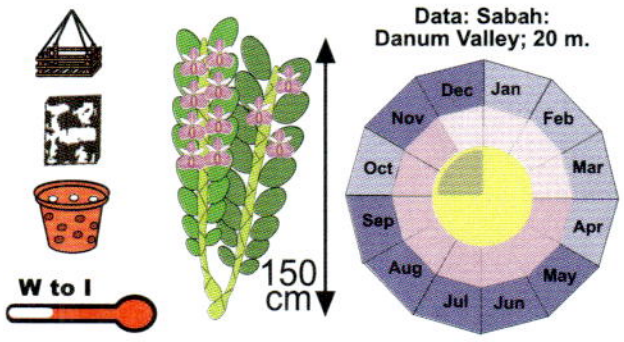


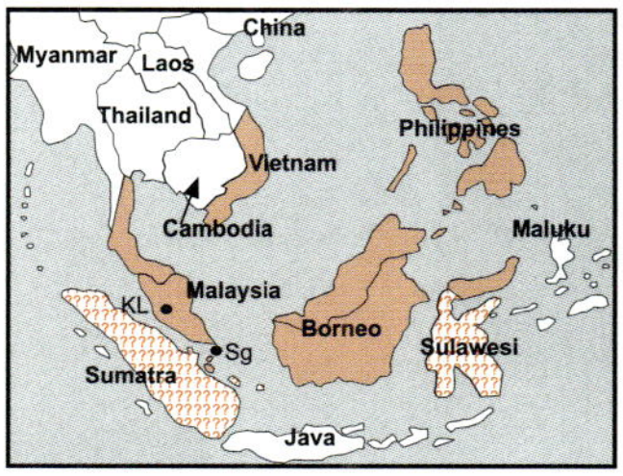

## *Dendrobium venustum* T. & B. Section: *Stachyobium*

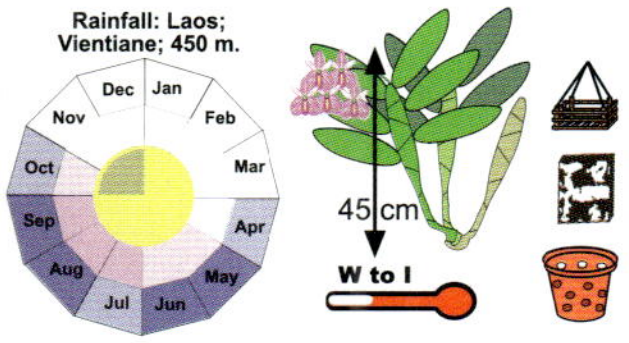


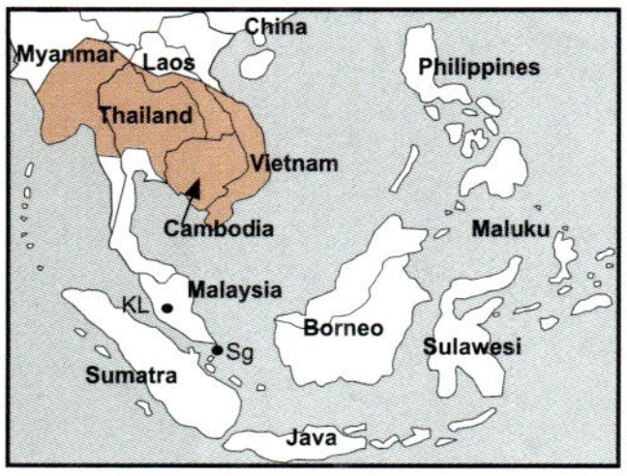


This fairly common lowland species grows on trees in deciduous forests. It is frequently confused with *D. delacourii*, which is a much smaller plant with shorter pseudobulbs and inflorescences. The pseudobulbs are usually about 10-20 cm tall, but sometimes much larger, with up to 7 persistent leathery leaves near the apex. Several arching inflorescences are produced from the pseudobulb apex, each with up to 25 flowers that are 25 mm wide and last 7-12 days.

# D

## Dendrochilum

This large genus has an estimated 260 - 280 species that are found in Myanmar, Thailand, Malaysia, Indonesia, the Philippines, Taiwan and New Guinea. The majority are found in Sumatra (about 75 species), Borneo (80 species), and the Philippines (89 species). A large proportion have a very restricted range; 94% of the Philippine species are found nowhere else, while 90% of the Borneo species are endemic. This is probably because most species are mountain plants; Mount Kinabalu in Sabah has 33 species, 12 of which are found nowhere else. *Dendrochilum* species are epiphytes or terrestrials with a creeping rhizome and pseudobulbs with a single leaf. The inflorescence usually emerges from a young shoot before the leaves develop fully, although some species flower from a mature pseudobulb while others produce a special flowering-shoot from the rhizome. There are usually 2 rows of small greenish, yellowish or brownish flowers. *Dendrochilum* species are steadily gaining in popularity amongst orchid growers, and a small number of species (mainly from the Philippines) are widely cultivated.

### *Dendrochilum carinatum* Carr

This unusual miniature species grows on tree trunks and branches in montane forest from 1300-1800 metres altitude. A short inflorescence of 6-12 tightly-packed 5 mm flowers is formed on a mature pseudobulb. It is easily identified by the two broad keel-like swellings that occupy most of the lip.

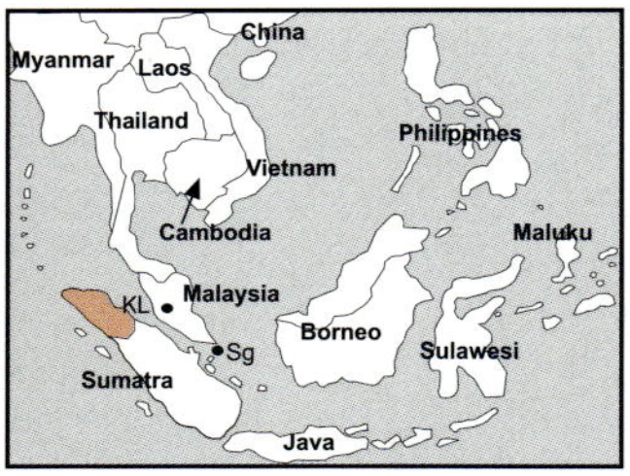


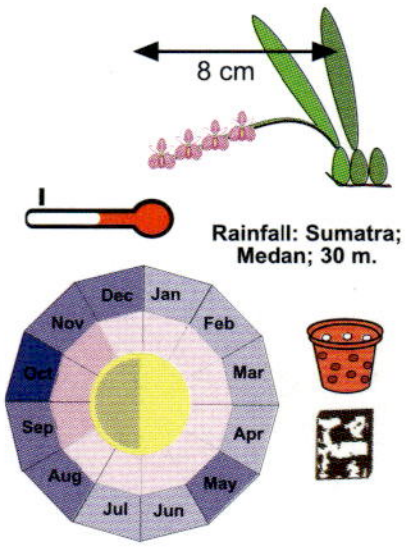


### *Dendrochilum cobbianum* Rchb.f.

This species is found in cultivation worldwide and is prized for its prolific floral displays. In the wild it can be found from 1400-2400 metres altitude, growing on trees or rocks in mossy forest. It has long yellow pseudobulbs and produces a 40 cm inflorescence on the young shoot. The flower colour varies from white to green or yellow, usually with a contrasting bright orange or yellow lip.

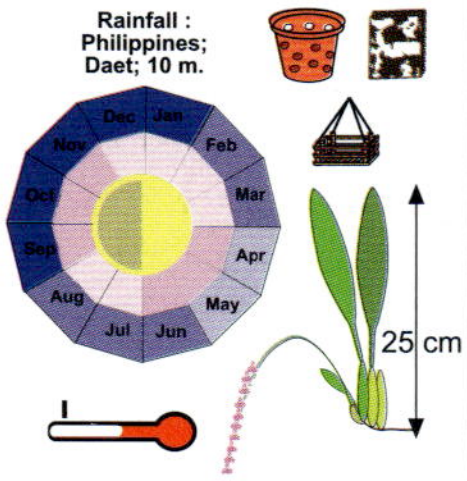


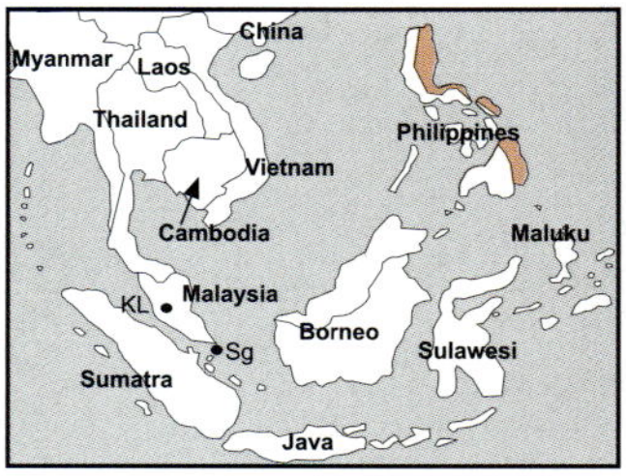

# D

## *Dendrochilum gracile* (Hk.f.) J.J.Sm.

This common species can be found on most mountains in Malaysia and Java from 760-1200 metres altitude; in Sumatra it is usually seen from 1400-1800 metres. It grows on tree trunks, branches and fallen logs, and is well known for liking exposed situations. The arching inflorescence has up to 35 flowers that are 1 cm across.

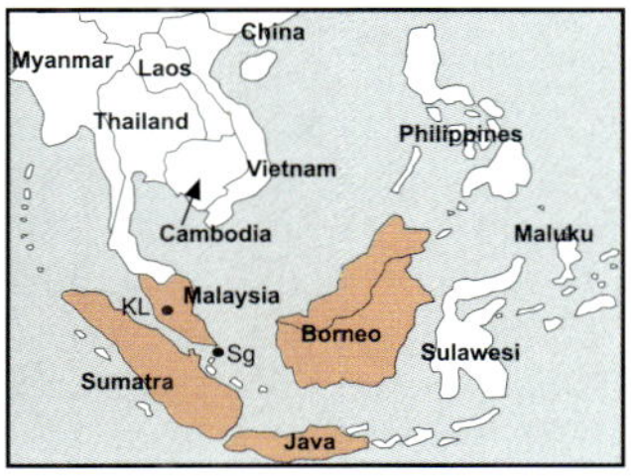


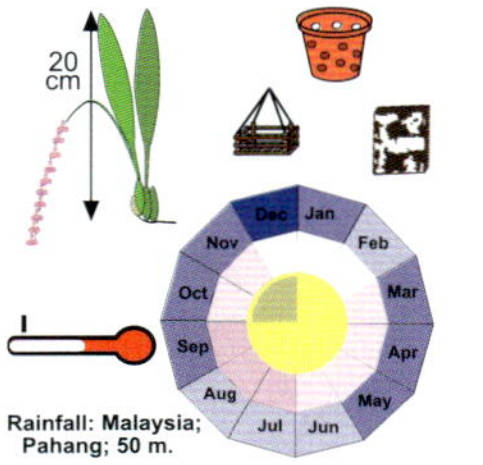


## *Dendrochilum megalanthum* Schltr.

This little-known species has the largest flowers in the genus; all of 2 cm in diameter ! It grows on tree trunks and branches from 1700-1800 metres altitude, preferring rather exposed situations. A 30 cm arching inflorescence is formed from a mature pseudobulb and carries up to 12 well-spaced flowers. The flower colour varies from yellowish to greenish or brownish.

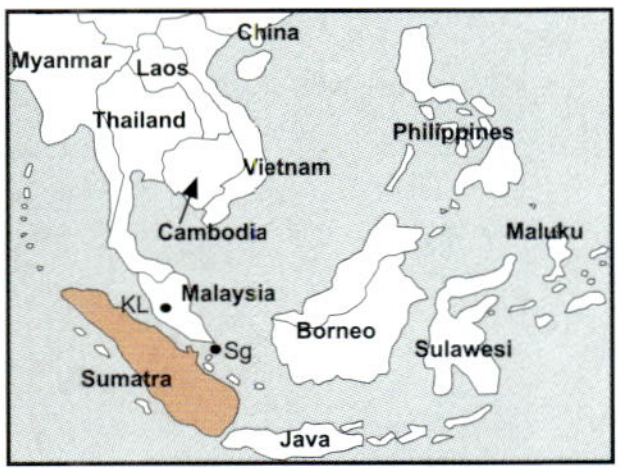


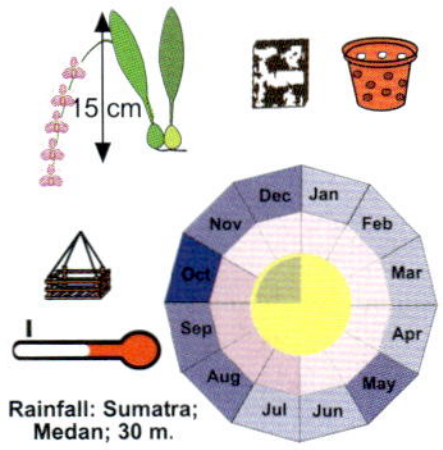


## *Dendrochilum pallidiflavens* Bl.

This widespread but variable species is quite common and can be found from sea-level to 1400 metres altitude, forming great masses that hang off tree branches or scramble over rocks. The illustrated specimen is an unusual colour; most plants have white or cream flowers, sometimes with orange tips and a green mark on the lip.

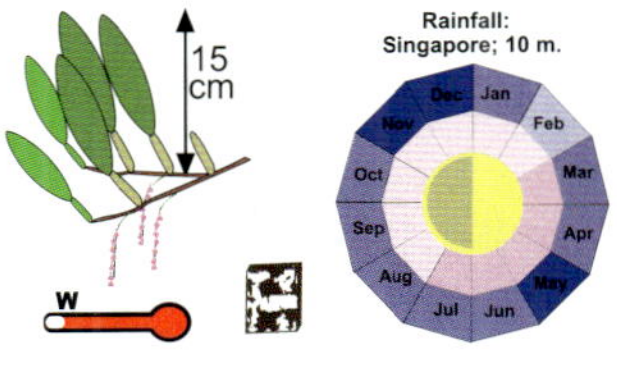


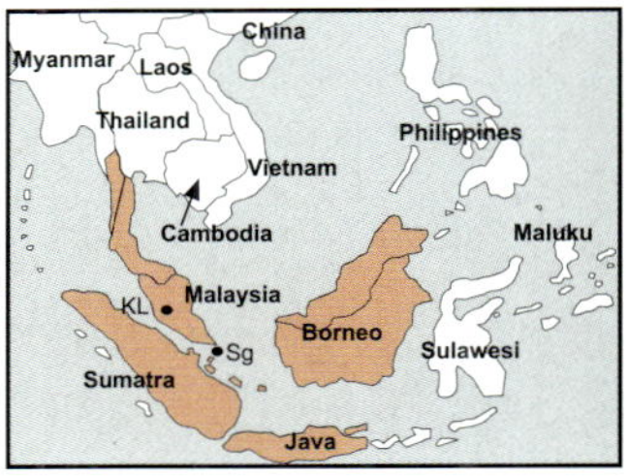

# D

## Doritis

This genus consists of a single widespread species, *D. pulcherrima*, and possibly a second species, *D. regneriana*, which apparently has only been found only once, in either Thailand or Vietnam. *D. pulcherrima* has been line-bred to the point where it bears little resemblance to the wild-type species; its progeny are popular houseplants and are cultivated worldwide. It has also been extensively hybridised with *Phalaenopsis* and the resulting hybrids are widely grown as pot plants and for the cut flower industry.

### *Doritis pulcherrima* Lindl.

This showy species grows from sea level to 800 metres altitude, but is usually found growing as a terrestrial or on mossy rocks in hilly country above 500 metres altitude. It has two compact rows of fleshy leaves and one to several inflorescences that extend slowly, reaching 40 cm tall, with several flowers open at a time. The flower size and colour are very variable.

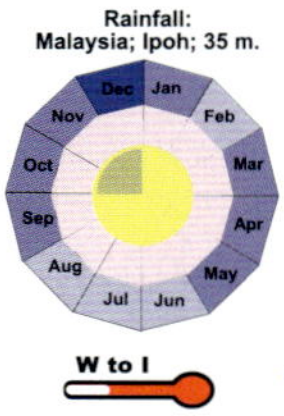


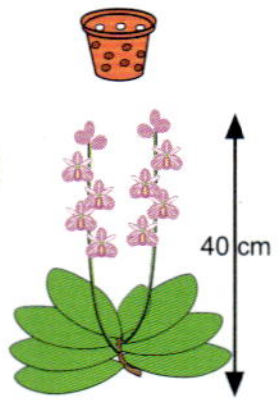


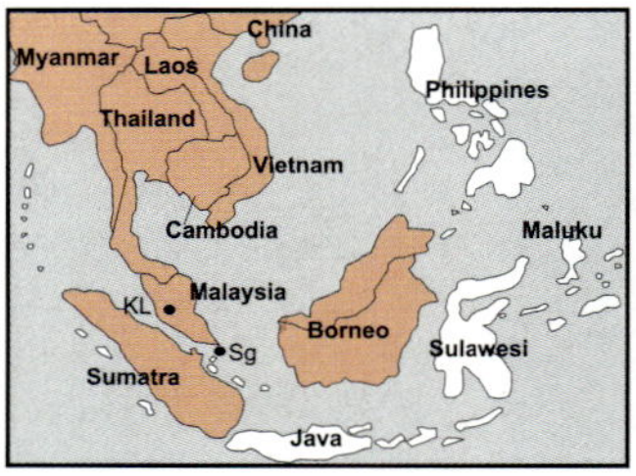

## Dyakia

This genus consists of a single species which was previously known as *Ascocentrum hendersonianum*. It is keenly sought after by orchid enthusiasts in S. E. Asia, but despite having huge potential as a pot-plant, it is hardly known outside the region.

### *Dyakia hendersoniana* (Rchb.f.) Christenson

This showy species has been found at sea level, but usually grows on trees in hilly country or the lower slopes of mountains from 500-700 metres altitude, in locations with constantly high humidity. It is now very rare and endangered in the wild, where it flowers once a year, with 1- 4 stiffly erect inflorescences. In cultivation it will flower twice a year, with up to 7 inflorescences, each carrying up to 40 brightly bicoloured flowers that are 6-7 mm tall.

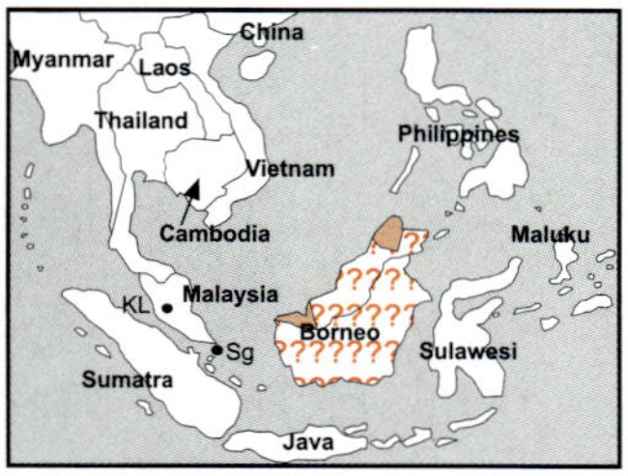


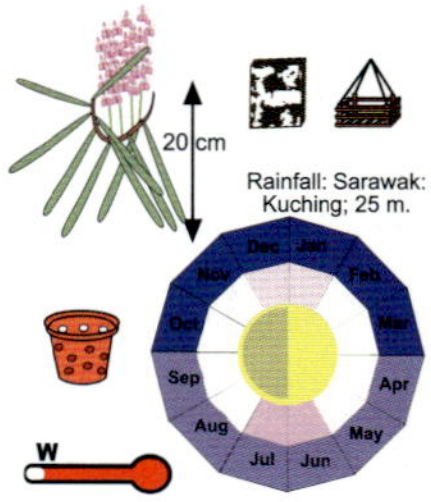

# E

## Epigeneium

This genus of about 35 epiphytic or terrestrial species is distributed from India and China to Indonesia and the Philippines. Most species are sun-loving mountain plants with long creeping or hanging bract-covered rhizomes and short pseudobulbs that have 1 to 3 leaves. Despite the fact that these free-flowering plants have long-lasting attractive flowers, they are hardly ever seen in collections and have not been used to create hybrids.

### *Epigeneium geminatum* (Bl.) Summerhayes

This distinctive species grows in Java from 700-1000 metres altitude and from 1300-1900 metres in Malaysia and Borneo. The rhizome forms roots from creeping sections with pseudobulbs 3-8 cm apart, and it forms aerial branches with pseudobulbs 10-30 cm apart. Young shoots produce 1-3 inflorescences before the leaves have developed properly; each carries 3-12 flowers.

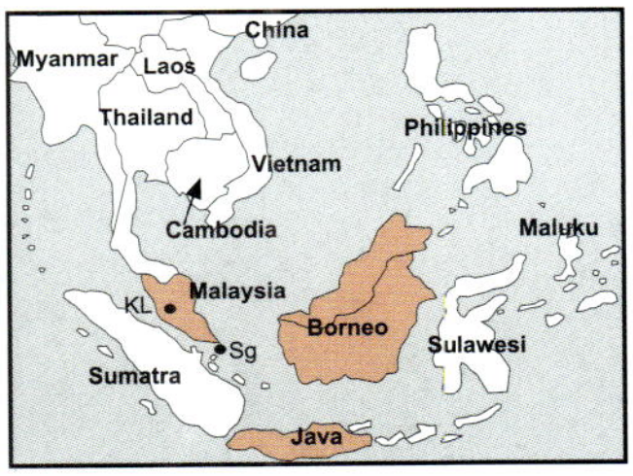


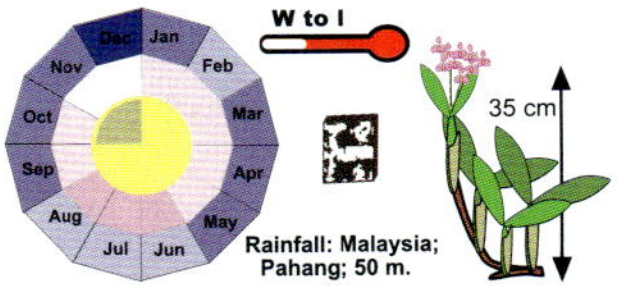


### *Epigeneium longipes* (Hk.f.) Summerhayes

In Malaysia this small species is confined to the highest mountains, where it grows on trees, rocks or terrestrial moss-pads on exposed situations. It is more common in Sumatra, where it grows above 1600 metres altitude. The creeping rhizome has small round pseudobulbs at 1-3 cm intervals. A long slender inflorescence carries 3-5 flowers that are flushed pink in some plants.

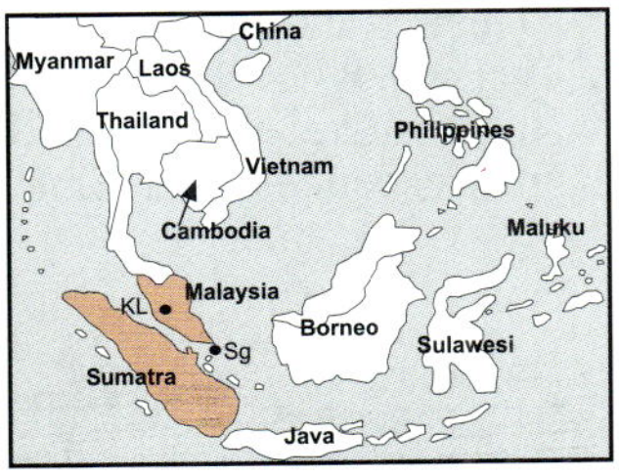


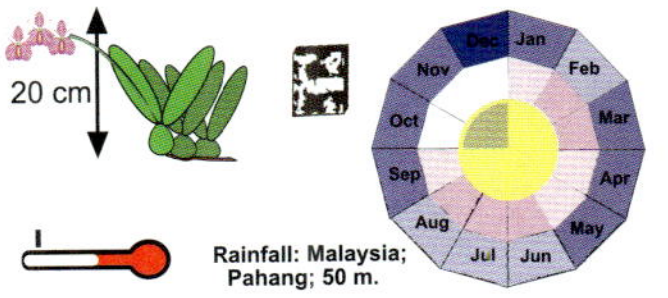

### *Epigeneium triflorum* (Bl.) Summerhayes **variety *orientale*** (J.J.Sm.) Comber

*E. triflorum* is a common epiphyte in Java, growing from 1000-1800 metres altitude. The three recognised varieties differ mostly in the sepal and lip details. Var. *orientale* has a slender drooping inflorescence with 6-17 flowers; a large plant in full bloom is an arresting sight.

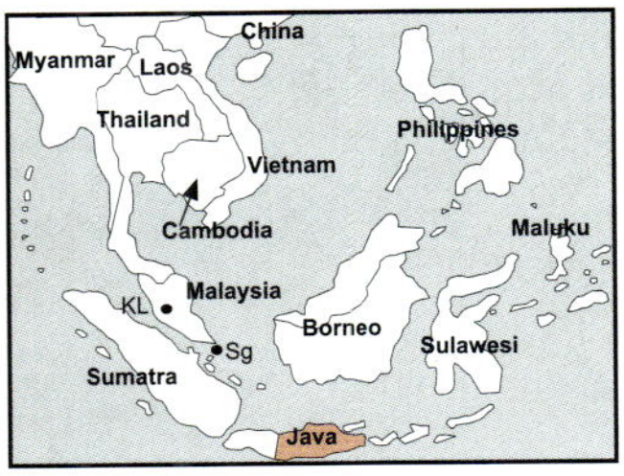


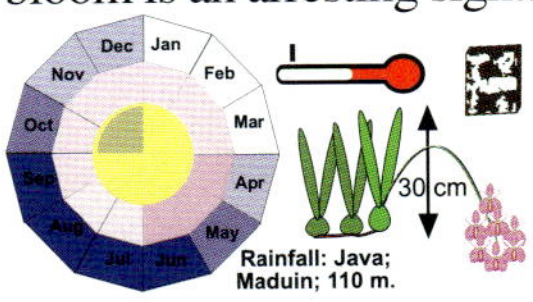

# E

## Eria

This is probably the third-largest genus in S. E. Asia, after *Bulbophyllum* and *Dendrobium*. It is distributed from India and the Himalayan foothills to New Guinea, Australia and the Pacific Islands. It has about 370-400 species which show a great diversity in habit. The genus is usually divided into sections in order to assist classification and identification. Despite the fact that many *Eria* species are highly attractive and easy to grow, their flowers are rather small and only last a few days, so they have never been popular with orchid enthusiasts, are seldom seen in collections outside S. E. Asia and have not made any impact on the hybrid industry.

### *Eria biglandulosa* J.J.Sm. Section: *Hymeneria*

This species is easy to spot in Kuching, where it grows in large numbers on the branches of mature trees in the city parks. It has large ovoid pseudobulbs with red-brown sheaths and 4-5 large leaves at the apex, and produces several hairy inflorescences that are longer than the leaves.

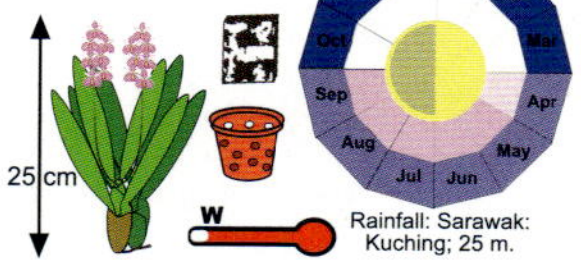


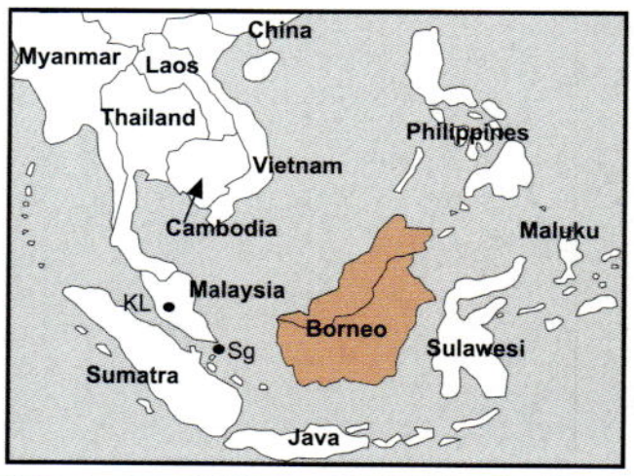

### *Eria cepifolia* Ridl. Section: *Urostachya*

This unusual small species is fairly common in the Borneo lowlands, growing on tree branches that overhang water. It has also been found in montane forests up to 1400 metres altitude. It is rare in Malaysia and Thailand. It has club-shaped pseudobulbs with semiterete leaves and produces 1-5 inflorescences at a time.

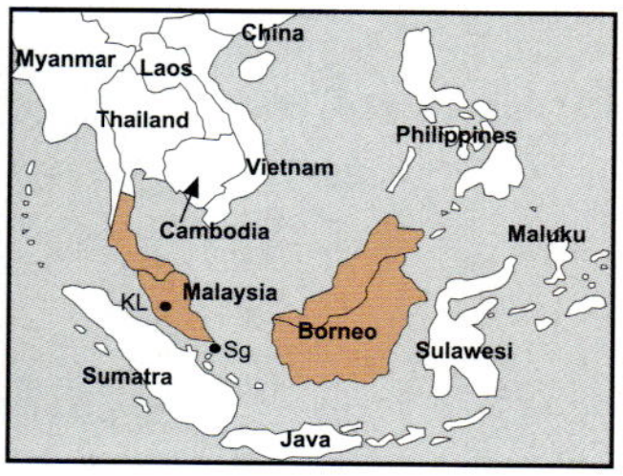

### *Eria citrina* Ridl. Section: *Mycaranthes*

This attractive epiphytic species is usually encountered in hilly country, but has been recorded in forests from sea-level to 1200 metres altitude. It has pencil-thick stems with a few well-spaced leaves that have a distinct stalk at the base. 2-3 inflorescences are formed at the top of the stem.

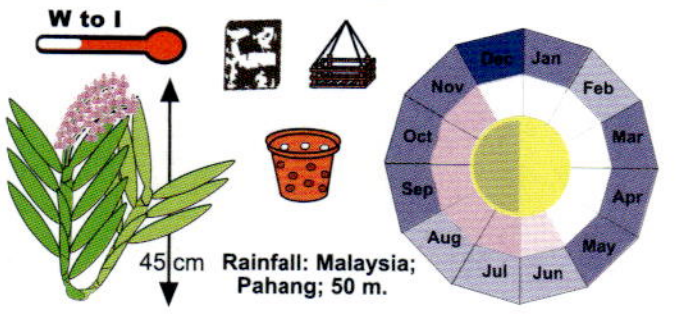


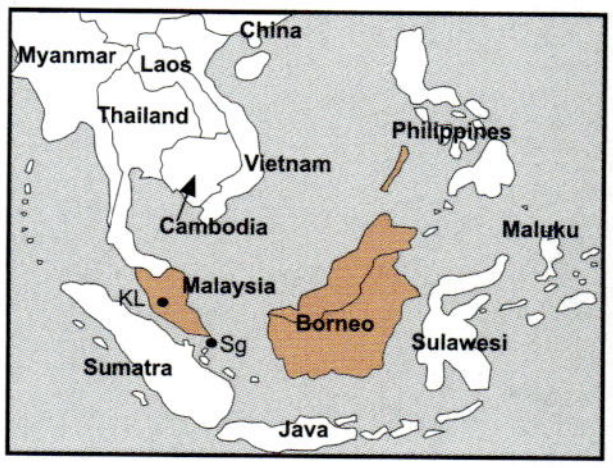

# E

## ***Eria cymbiformis*** J.J.Sm. Section: *Cymboglossum*

This imposingly large epiphytic species is common in mountain areas from 1100-2500 metres altitude. It has not yet been found in Borneo. It has a short rhizome with small pseudobulbs that are totally concealed by the 60-70 cm high *Cymbidium*-like leaves. It usually forms 2 slightly arched inflorescences at a time. These are shorter than the leaves and are crowded with 12 mm flowers.

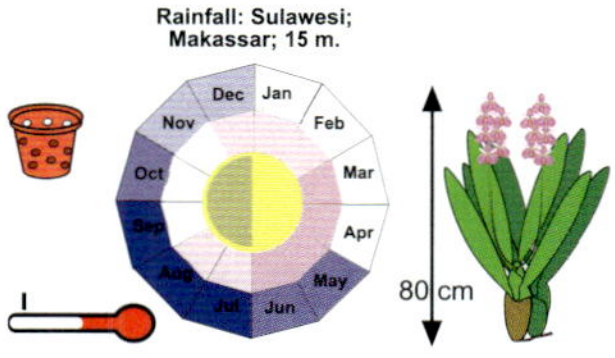


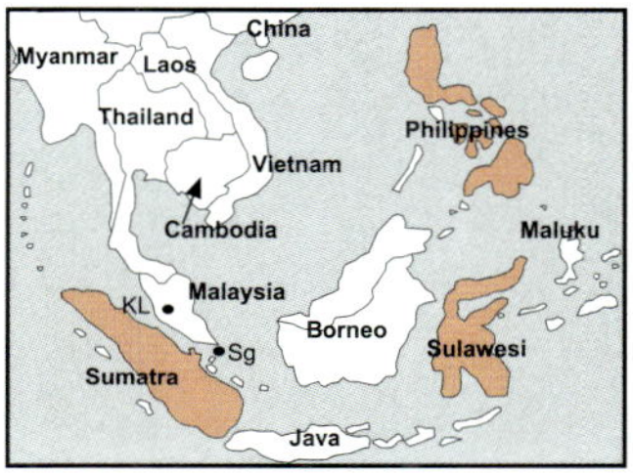

## ***Eria densa*** Ridl. Section: *Urostachya*

This large epiphytic species is common in the mountains of Malaysia from 1300-2000 metres altitude. In Borneo it appears to grow in the lowlands. It has tall thick stems with a few leaves near the apex. Several 10 cm arching inflorescences are formed from nodes near the stem apex. It is easy to confuse with the closely-related *E. floribunda*, which has thinner stems.

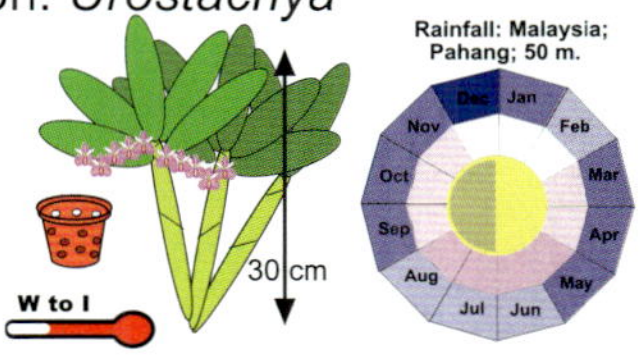


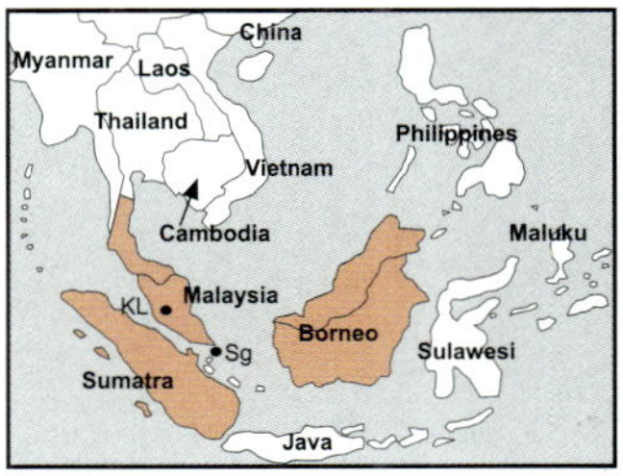

## ***Eria elisheae*** O'Byrne Section: *Cylindrolobus*

This species from Quezon Province was described as recently as 2000. It has a creeping rhizome with many slender erect leafy stems that grow and flower for several years. Each stem produces several single-flowered inflorescences that only last for one day. The lip has 3 keels and a thick hairy midlobe. Flowering is intermittent throughout much of the year and is probably triggered by heavy rain.

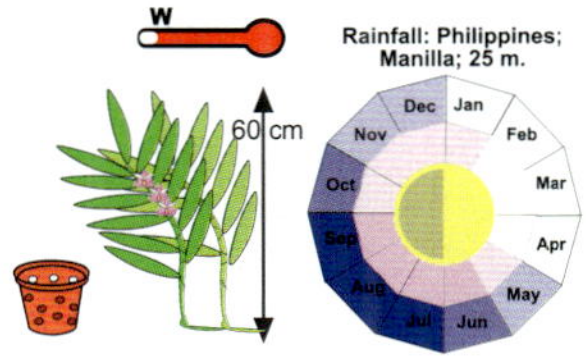


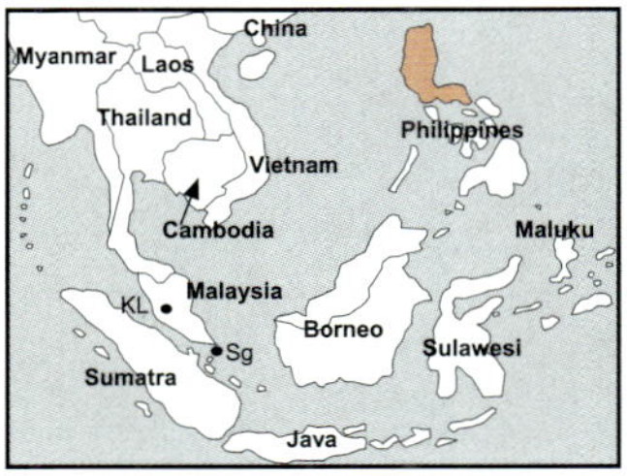

# E

## *Eria flavescens* (Bl.) Lindl. Section: *Hymeneria*

There is some doubt about this epiphytic species because the Malaysian plants (2 colour forms illustrated here) differ so much from the Javanese plants. In Malaysia, plants are found in the central mountains from 700-1500 metres altitude. The Javanese form occurs at higher altitudes and has brown flowers with darker brown stripes.

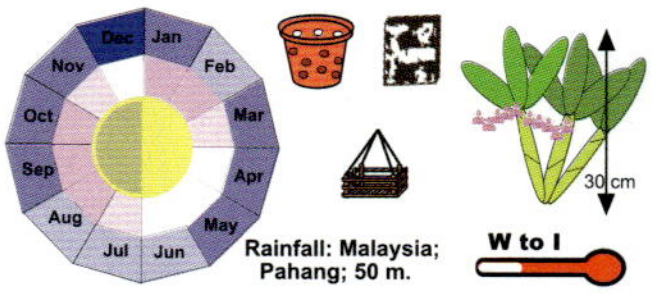


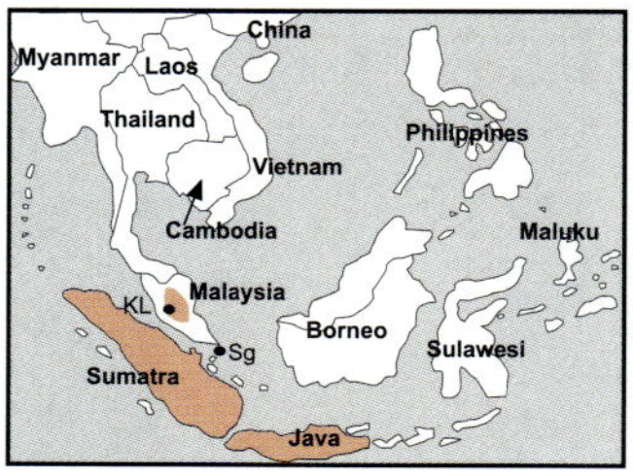

## *Eria floribunda* Lindl. Section: *Urostachya*

This epiphytic species has more slender stems and a slightly different lip to *E. densa*. It is a common lowland species that grows in a variety of forest types and also quickly establishes itself in rubber and fruit plantations. It has been recorded growing up to 1200 metres altitude.

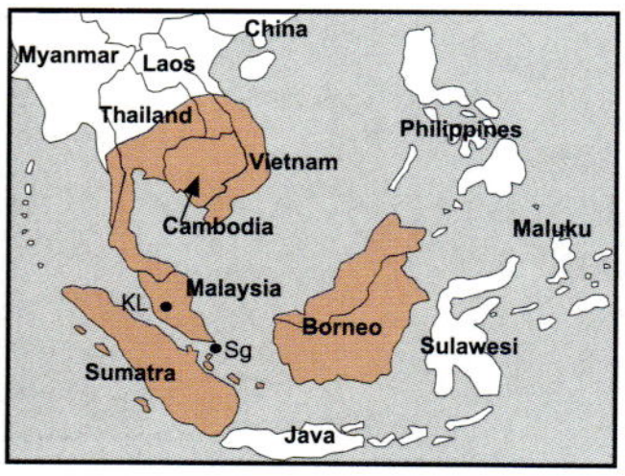


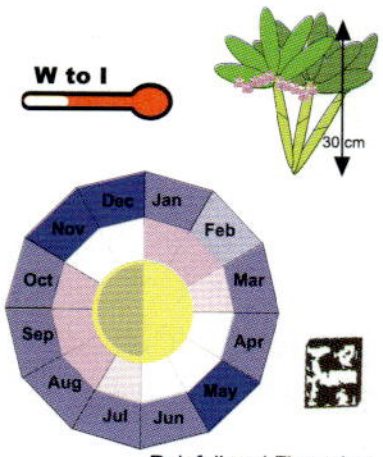


Rainfall and Flowering: Singapore; 10 m.

## *Eria hyacinthoides* (Bl.) Lindl. Section: *Hymeneria*

This species grows on tree branches and rocks from 500-1500 metres altitude. The conical pseudobulbs are hidden behind several thick green sheaths, and have 2-3 leathery leaves at the tip. There are usually two 25 cm long inflorescences at a time, each with 30-40 *Hyacinth*-like flowers, most of which are open at the same time.

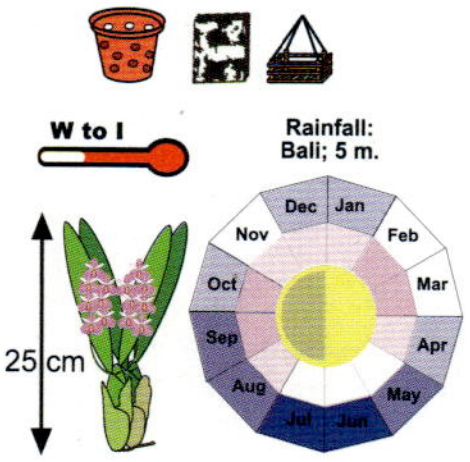


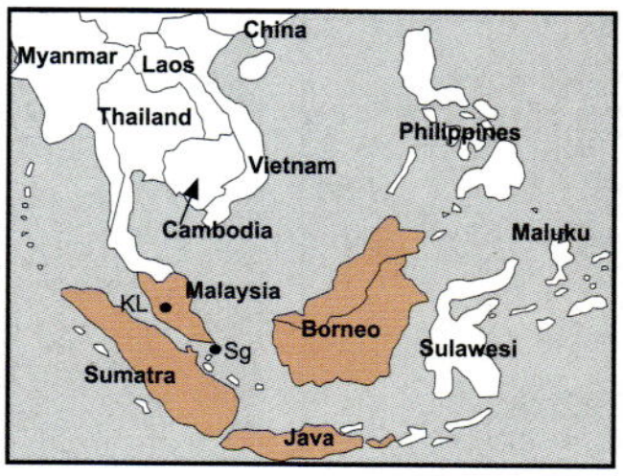

# E

## *Eria iridifolia* Hk.f. Section: *Mycaranthes*

This species is extremely common in the mountains of Sumatra and Malaysia, where it is found on almost every mountain above 1000 metres altitude, growing in tree forks and on moss-covered rocks. The rather short stems are hidden by the overlapping leaf bases. The stem apex produces three or four 30 cm inflorescences at a time.

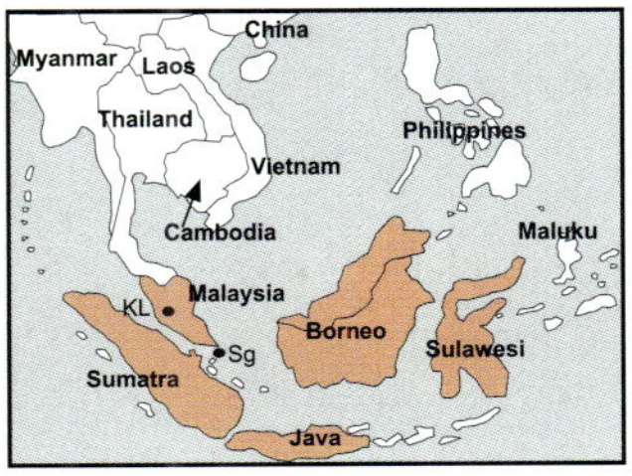


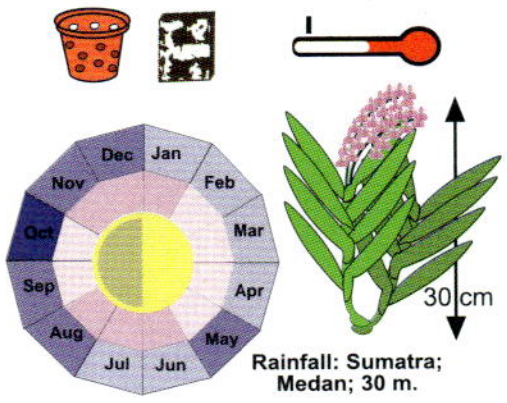


Rainfall: Sumatra; Medan; 30 m.

## *Eria javanica* (Sw.) Bl. Section: *Eria*

This fragrant species grows in tree forks and on rocks from sea-level to at least 2400 metres altitude, sometimes forming extremely large colonies. The pseudobulbs form 1-4 inflorescences at a time; these are usually 50 cm long but may be much longer. There is a lot of variation in flower colour and size.

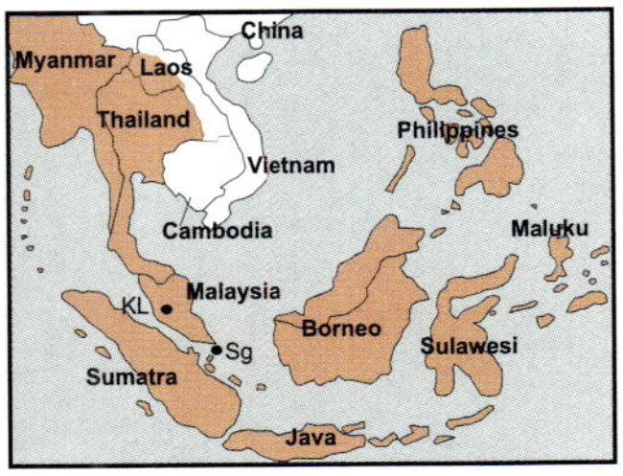


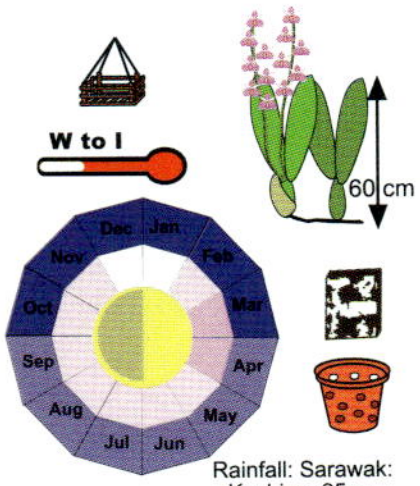


Rainfall: Sarawak: Kuching; 25 m.

## *Eria jenseniana* J.J.Sm. Section: *Cylindrolobus*

This rare species was first found in Sumatra, but seems to be slightly more common in Borneo, growing on rainforest tree branches from 100-600 metres altitude. Several inflorescences of 1-2 flowers are formed at a time, probably in response to heavy rain. The flowers only last 1 day. Some Sumatran plants lack brown spots on the lip sidelobes.

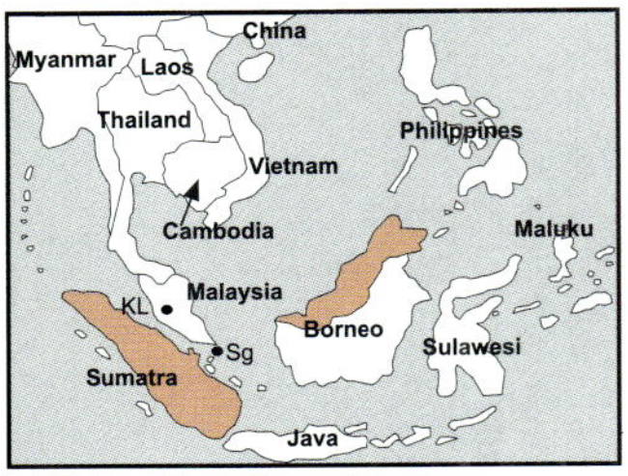


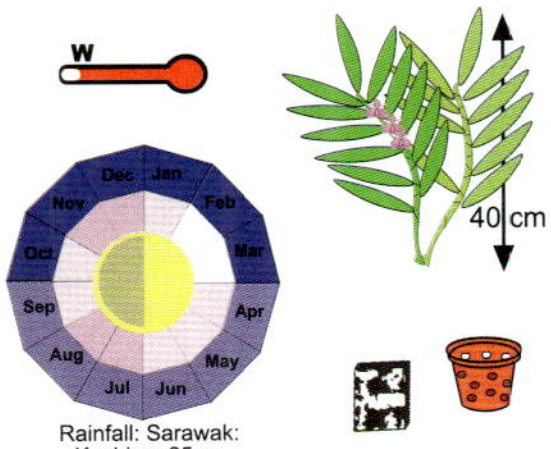


Rainfall: Sarawak: Kuching; 25 m.

# E

## *Eria kingii* F. Muell. Section: *Hymeneria*

This large showy species occurs from New Guinea to Sulawesi, where it is often seen in coffee plantations, growing on the branches of old bushes at about 1000 metres altitude. It has large egg-shaped pseudobulbs which usually produce 2-3 30 cm long crowded inflorescences that last for about 3 weeks.

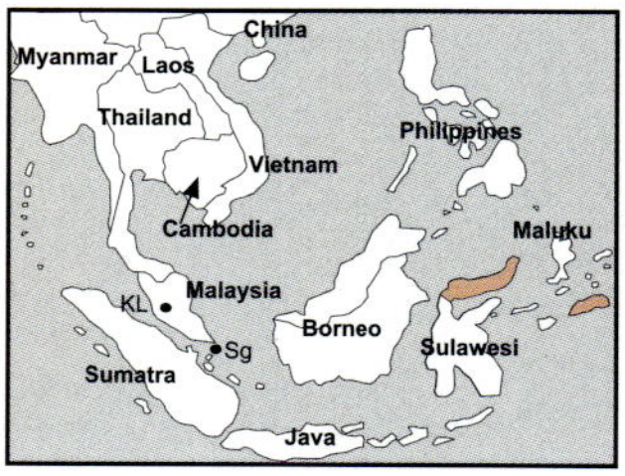


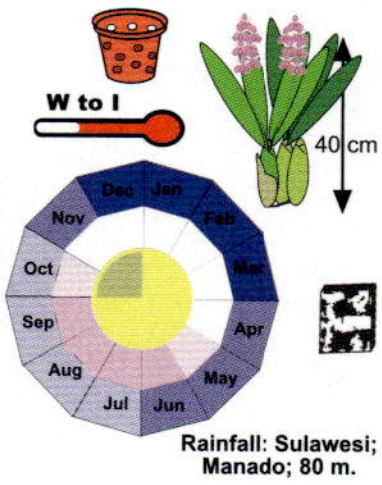


## *Eria lamonganensis* Rchb.f. Section: *Hymeneria*

This species is common in Java from 420-1500 metres altitude, growing on trees or rocks. In Malaysia it is found only in the mountains. It has only recently been found in Sumatra, growing on tree branches from 600-1000 metres altitude. In some of these colonies an *alba*-form was growing alongside plants of the normal colour.

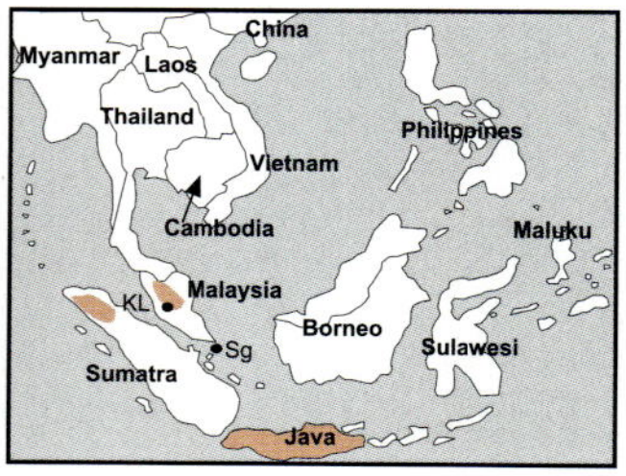


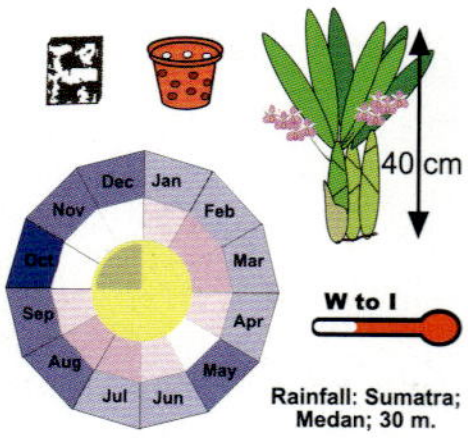


## *Eria leiophylla* Lindl. Section: *Strongyleria*

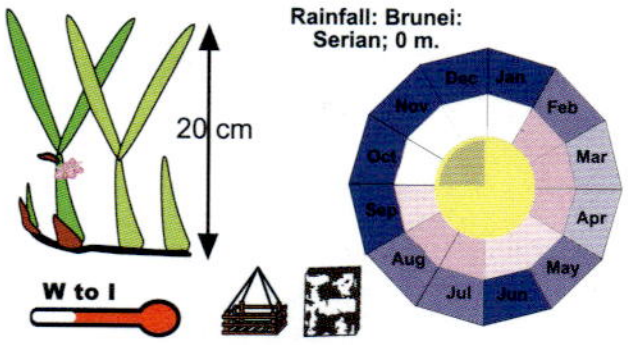


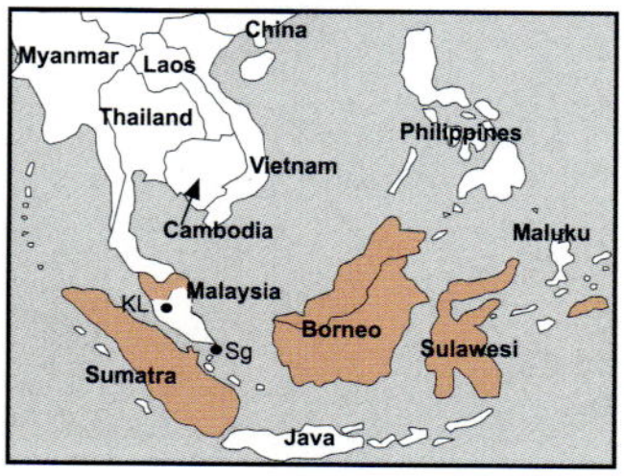


This epiphytic species is common in some habitats in Borneo from sea level to 2000 metres altitude, but only occurs infrequently elsewhere. It prefers rather exposed situations such as high tree branches in heath forest or the crowns of trees on ridgetops. It has a long creeping rhizome with rather attenuated pseudobulbs. Several 1-3 flowered inflorescences are produced just below the twin leaves.

# E

## *Eria nutans* Lindl. Section: *Cylindrolobus*

This common species can be found growing as an epiphyte in a range of lowland forest types and also grows on trees and rocks in the mountains up to 1500 metres altitude, often forming big clumps. Mountain plants are usually much larger than lowland plants. 1 or 2 red-brown sheaths in the leaf axil each produce a succession of short-lived single flowers.

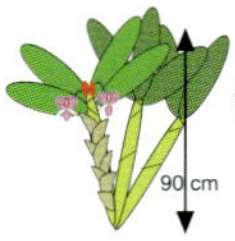


Rainfall and Flowering: Singapore; 10 m.

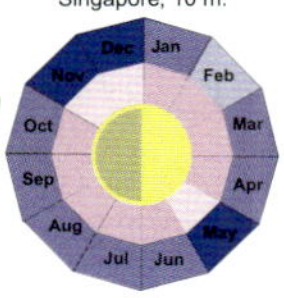


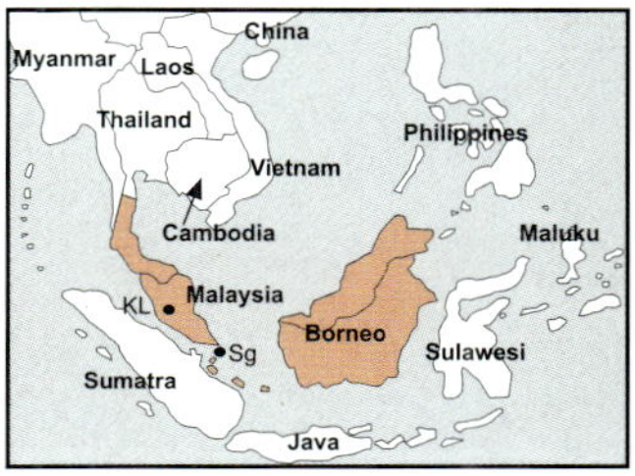


## *Eria obliqua* (Lindl.) Lindl. Section: *Mycaranthes*

Rainfall : Sarawak: Kuching; 25 m.

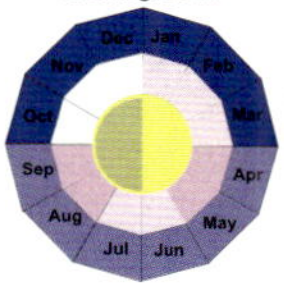

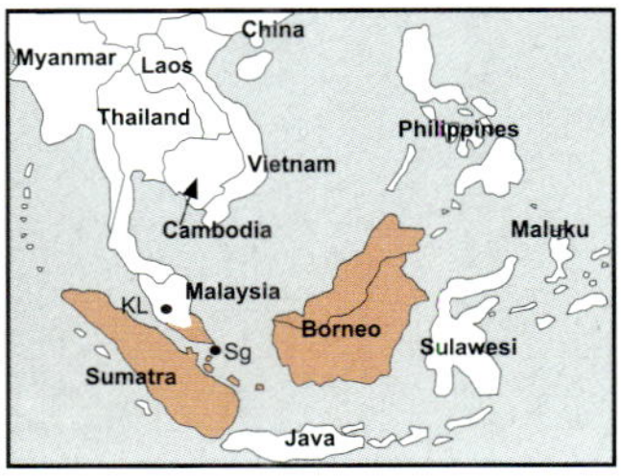


In Malaysia this small epiphytic species is usually found at sea-level on mangroves, but elsewhere it grows in a range of forest types up to 1500 metres altitude. The short stems have 4-5 rather fleshy leaves and usually produce 2 inflorescences at a time. The sepals and translucent petals are usually white, but in some specimens they are suffused a pale salmon-pink colour (see photo).

## *Eria oblitterata* (Bl.) Rchb.f. Section: *Mycaranthes*

This large species is common in most mountainous areas in the region, growing on tree forks and branches and on mossy rocks. The usual colour form is shown on the left of the photo; the right side shows a variety from Sumatra which has darker colours and pea-green calli on the lip.

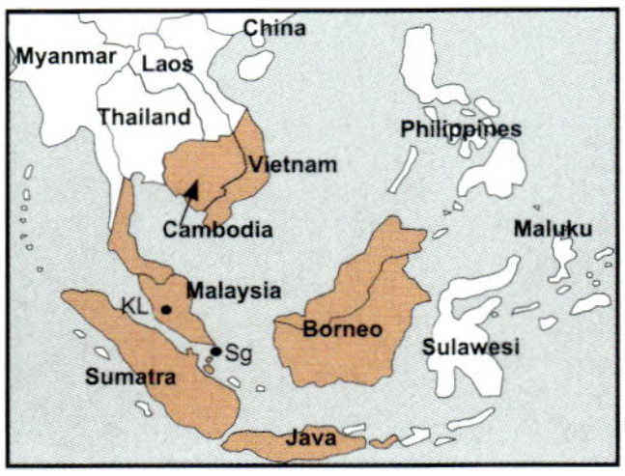

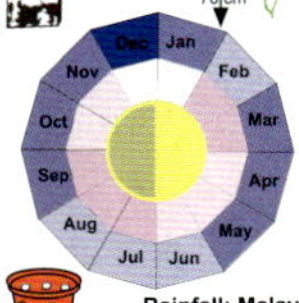

Rainfall: Malaysia; Pahang; 50 m.

# E

## *Eria ornata* (Bl.) Lindl. Section: *Dendrolirium*

This well-known species is found from sea-level to about 1500 metres altitude, usually growing as an epiphyte on isolated trees or in the crowns of emergent forest trees. It has a creeping rhizome with large leathery pseudobulbs at 5-10 cm intervals, each bearing 3-5 large tough leaves. The 40 cm tall inflorescence emerges from the base of the pseudobulb. Although the flowers are rather dull and inconspicuous, the bright orange bracts attract a lot of attention.

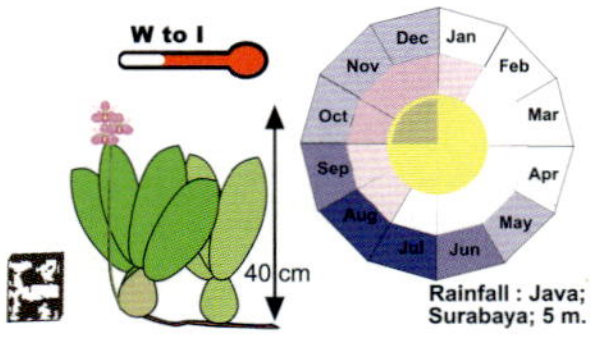


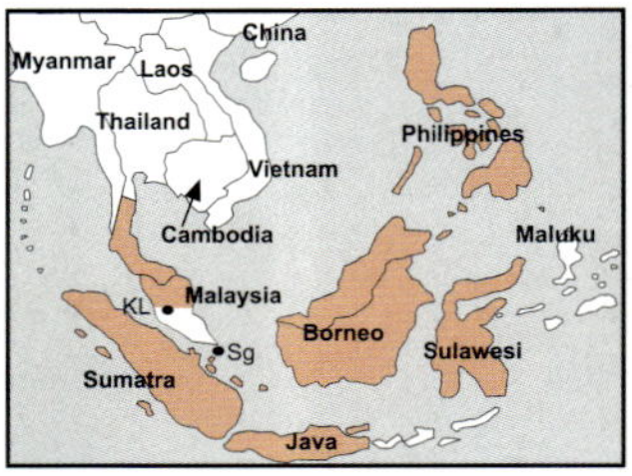


## *Eria pannea* Lindl. Section: *Strongyleria*

In Malaysia this creeping epiphytic species is common in lowland forests in hilly country. Elsewhere it can be found in swamp forests and rainforest up to 1200 metres altitude. The short unswollen stems have 2-3 narrow leaves and produce several 1-3 flowered inflorescences from near the apex. The attractive 2 cm wide flowers smell strongly of vanilla.

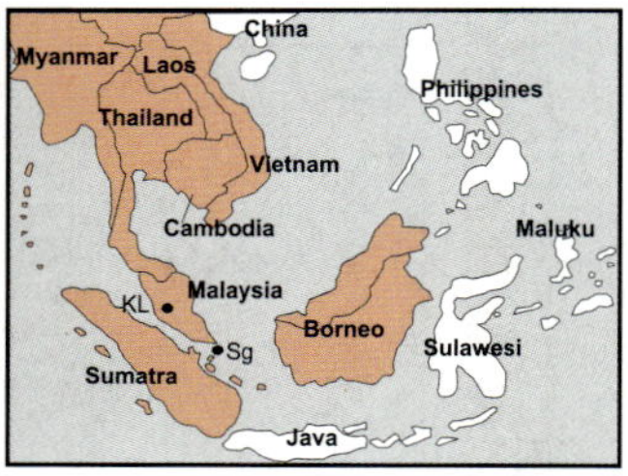


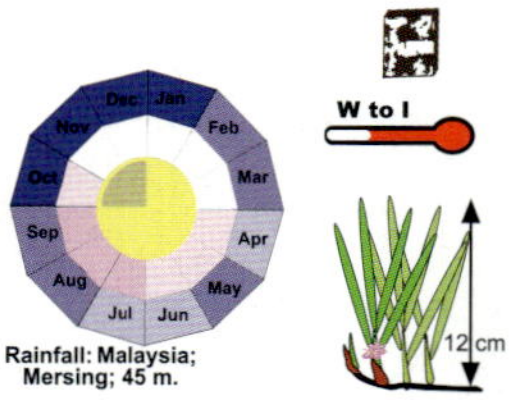


## *Eria pilifera* Ridl. Section: *Cylindrolobus*

In Java this species grows as a terrestrial on lava rocks in full sun, but elsewhere it is usually epiphytic, growing on tree branches in forests and plantations from 600-1300 metres altitude. The height and width of the stems varies according to the light levels, and the floral bracts vary in colour from green (Sumatra and Java) to orange-yellow (Malaysia). The lip midlobe is covered in fine yellow hairs.

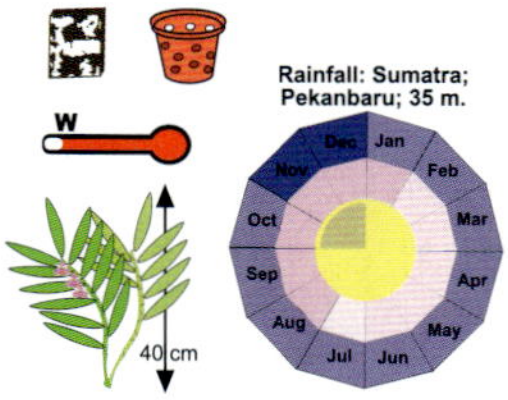


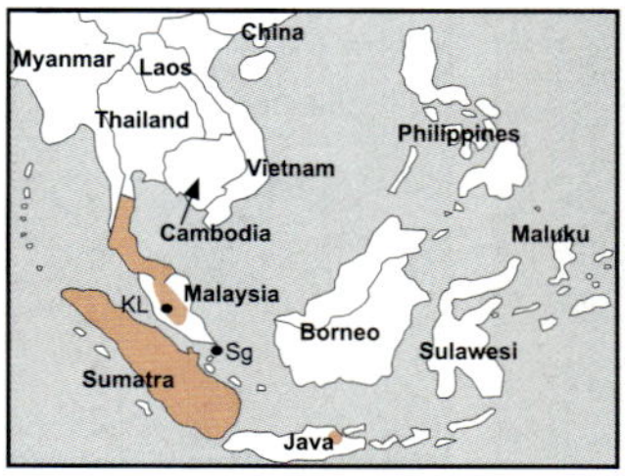

# E

## *Eria pseudoleiophylla* Lindl. Section: *Strongyleria*

This species differs from others in the section in having stems that may contain several nodes, and producing flowers from all parts of the stem. It is found from 800-1700 metres altitude, forming large masses that completely cover trunks and branches of forest trees. The rhizome and bracts are covered in woolly brown-orange hairs.

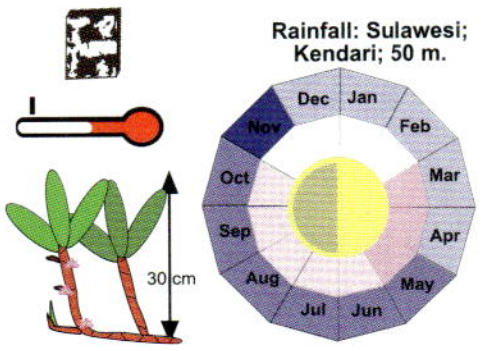


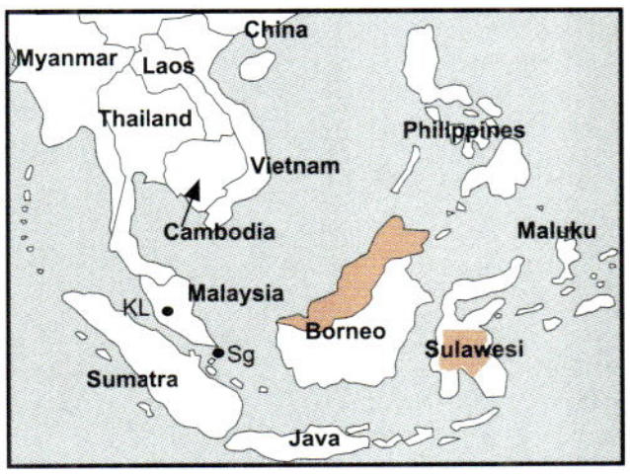


## *Eria pubescens* (Hk.f.) Steudel Section: *Dendrolirium*

This widespread but variable species is found from sea-level to 1600 metres altitude, growing as an epiphyte or (in Sumatra and Thailand) as a terrestrial. Plants from the southern part of the range have deep purple-brown lips and are widely known as *E. albidotomentosa*, a name that refers to the downy white hairs which cover the inflorescence.

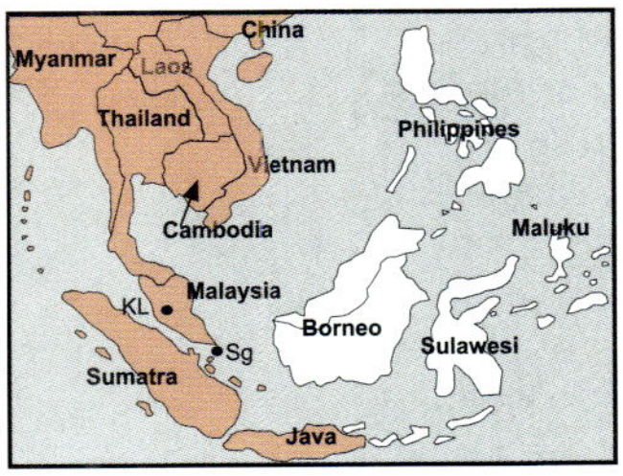


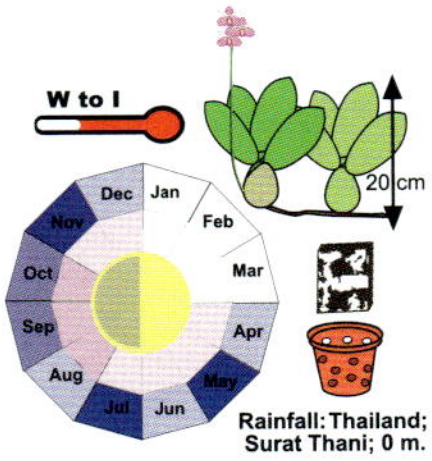


## *Eria pudica* Ridl. Section: *Hymeneria*

This small epiphytic species is usually seen on old mangroves and tree branches that overhang rivers, but it is also quite common in lowland forest up to 400 metres altitude. It has small swollen pseudobulbs with 1-2 leaves and a 3-5 cm tall inflorescence that is covered in short stiff white hairs. It forms up to 10 flowers that are about 6 mm across and do not open very wide. The lip has a swelling at the base of each sidelobe.

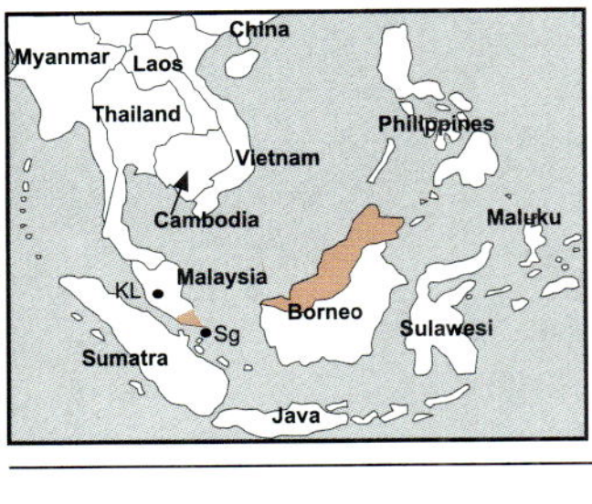


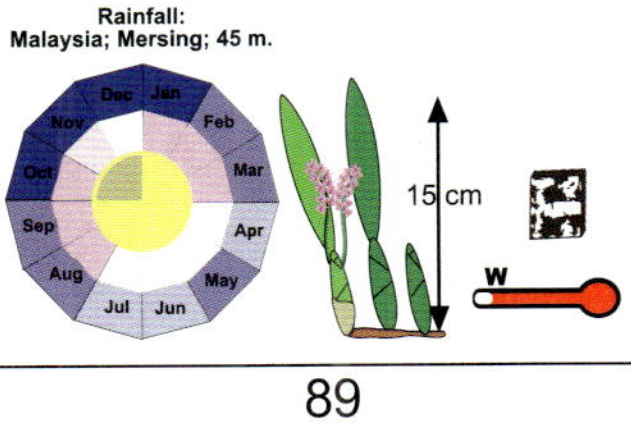

# E

## *Eria pulchella* Lindl. Section: *Tylostylis*

This epiphytic species is very common in old mangroves in southern Malaysia and Sumatra, and is also found in a variety of moist forest types from sea-level to 1400 metres altitude. It has a long creeping rhizome with 2-4 leaved flattened oval pseudobulbs at 5-15 cm intervals. There are 1-3 terminal white-haired inflorescences which extend slowly to about 10 cm long, bearing up to 20 flowers which open one at a time in quick succession.

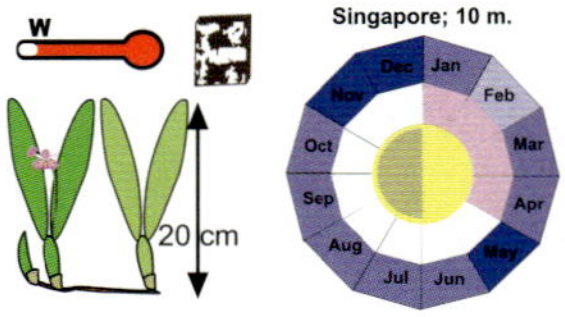


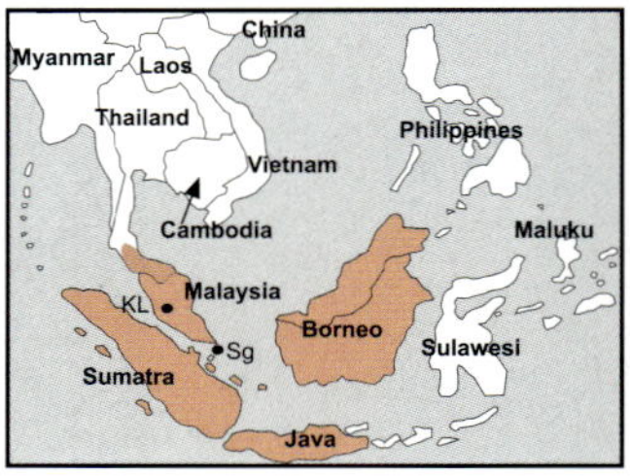

## *Eria pulla* Schltr. Section: *Urostachya*

This little-known but attractive species is widespread in the mountains of Sulawesi between 800-1200 metres altitude, but appears to be uncommon. It grows on the trunks and branches of trees in montane forest and coffee plantations. The bract-covered 3-6 cm tall pseudobulbs have two leaves and produce 1-2 lateral inflorescences which are about 12 cm long and bear up to 30 flowers. The flowers are pale pink with rose-red sidelobes and marks on the column.

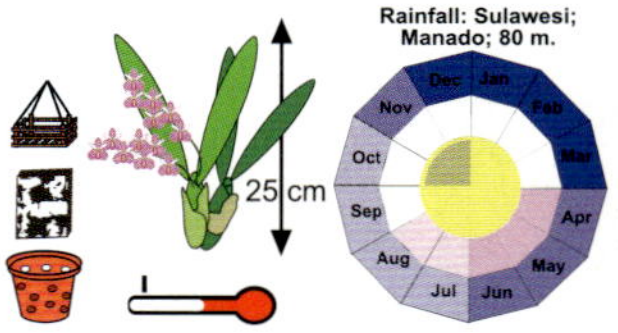


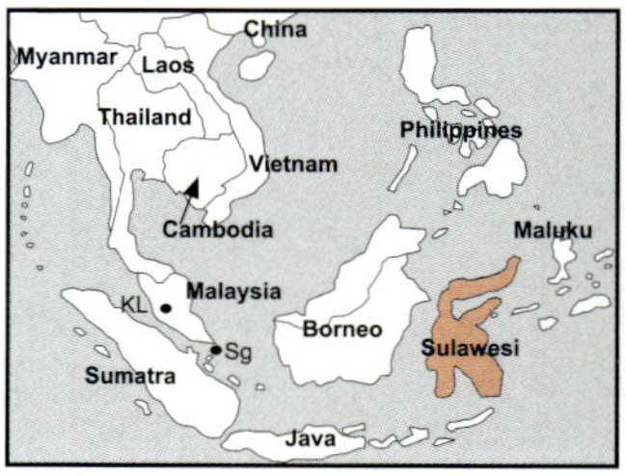

## *Eria punctata* J.J.Sm. Section: *Hymeneria*

This uncommon species is confined to a few scattered locations in the mountains, where it grows in mossy trees from 1000-1800 metres altitude. The rhizome has close-set 3 cm tall bract-covered pseudobulbs that bear a single large thick leaf. There is usually a single 10-15 cm short-haired inflorescence that carries 8-11 well-spaced flowers which are all open at the same time. The 10 mm high flowers do not open widely. The lip has two large keels.

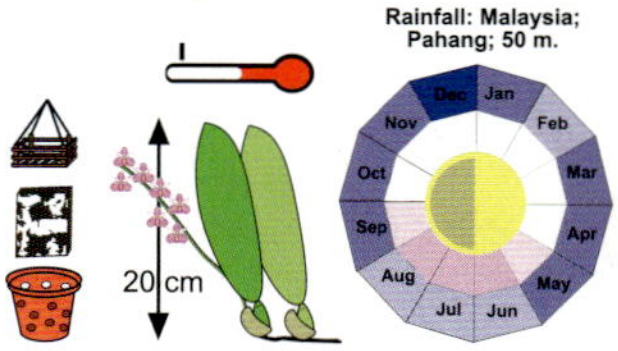


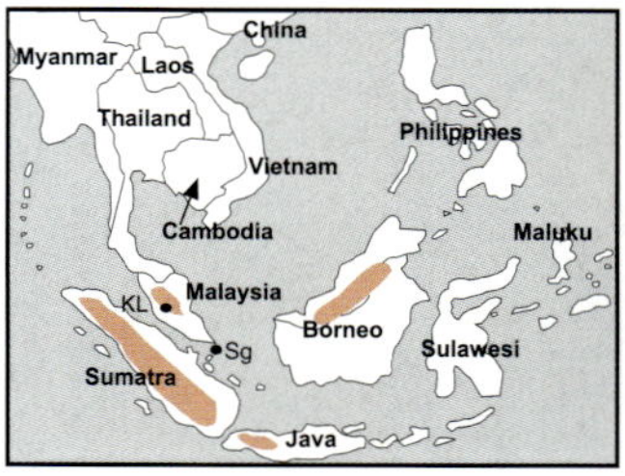

# E

## *Eria rhychostyloides* O'Brien Section: *Urostachya*

This attractive but uncommon epiphyte is a favourite with the few orchid enthusiasts who grow *Eria* species. It is found only in Java, where it grows on the forested slopes of several volcanoes, usually from 500-800 metres altitude. The 10 cm tall fat pseudobulbs have 4-5 leaves near the apex, and usually produce 3 arching inflorescences that are 20 cm long and carry up to 150 crowded flowers which are all open at the same time.

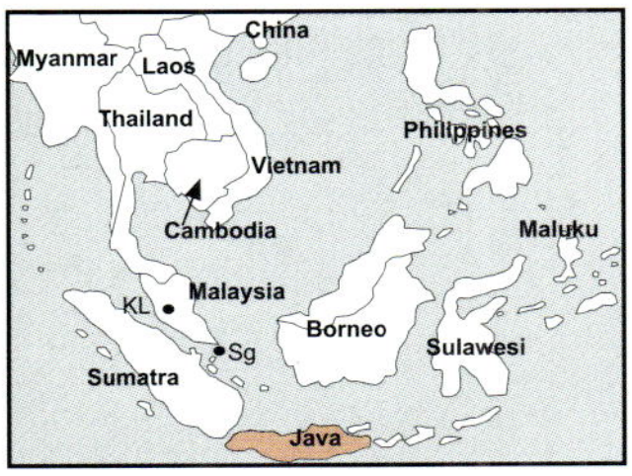


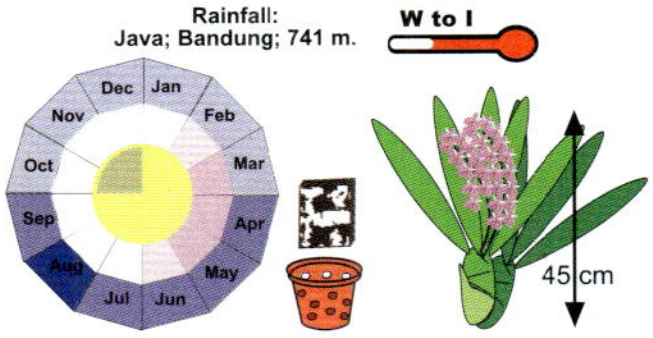


## *Eria rubifera* J.J.Sm. Section: *Hymeneria*

This small epiphytic species was thought to be restricted to a few localities in West Sumatra, but has recently been found in Aceh Province, where it is quite common from 700-1000 metres altitude. It grows on the branches of large forest trees and on isolated trees in villages and farmland. It is fast growing and often covers a whole branch. The short pseudobulbs have 2 leaves and produce several 10 mm high inflorescences that bear 2-4 crowded flowers.

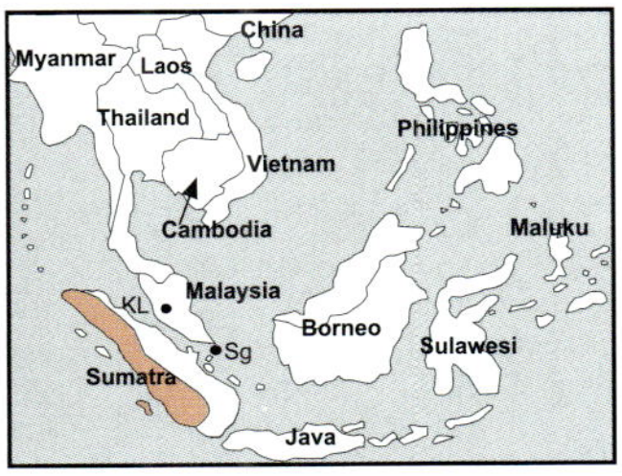


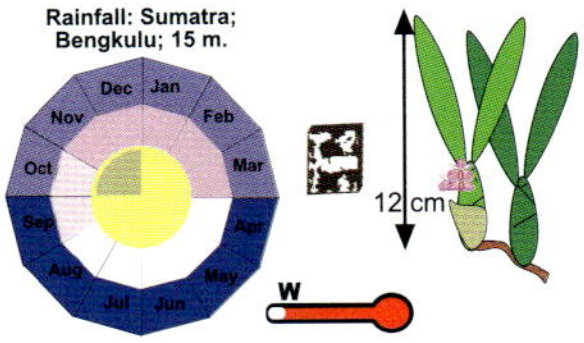


## *Eria tricuspidata* Rolfe Section: *Mycaranthes*

This species is superficially similar to *E. iridifolia*, but has much longer cane-like stems and the flowers have a broader midlobe and a different pattern of keels. It is confined to the mountains of Sulawesi, where it is quite common from 900-1500 metres altitude. It grows in the forks of large montane forest trees and on old coffee bushes. The stem produces 7-8 inflorescences which extend to about 20 cm long, each with about 5-10 flowers open at a time.

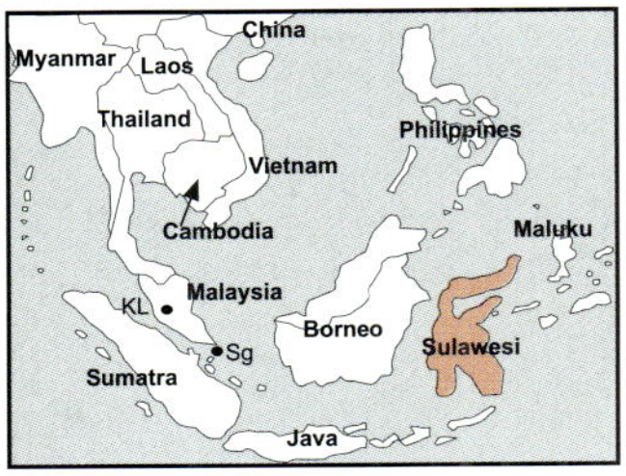


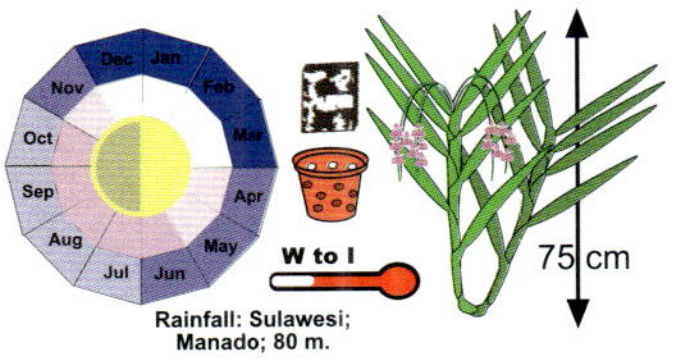

# E

## *Eria vanoverberghii* Ames Section: *Mycaranthes*

This unusual epiphytic species is found on Mindoro Island and the Cordillera range of Luzon, where it grows from 300-1500 metres altitude. It has leafy stems that often droop towards the apex and usually produce up to six horizontal inflorescences which grow to about 20 cm long. Although the flowers are rather small, the snow-white bracts form a spectacular display.

Rainfall: Philippines; Baguio; 1512 m.

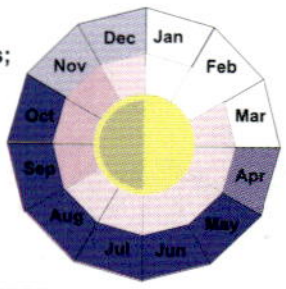


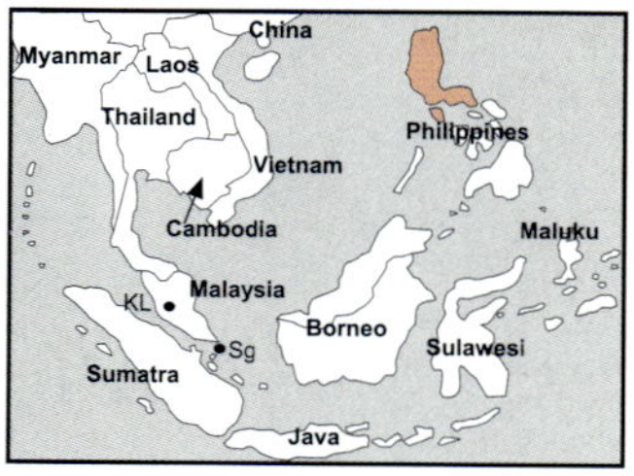

## *Eria xanthocheila* Ridl. Section: *Hymeneria*

This striking epiphytic or lithophytic species is usually found in lowland forests near lakes and rivers, but it is known to occur up to 1600 metres altitude. Each stem produces several inflorescences that have up to 25 brightly coloured flowers which unfortunately only last a few days. A small form of this species grows on granite rocks at Pedu Lake in Malaysia. The stems are only 10 cm tall and the flowers, which are about two-thirds of the normal size, have greenish-white sepals and petals (see bottom right).

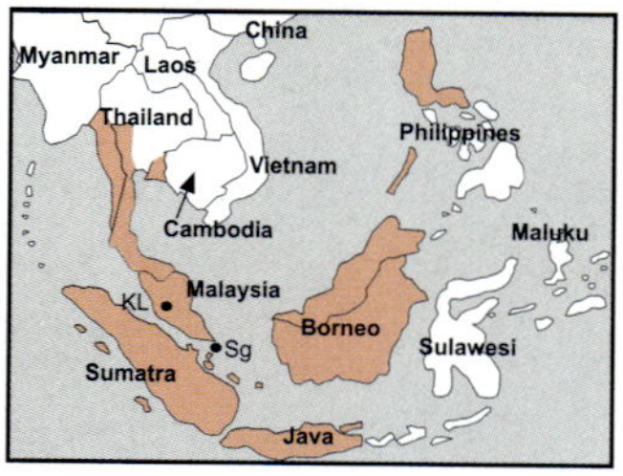

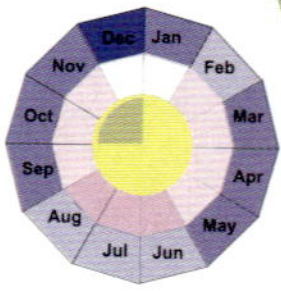


Rainfall: Malaysia; Ipoh; 35 m.

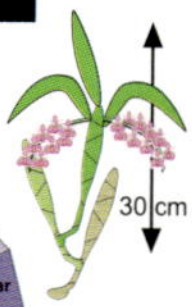

# E

### ***Eria* unidentified species** Section: *Polyura*

This dramatic species was photographed in the Singapore Botanic Gardens. Unfortunately, there is no record of who donated the plant to the SBG, nor any indication of where it originally came from. The majority of species in section *Polyura* are from the Philippines or New Guinea. This unidentified species has a stout creeping rhizome with 15 cm tall club-shaped pseudobulbs at 4-5 cm intervals. Each pseudobulb has 5-6 leaves near the apex and produces 4-6 drooping 12 cm inflorescences which each bear up to 90 crowded small flowers, all of which are open at the same time. It is similar to *E. ovata*, but it has a simple oval lip with three small flap-like keels instead of two.

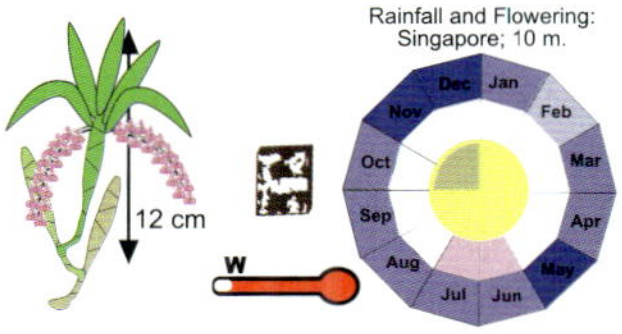


## *Eulophia*

This huge genus of over 200 species has a worldwide distribution, with the majority of species (and the most beautiful) being found in Africa. Although every country in S. E. Asia has a few *Eulophia* species, nowhere in the region is particularly rich in these species. Most *Eulophia* species are deciduous terrestrials with large underground storage organs and are thus particularly well-adapted to scrub or grassland that has a severe dry season during which there are occasional fires; this environment is not common in our region. Asian *Eulophia* species usually have large pseudobulbs, either at ground level or just under it, with the inflorescence arising from a separate shoot. Although most are terrestrial, some (including the widespread *E. zollingeri*) are leafless saprophytes. Asian *Eulophia* species are seldom seen in collections.

### ***Eulophia andamanensis*** Rchb.f.

This deciduous lowland species grows in sandy soil, open grassland and scrubland in regions with a pronounced dry season. It readily colonises disturbed soil and can appear on paths, in farmland, in village gardens, and in parkland and wasteland in towns. It has large green pseudobulbs with 4-5 narrow leaves. At the start of the dry season the leaves drop off, after which a 40 cm tall inflorescence is formed that bears up to 15 well-spaced 3 cm wide flowers.

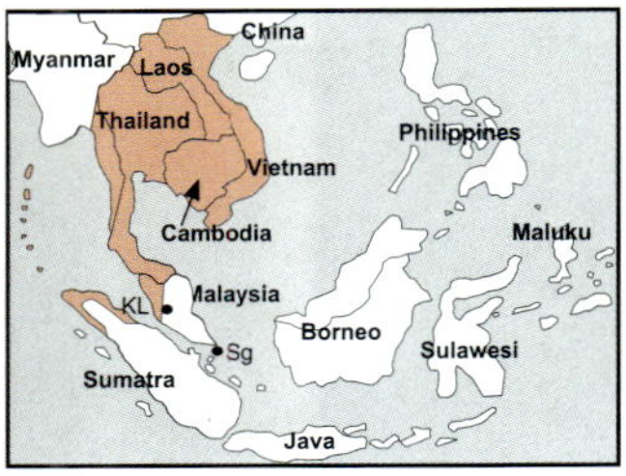


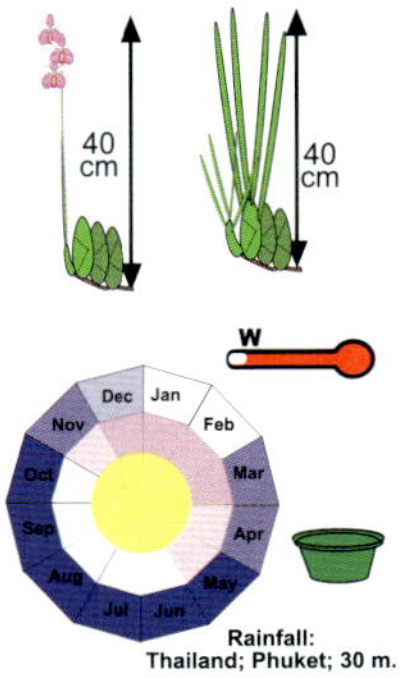

# E

## *Eulophia graminea* Lindl.

This deciduous lowland species can be found almost anywhere in our region where the habitat is suitable. It grows on sandy beaches, open grassland, farmland and wasteland. It spreads wherever humans transport soil, so it can appear on roadsides, playing fields and building sites. The large pseudobulbs have several grass-like leaves that fall at the end of the growing season, after which a tall branched inflorescence is formed that bears many 3 cm wide flowers. A form with a white lip is known from Java.

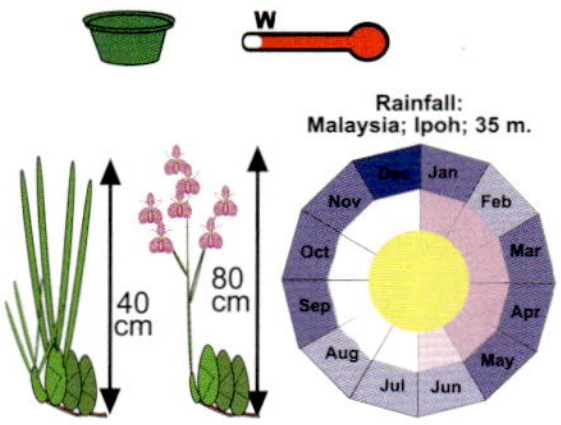


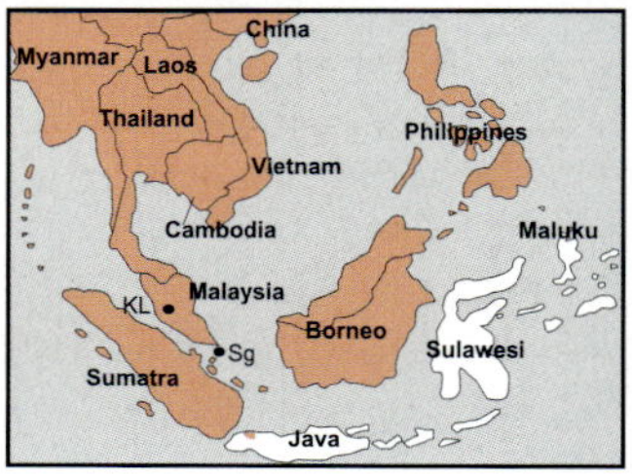


## *Eulophia spectabilis* (Dennst.) Suresh.

This widespread species is often called by the old names *Eulophia nuda* and *Eulophia squalida*. It can be found from sea-level to about 900 metres altitude, growing in open grassland and swamplands. It is quick to colonise disturbed soil, roadside cuttings, paths and farmland. The 30-100 cm tall inflorescence is produced shortly after the leafy shoot matures, and bears up to 15 well-spaced flowers, with up to 5 open at a time. Flower colour is highly variable; the illustrated plant (from Sulawesi) has a yellow lip centre instead of the usual white; in Thailand flowers are often bright yellow or brilliant scarlet.

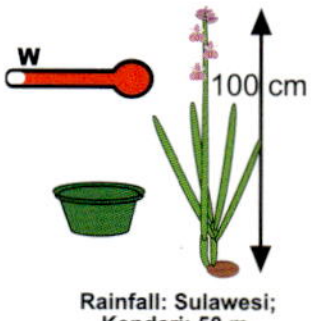


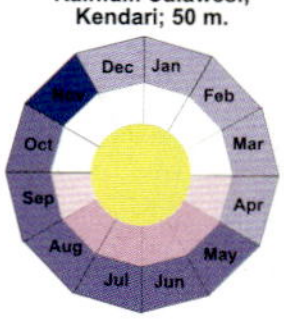


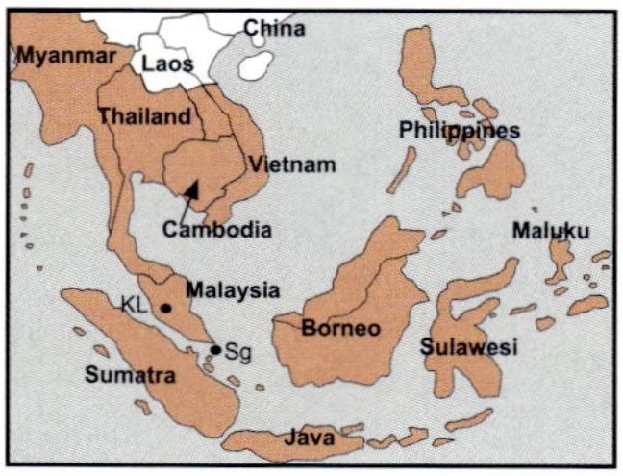

# F

## Flickingeria

Although hardly known by orchid growers on account of its extremely short-lived flowers, this genus is very common in S. E. Asia and is easy to spot in the wild. It has an estimated 70 epiphytic or lithophytic species and occurs throughout the region, from India to the Pacific Islands. With the exception of *F. bicostata* (below) the species have an unusual growth habit; the tough creeping rhizome produces aerial shoots that bear single-leafed pseudobulbs at diminishing distances. In many species the pseudobulbs progressively diminish in size along each aerial shoot. Flowers are usually produced from bracts at the tip of the pseudobulb, and are usually only open for less than one day. As is the case with most fugacious orchids, they flower frequently throughout the year.

---

### ***Flickingeria bancana*** Hawkes

This small lowland species is quite common in southern Malaysia. The aerial branches are quite short and the largest pseudobulbs are only 4-6 cm long. The brightly-coloured flowers emerge behind the leaf axil, are only about 12 mm across, and are quite thick and waxy.

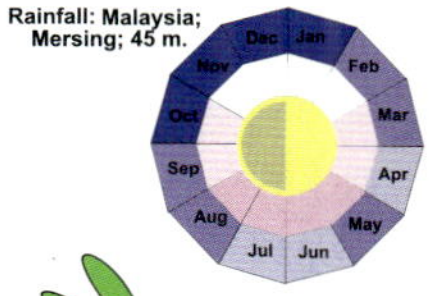


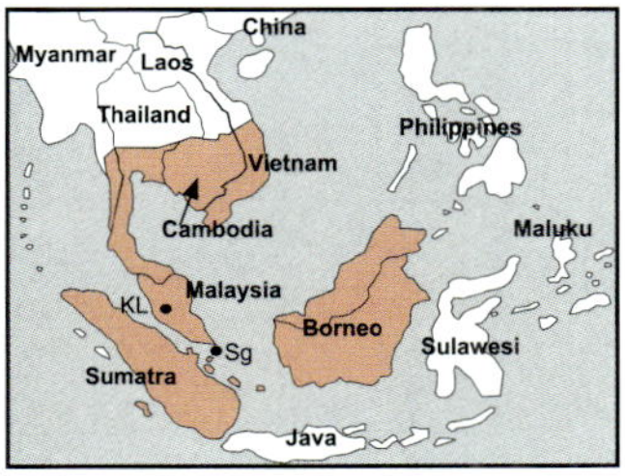

---

### ***Flickingeria bicostata*** (J.J.Sm.) Hawkes

In many ways this is an unusual *Flickingeria*. It does not have aerial branches on the rhizome; the razor-shaped pseudobulbs are very large and all approximately the same size. The strongly fragrant flowers are the largest in the genus, last 2-3 days, and are formed intermittently throughout the year. The species is quite rare, growing in forests from 500-1200 metres altitude, preferring locations where there is plenty of moss.

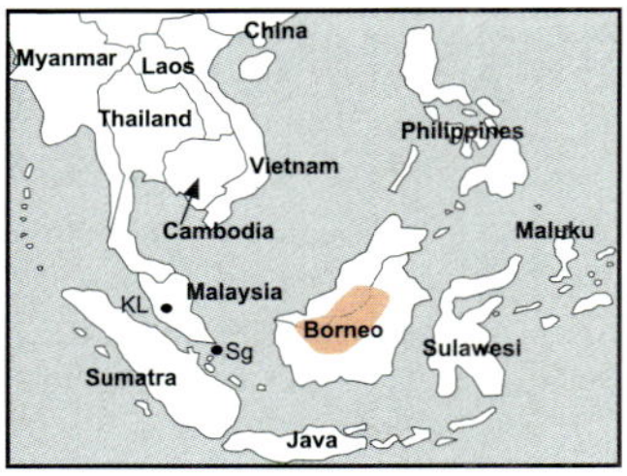


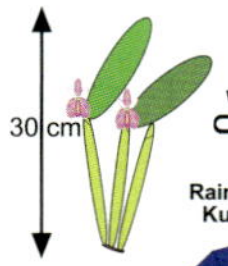

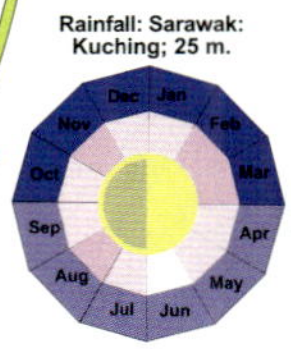

# F

## *Flickingeria comata* (Bl.) Hawkes

This large species can be spotted throughout much of the region on trees in plantations and parkland. On the flowering days it produces more blooms than most *Flickingeria* species as the multi-flowered inflorescences are formed from the aerial sections of the rhizome as well as from nodes on the pseudobulb.

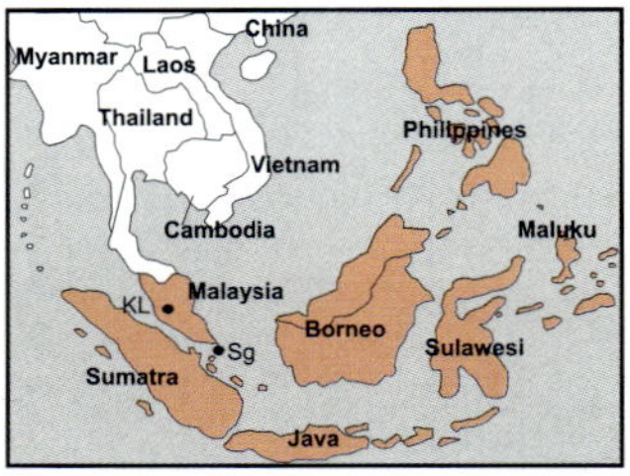


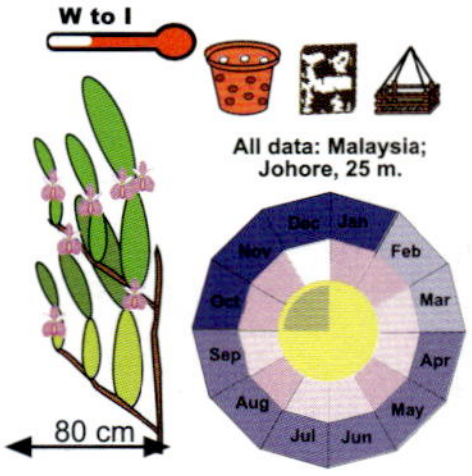

## *Flickingeria fimbriata* (Bl.) Hawkes

This is a large species and can form huge masses that spread outwards and hang downwards from a tree branch. The pseudobulbs are much flattened, up to 15 cm apart, and up to 8 cm long. The flowers are lightly scented and vary in colour from white to yellow with red, brown or purple spots.

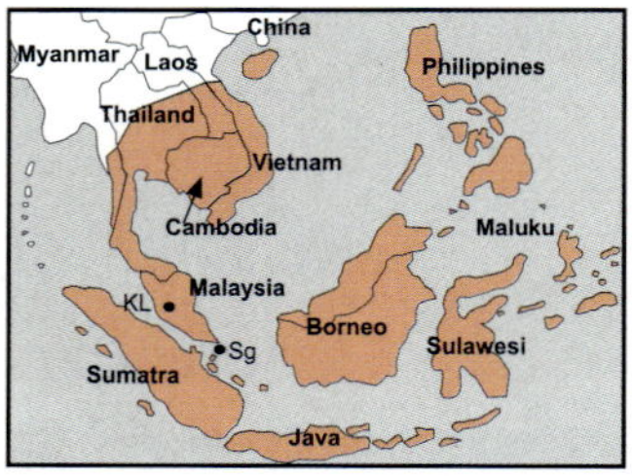


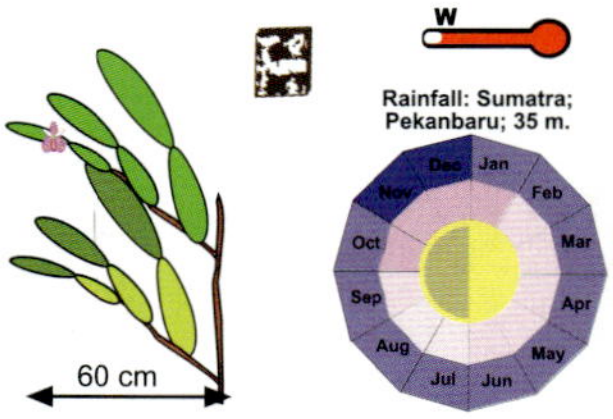

## *Flickingeria xantholeuca* (Rchb.f.) Hawkes

Although the branches of this plant are fairly short, it spreads rapidly and can quickly cover a branch or a rock. The species is very common in parts of Borneo and Java from sea-level to 1200 metres altitude, but is usually found in dense moist forest. The largest pseudobulbs are up to 5 cm long and are rather spindle-shaped. The 1 cm wide flowers emerge behind the leaf axil.

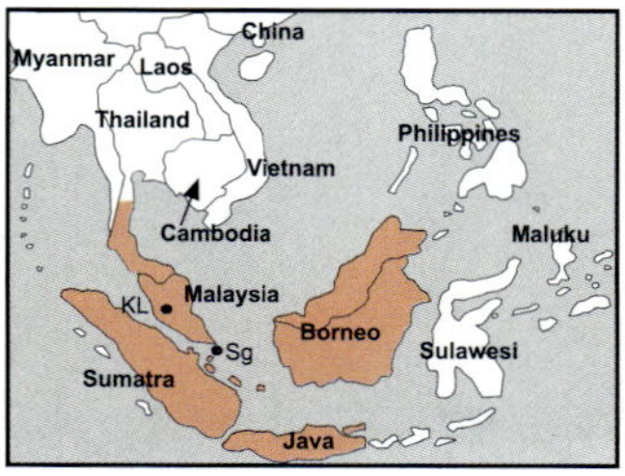


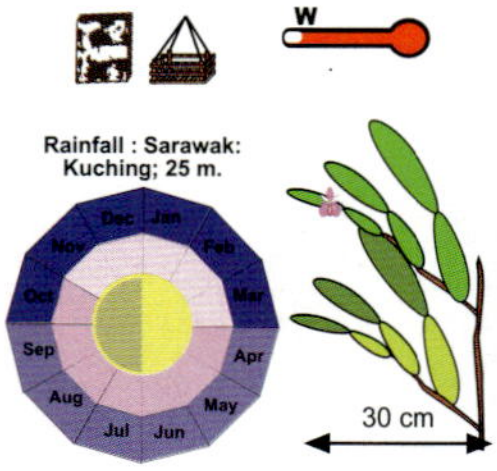

# G

## Gastrochilus

This genus of about 40 epiphytic species is distributed from Sri Lanka and the north-west Himalayas to Japan and the Philippines. Most species have short erect stems and close-set leaves although one group of species have long pendulous stems. The inflorescence is usually short with up to 10 flowers in a tight cluster. The flowers all have a large saccate lip with a fan-shaped mid-lobe. *Gastrochilus* species are becoming popular with orchid enthusiasts because of their cheerful colours and frequent flowering.

### *Gastrochilus calceolaris* (Buch.-Ham. ex J.J.Sm.) D.Don

This small epiphytic species is quite common in mountainous regions at altitudes above 1000 metres. The narrow leaves are thick and leathery and have a deep V-shaped notch in the tip. The plant usually produces several short inflorescences at a time, each bearing a crowded cluster of 5 or 6 flowers. The lip midlobe is covered with long flat square-edged hairs.

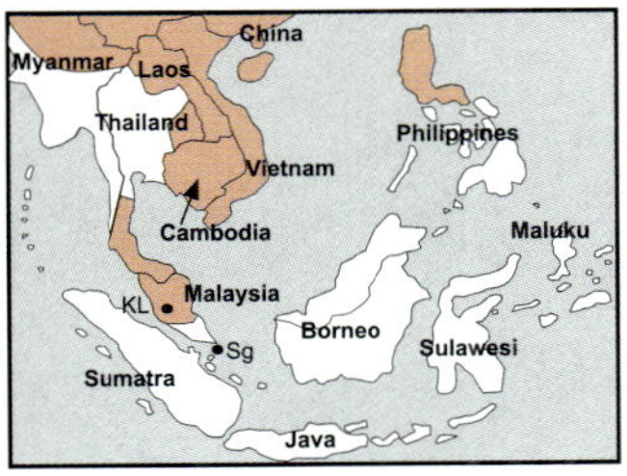


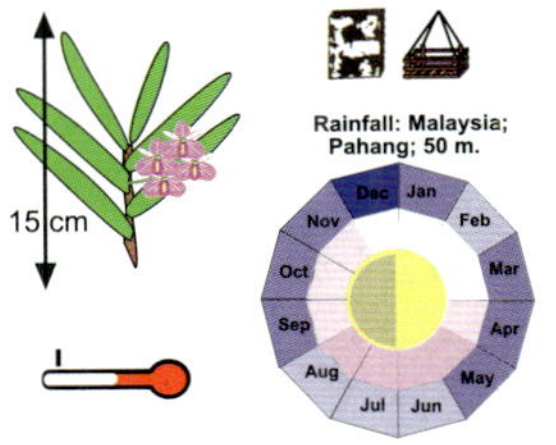

## Goodyera

This genus of about 40 terrestrial species occurs worldwide in all tropical and temperate regions except in Africa. S. E. Asia has at least 25 species which are found growing in deep shade in humus on the forest floor. They have a creeping rhizome that produces erect leafy shoots at intervals, with an inflorescence on the top of the shoot. Some *Goodyera* species have ornate leaves and are occasionally seen in cultivation; these (along with other variegated terrestrial orchids) are known as "Jewel Orchids".

### *Goodyera reticulata* (Bl.) Bl.

This attractive little species is found in the hills of Java, usually in the moister forests, above 500 metres altitude. Each shoot forms a low rosette of 3-5 leaves, the largest of which is about 6 cm long. A 15 cm tall erect inflorescence emerges from the centre of the rosette and bears up to 35 flowers, about half of which are open at the same time. The flowers do not open widely; the dorsal sepal (which is uppermost) and the petals form a "hood" over the column and lip. In cultivation plants flower throughout the year.

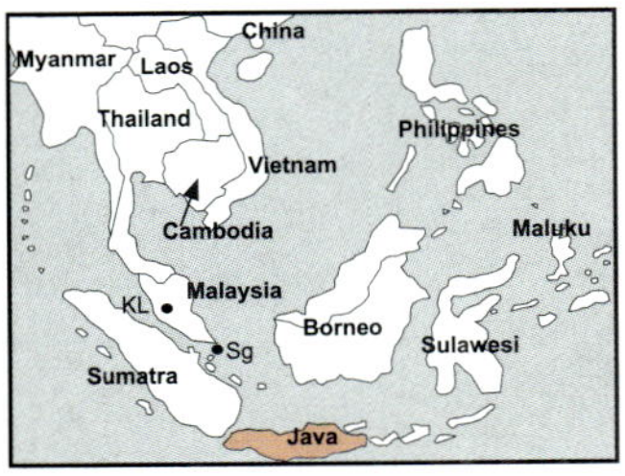


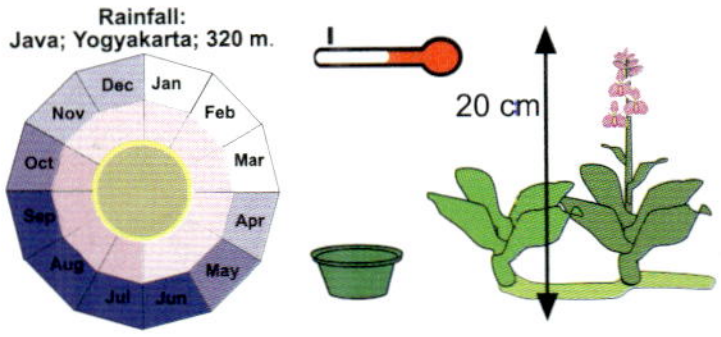

# G

## Grammatophyllum

This is a small genus of only 10-12 epiphytic species, found from S. E. Asia to the Pacific Islands, and usually considered to be related to *Cymbidium*. Three species are often seen in cultivation; *G. scriptum* ( Sulawesi to the Pacific Islands), which has been "improved" by line breeding, has yellowish-flowered arching inflorescences that may exceed 2 metres in length; and the two species below, *G. speciosum* and *G. stapeliiflorum*.

### *Grammatophyllum speciosum* Bl.

The "Tiger Orchid" is often cited as "The World's Largest Orchid", (although the "heaviest orchid" would be a better description). It is too large to cultivate in pots and is usually grown in a tree fork or on the ground. Flowering is infrequent; some plants only flower every third or fourth year !

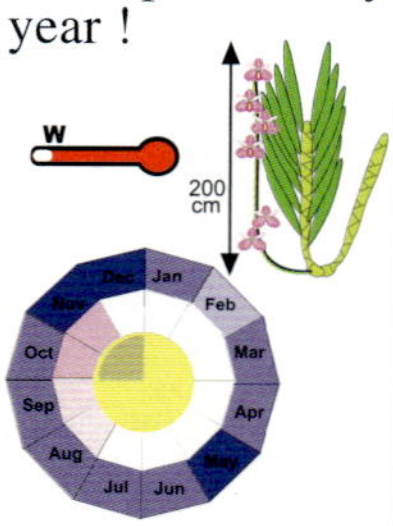

Rainfall and Flowering: Singapore; 10 m.

### *Grammatophyllum stapeliiflorum* (T.& B.) J.J.Sm.

This species, the famous "Black Orchid" of Sulawesi, is found in hilly country from Malaysia to New Guinea, usually at around 500-1200 metres altitude. It has large heavy egg-shaped pseudobulbs and a descending inflorescence. The flower colour is variable and can be very dark, although a true black has not yet been seen; they are usually brown to dark maroon, grey or blue-grey with dark brown to maroon-black markings; some plants from New Guinea have off-white flowers with dark lavender to purple spots and petals.

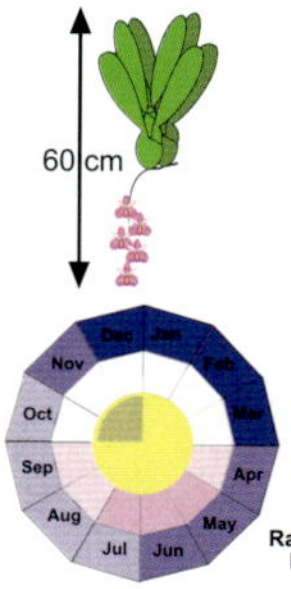

Rainfall: Sulawesi; Manado; 80 m.

# G

## Grosourdya

A genus of about 10 small species, mostly twig epiphytes, found from the Andaman Islands and Myanmar to Sulawesi and the Philippines. Although *G. appendiculata* occasionally turns up in collections in S. E. Asia, these species are not sought after because of their small size and short-lived flowers. Although the brightly-coloured flowers usually only last one day they are produced intermittently throughout the year and have a fascinating structure, with 4 erect hornlike lobes on the lip.

---

### *Grosourdya appendiculata* (Bl.) Rchb.f.

This miniature species is widespread throughout most of the wet tropical lowland regions of S. E. Asia. It is a twig epiphyte, often occurring in very large numbers on the outer branches and twigs of a particular tree. It flowers from a very small size, possibly as little as three years after germination. The colours and spur shape are quite variable.

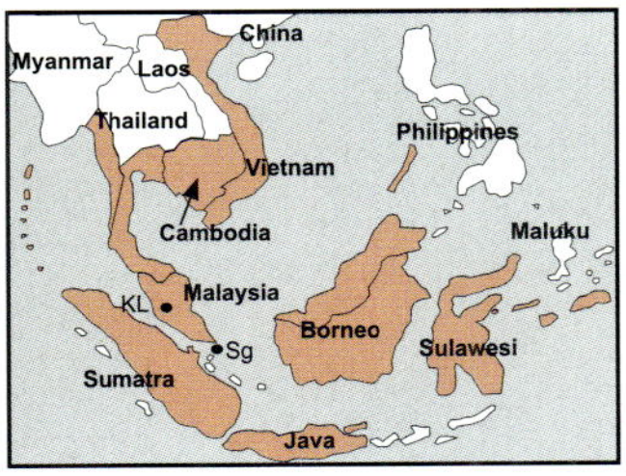


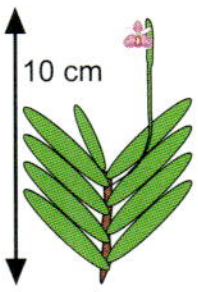

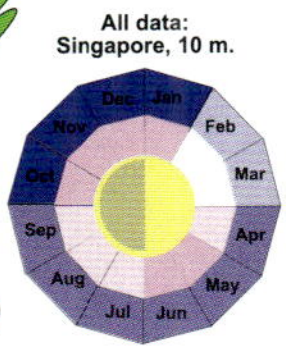

---

### *Grosourdya pulvinifera* (Schltr.) Garay

This little-known miniature species grows on forested hillsides from 150-650 metres altitude. It is a twig epiphyte, often occurring in very large numbers on the outer branches and twigs of a particular tree. A full sized plant is only 5 cm long, but it is a prolific producer of flowers, and can have 8-10 active inflorescences, each producing up to 3 flowers at each flowering session. Unlike other *Grosourdya* species, the flowers last up to 5 days. There are two colour variants; both are shown in the illustration.

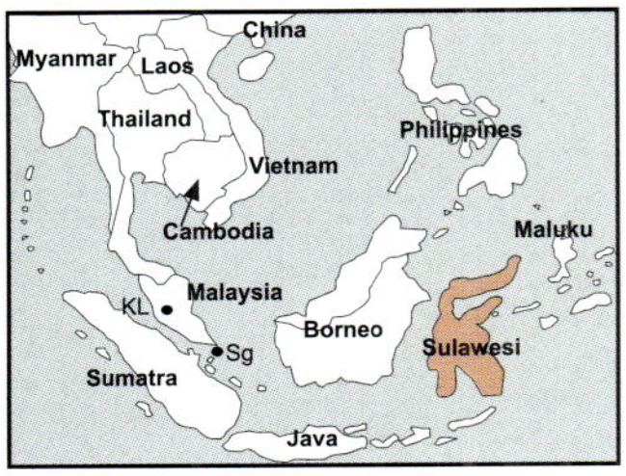


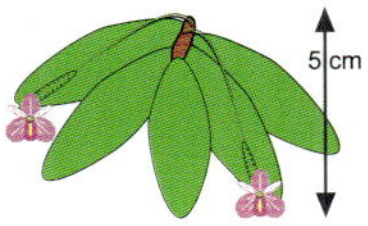


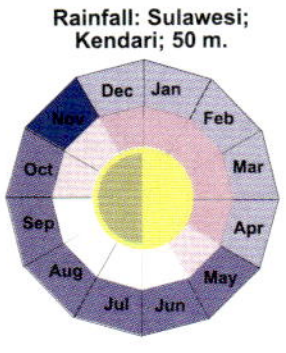

# H

## Habenaria

*Habenaria* is a terrestrial genus found on every continent except Antarctica; there are probably 700-800 species altogether, of which about 100 are from S. E. Asia. Because the majority of *Habenaria* species prefer a temperate seasonal climate, few are found in the wet tropical lowlands, and most Asian species come from mountainous regions with a distinct wet-dry climatic cycle. Although this genus contains some species with spectacular flowers, they can be difficult to maintain in cultivation and are not often grown; occasionally individual specimens are seen in orchid collections. Many species have large white or pale green flowers with elaborate lip structures and long spurs.

### ***Habenaria marmorophylla*** Ridl.

This tall species grows on the forest floor in limestone regions. It has up to 10 greyish-green leaves and a 100 cm-long inflorescence that bears up to 30 flowers, with 2-6 open at a time. The 3 cm long flowers are not particularly beautiful until about an hour after sunset when they suddenly emit large quantities of a strong vanilla-like fragrance.

Needs limestone in the media.

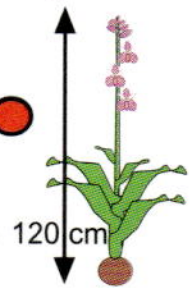


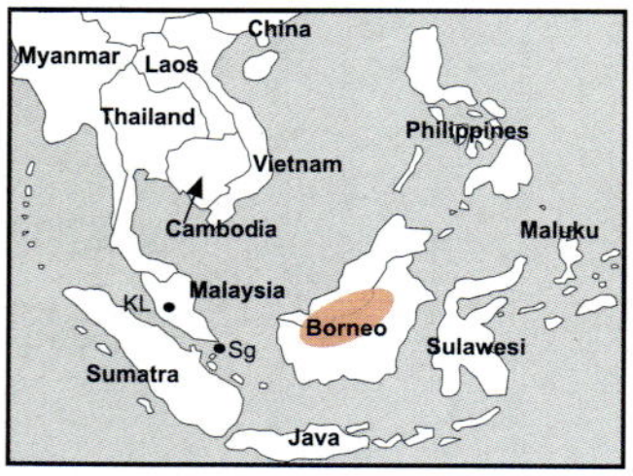


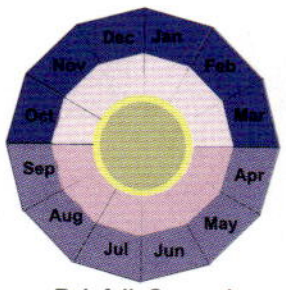


Rainfall: Sarawak: Kuching; 25 m.

### ***Habenaria undulata*** (J.J.Sm.) J.J.Sm.

This 60 cm tall species grows on the floor of forests from 500-1000 metres altitude. It has up to 6 rather soft green leaves near the base of the shoot, then a 50 cm long inflorescence that bears up to 40 flowers, with up to half of them open at a time. The flowers are about 15 mm long and have a straight slender spur that is longer than the pedicel and ovary. In most *Habenaria* species the lip is divided into 3 lobes, and in this species the midlobe is shorter and narrower than the outer lobes.

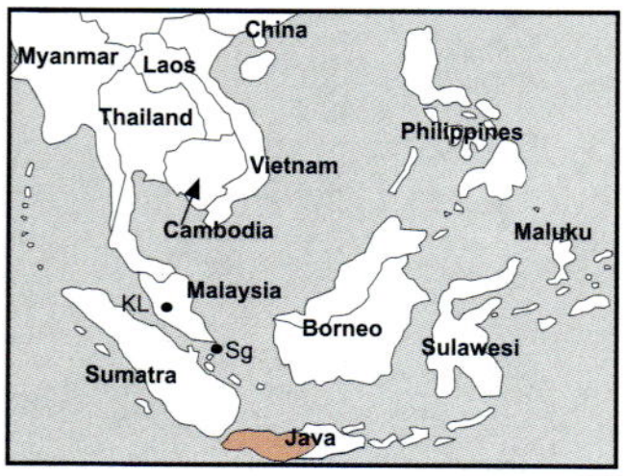

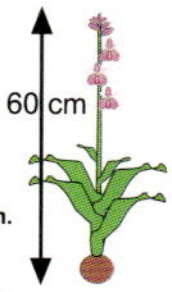


Rainfall: Java; Bandung; 741 m.

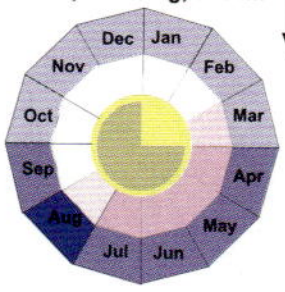

# K

## Kingidium

This genus of 5 epiphytic species, which some botanists consider to be part of *Phalaenopsis*, is distributed from the Himalayas to Sri Lanka, China and the Philippines. There are several species in cultivation, of which *K. deliciosum* is probably the best known. *Kingidium* plants look like miniature *Phalaenopsis* plants and require similar cultivation conditions. The inflorescences lengthen continuously for several months, usually with a few flowers open at any time.

### *Kingidium deliciosum* (Rchb.f.) Sweet

This widespread species is sometimes known by the old names *Phalaenopsis deliciosa* or *Phalaenopsis bella*. It is quite common from sea-level to 500 metres altitude in some areas, growing on trees in forest, plantations and villages. It is frequently seen in cultivation, where it can produce flowers throughout the year; the illustrated inflorescence flowered continuously for 18 months. An *alba*-form occurs in Java.

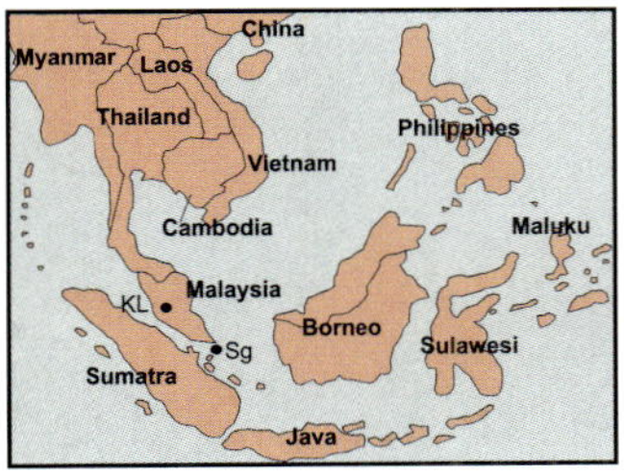


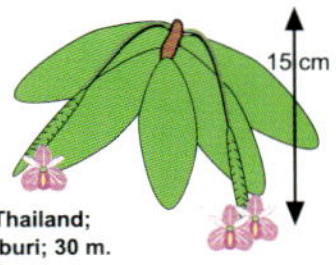


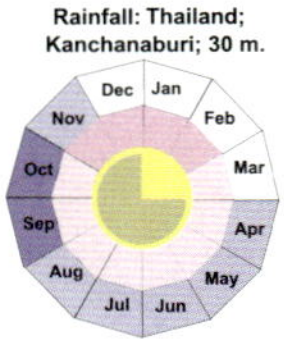

### *Kingidium minus* Seidenfaden

This uncommon species was only discovered in 1988 when the Thai orchid nursery which was growing the plant sent in some flowers for identification. *K. minus* appears to be confined to north-eastern Thailand. It has a short stem with rather fleshy broad leaves. The inflorescences continue to grow for several months, intermittently producing 18 mm long flowers. This species can be identified by the rather square sidelobes on the lip.

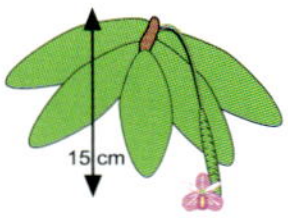


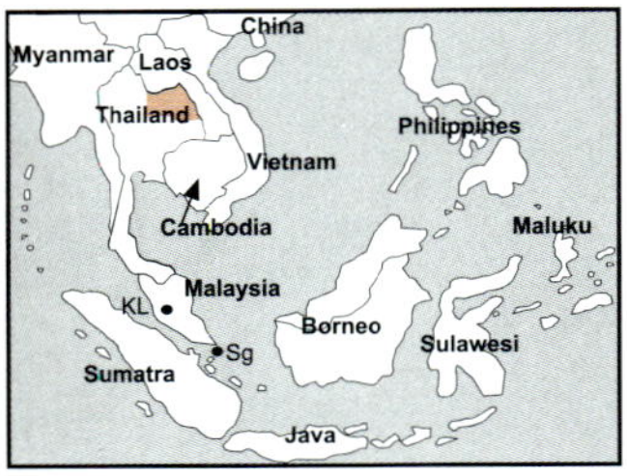


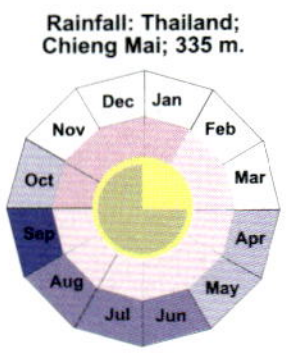

# L

## Liparis

This huge genus of at least 350 species has a worldwide distribution, with the majority of species found in New Guinea and S. E. Asia. Although some tropical *Liparis* species are terrestrials with underground storage organs (like their temperate relatives), many are epiphytes. Some tropical species have pseudobulbs while others lack them. Despite the fact that some species are extremely beautiful, none are in widespread cultivation although individual specimens are occasionally seen in orchid collections.

---

### ***Liparis caespitosa*** (Thou.) Lindl.

This quick-growing species is distributed from eastern Africa to the Himalayas and through S. E. Asia to the Pacific Islands. In the wild it is an epiphyte or lithophyte, growing from sea-level to 1500 metres altitude, but it has proved very versatile and highly adaptable and has gained a foothold in some temperate countries as a weed that infests greenhouse orchid collections. The 15 cm long inflorescence appears from the top of a new shoot before the pseudobulb fully develops and can bear up to 25 flowers which are only 3 mm wide.

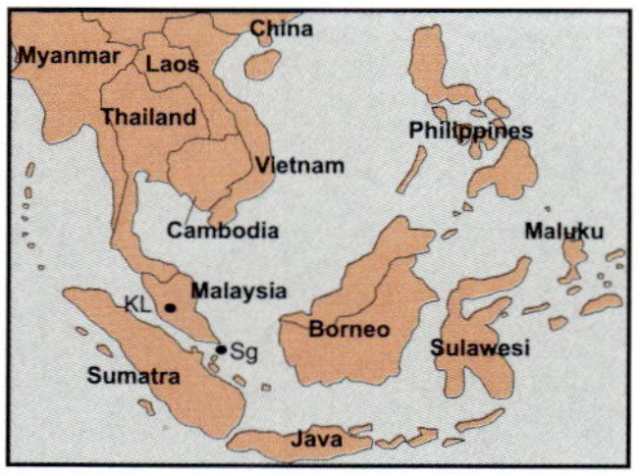


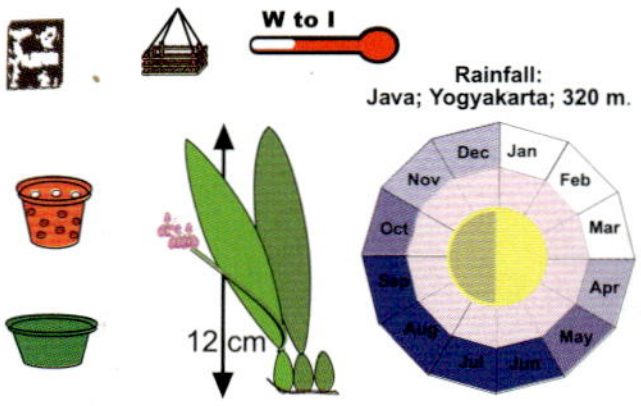


---

### ***Liparis condylobulbon*** Rchb.f.

Distributed through much of S. E. Asia to the Pacific Islands, this fast-growing epiphytic species is common in Sumatra and Java from 500- 1500 metres altitude, but in New Guinea is usually found below 100 metres. It can form huge clumps that totally cover a tree trunk or branch. The flower colours are variable, as the photograph shows.

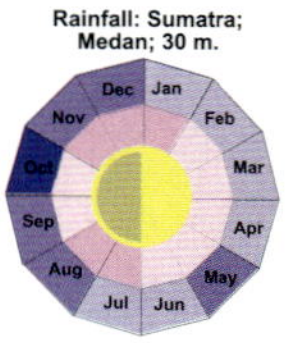

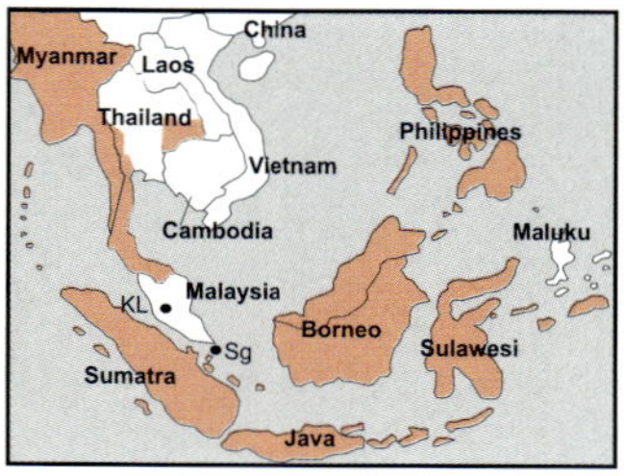

# L

## *Liparis elliptica* Wright

This small epiphytic mountain plant is also found in Sri Lanka, India and Taiwan; a strange distribution ! It gained its name from the elliptic 2 cm tall pseudobulbs that are strongly flattened bilaterally.

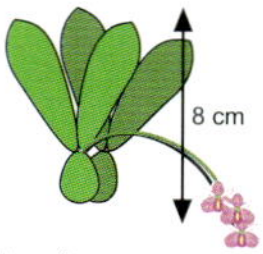


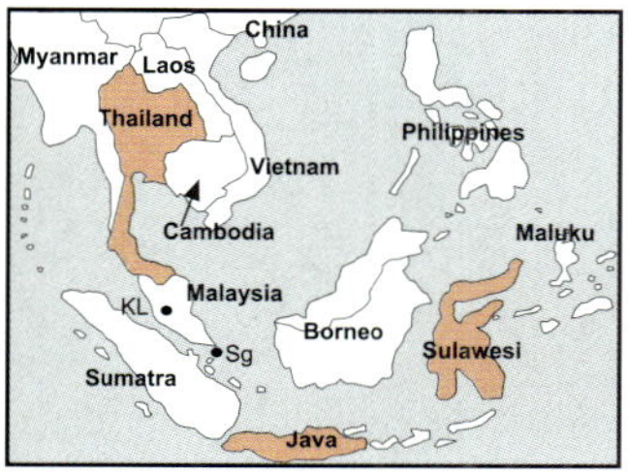


Rainfall: Sulawesi; Kendari; 50 m.

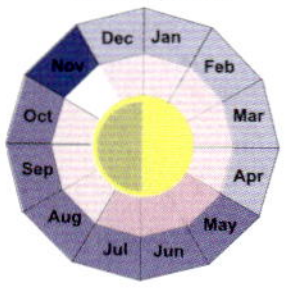

## *Liparis gibbosa* Finet.

This striking epiphytic or lithophytic species is found from 100-1500 metres altitude and is quite common at around 1000 metres across much of the area. The inflorescence produces 1-2 flowers at a time over a period of several months. There is considerable variation in flower colour.

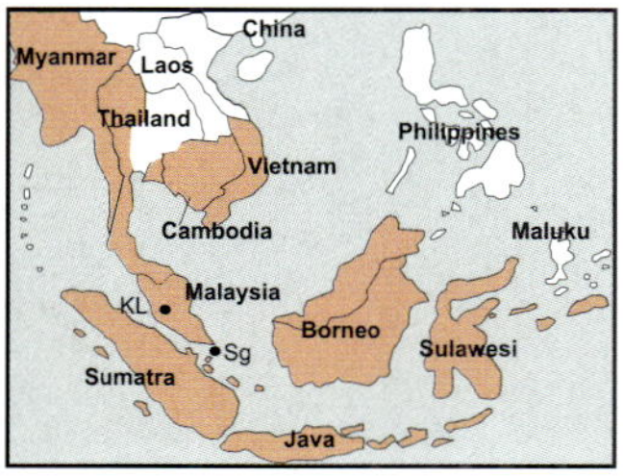


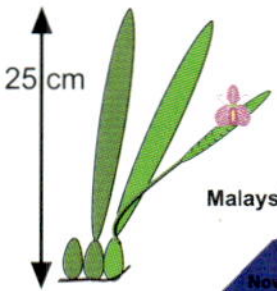


Rainfall: Malaysia; Kuantan; 20 m.

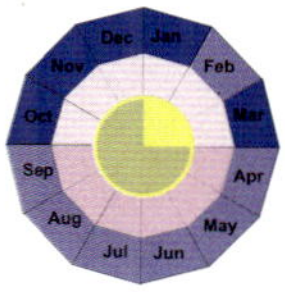

## *Liparis lacerata* Ridl.

This beautiful but uncommon epiphytic species is found from sea-level to 600 metres altitude. It forms clumps of 2-3 cm tall ovoid pseudobulbs that have two 18 cm lanceolate leaves. The inflorescence hangs freely from a very thin peduncle and can extend to about 40 cm long. The flowers open gradually from the base to the apex of the inflorescence with about a 10 cm length open at once. The flowers are about 15 mm across and have a lacerated lip apex.

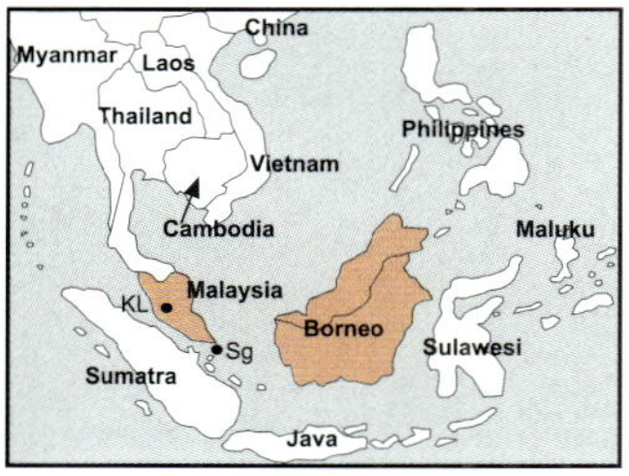

All data: Malaysia; Johore

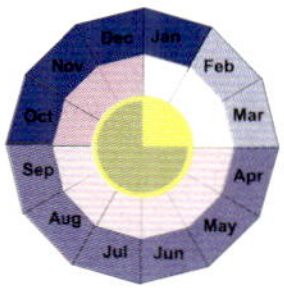


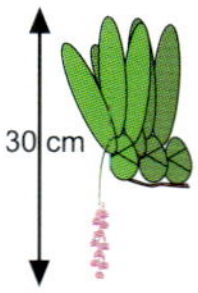

# L

## *Liparis latifolia* (Bl.) Lindl.

This epiphytic species grows in the forks of trees in forests from 250-1700 metres altitude and often forms fairly large plants that are spectacular when in full bloom. The pseudobulbs produce several 25 cm long inflorescences each with 20-30 flowers. Flower colour is variable.

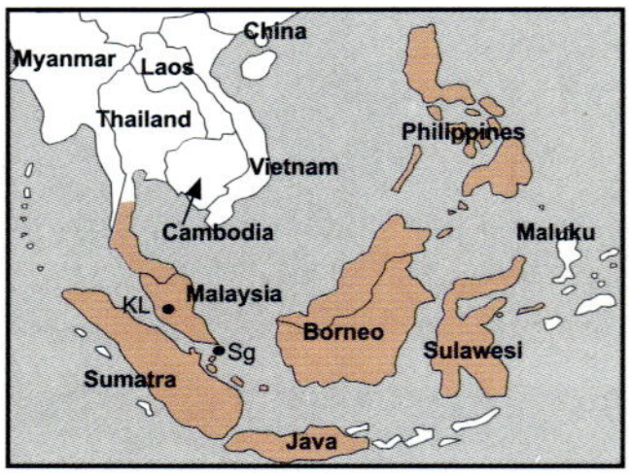


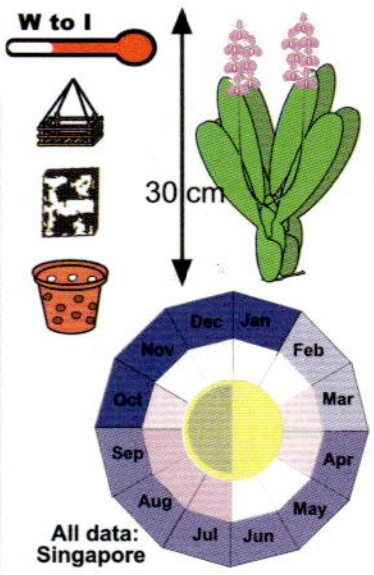

## *Liparis parviflora* (Bl.) Lindl.

This species is common across the region from 100-1500 metres altitude. It grows on the trunks of forest and roadside trees and quickly colonises teak and rubber plantations. It has rather soft pseudobulbs that turn yellow-orange if the plant is growing in direct sunlight. The 50-70 cm long inflorescence bears several hundred tiny flowers which are all open at once.

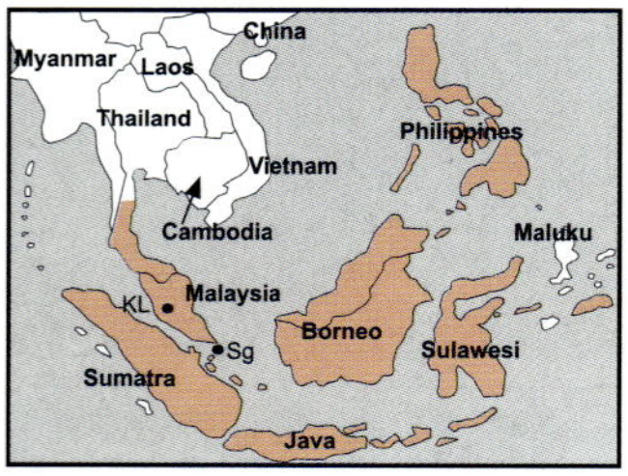


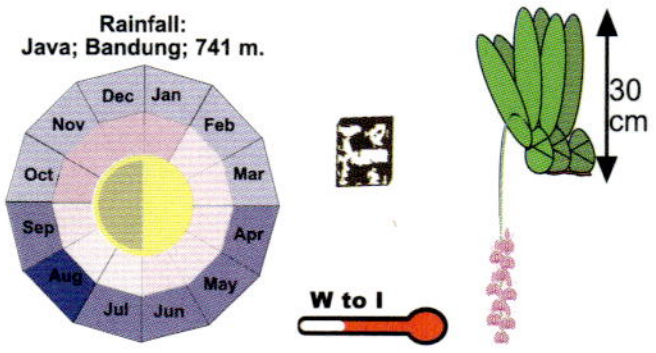

## *Liparis pilifera* J.J.Smith

This uncommon terrestrial species is found in the mountains of Sumatra, growing on the forest floor in deep shade. It is very closely related to *L. purpureovirdis*. It has a short swollen shoot and a single broad green leaf. The suberect inflorescence extends slowly with 1-3 flowers open at a time. The lip has hairs only at the blunt apex, unlike *L. purpureovirdis* which has hairs along the side margins.

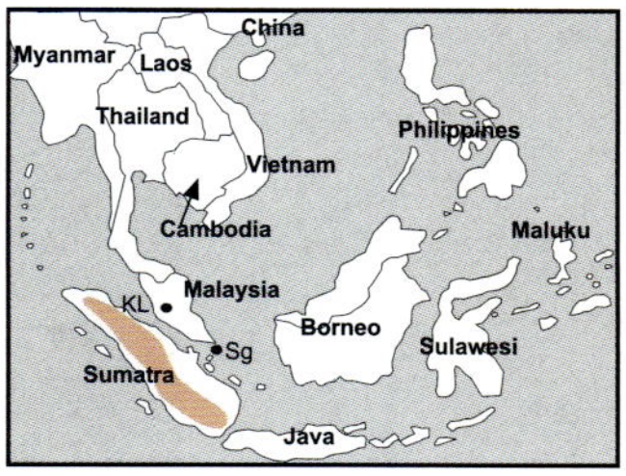


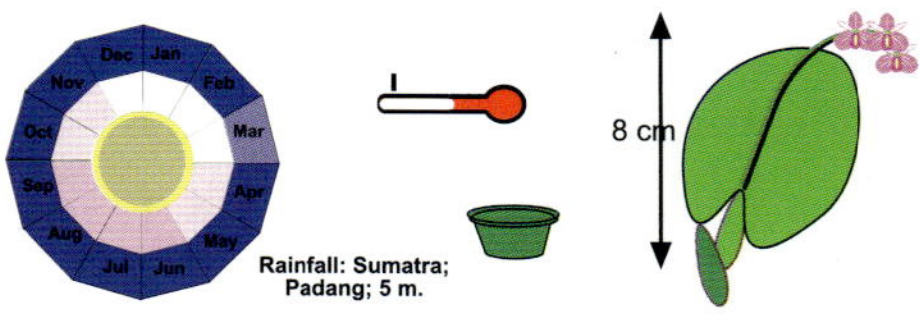

# L

## *Liparis purpureoviridis* Burkill ex Ridl.

This uncommon terrestrial species is found on the forest floor above 1300 metres altitude. It has a short stem with a broad greenish- purple heart-shaped leaf. The inflorescence extends slowly to 15-20 cm long, and has 1-5 flowers open at a time.

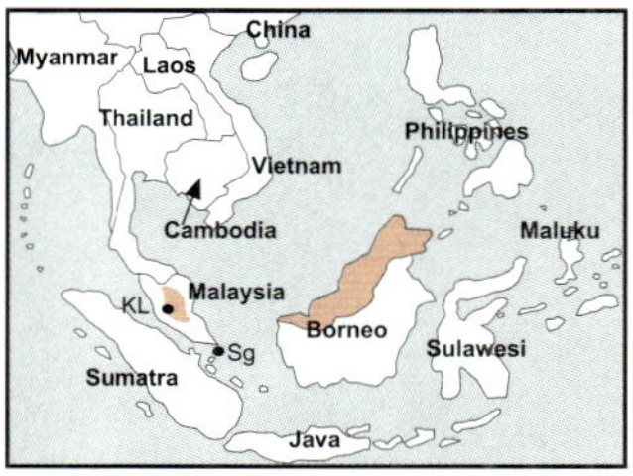


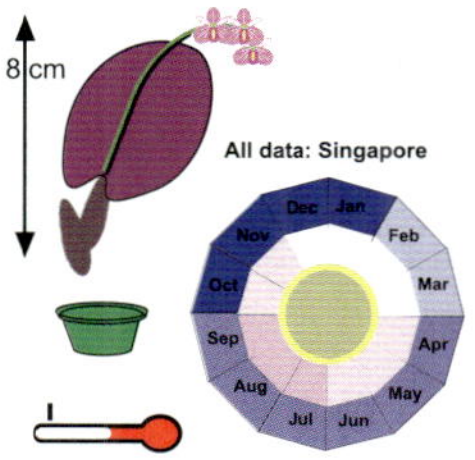

## *Liparis rhombea* J.J.Smith

This uncommon epiphytic species is unmistakable when in flower. It is found from 400-1100 metres altitude in Malaysia and Java, but in Borneo it is found in montane forest from 1200-1300 metres. It has rather flattened 3 cm long pseudobulbs and 1-2 leaves. The inflorescence is a bit longer than the leaves and has up to 30 brilliantly coloured flowers that are all open at once.

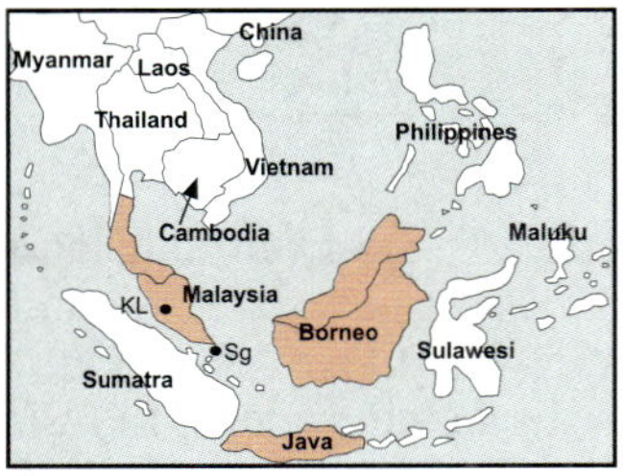


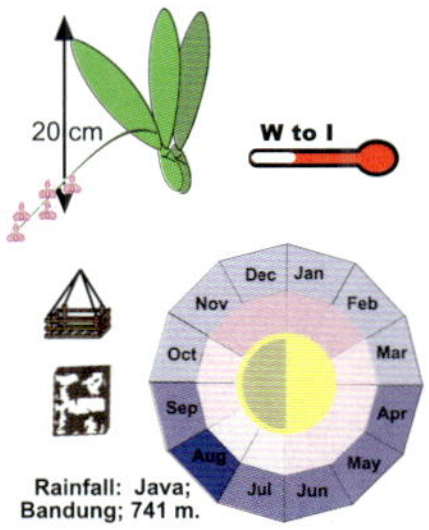

## *Liparis tricallosa* Rchb.f.

This uncommon terrestrial species is found on the floor of swamp forest and the wetter parts of primary rainforest from sea-level to about 700 metres altitude. The 40-50 cm tall inflorescence bears up to 30 flowers, about half of which are open at the same time. The flowers are about 3 cm high, which is very large for this genus.

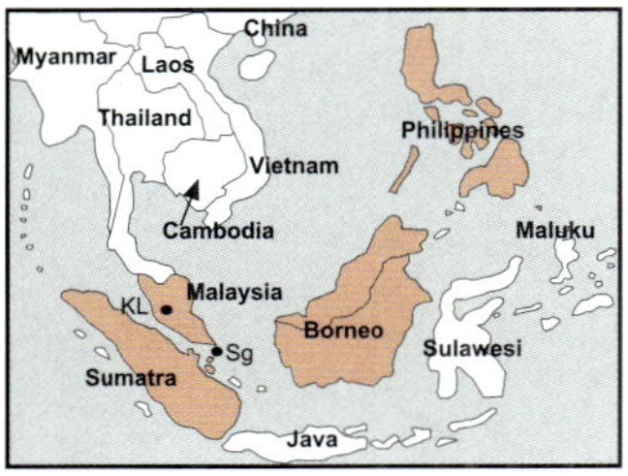


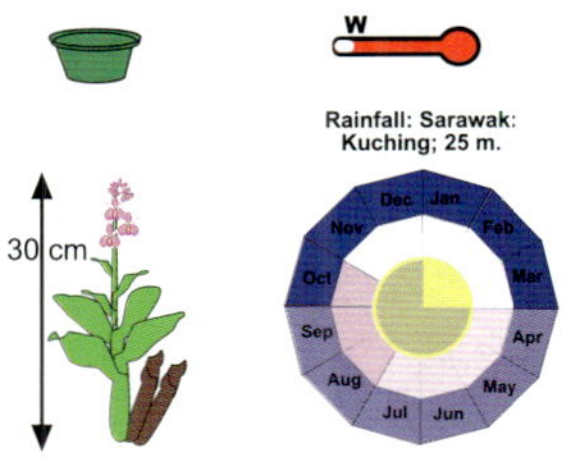

# L

## *Ludisia*

A genus of only one species, although it is highly variable and many varietal and species names have been proposed for the variants. *L. discolor* belongs to a group of terrestrial orchids known as the "Jewel Orchids" which are cultivated for their attractive multicoloured foliage rather than their flowers.

### ***Ludisia discolor*** (Ker Gawler) A.Rich.

This species is widely available from nurseries and garden centres, often being sold under the synonym *Haemaria discolor*. The illustrated plant is one of the "improved varieties" which have become available in recent years; these are much larger than the wild type. *L. discolor* has a creeping rhizome which produces erect leafy shoots at intervals. If kept warm and moist they will grow and flower continuously throughout the year, although in the wild they respond to dry spells by going through a dormant phase.

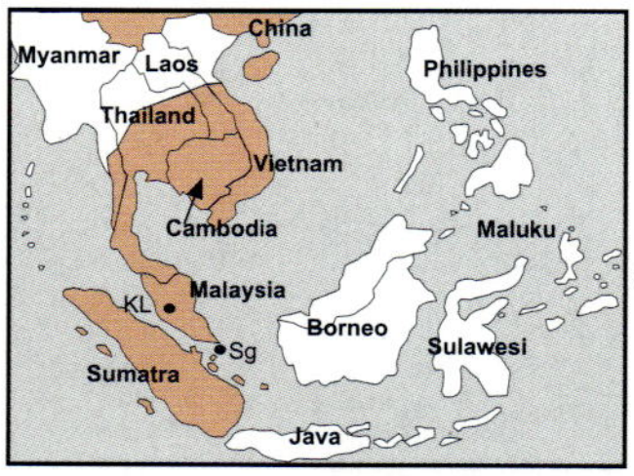


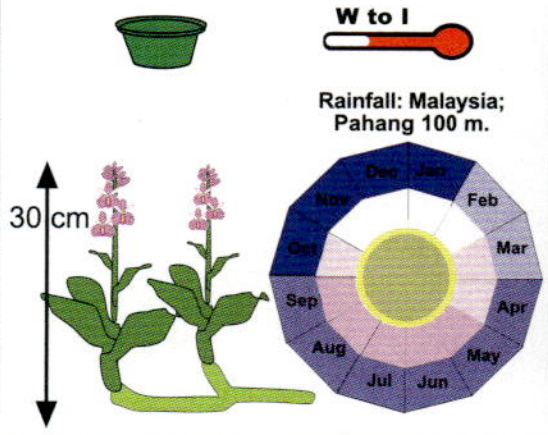


## *Luisia*

A genus of up to 40 terete-leaved epiphytic or lithophytic species, found from India and Sri Lanka to the Pacific Islands. *L. jonesii* and *L. curtisii* are grown in S. E. Asia, and other species are occasionally seen in collections. All species have very short club-shaped inflorescences and flowers with large fleshy lips. In the last decade *L. jonesii* has been utilized in hybrid programs with other monopodial orchids, resulting in offspring with striking flower shapes and colours, although the inflorescence remains short.

### ***Luisia celebica*** Schltr.

This species is quite common in Sulawesi where it is found growing on forest trees from sea-level to at least 1200 metres altitude. Some botanists think it is the same species as the closely related *L. javanica*. The very short inflorescence bears 2-5 long-lasting flowers.

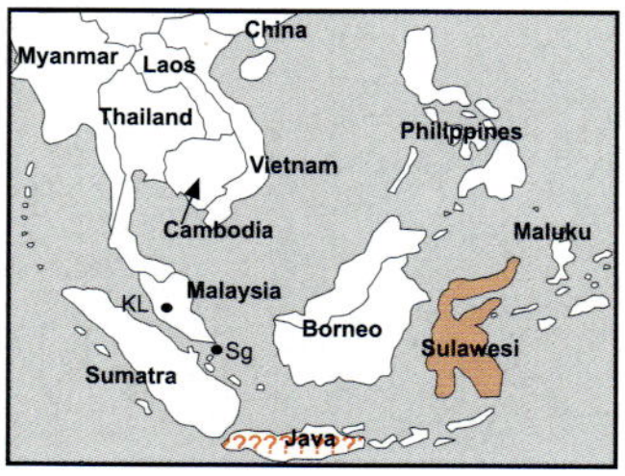


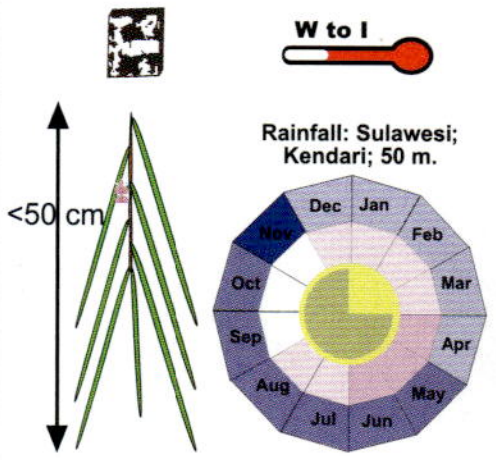

# L

## *Luisia curtisii* Seidenfaden

This species is found on trees in moist lowland forests. Like most *Luisia* species it is rather slow-growing but it can eventually develop into quite a large plant with leaves up to 20 cm long. The 1-2 cm long inflorescence can form up to 8-10 flowers, though only 1-2 are open at a time.

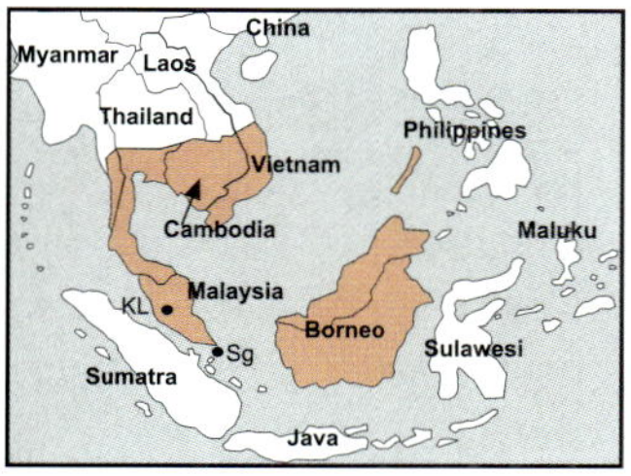


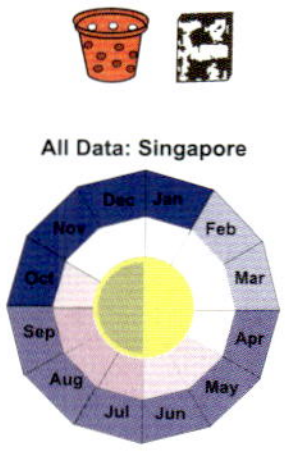

## *Luisia taurina* J.J.Smith

This species is quite common in eastern Java from 400-800 metres altitude, where it grows on the trunks of isolated trees along roads and in villages, pastureland and parkland. The 1-2 cm long inflorescence can form from 6-10 flowers, though only 1-2 are open at a time. The swollen lip midlobe has a beautiful pattern of evenly spaced furrows.

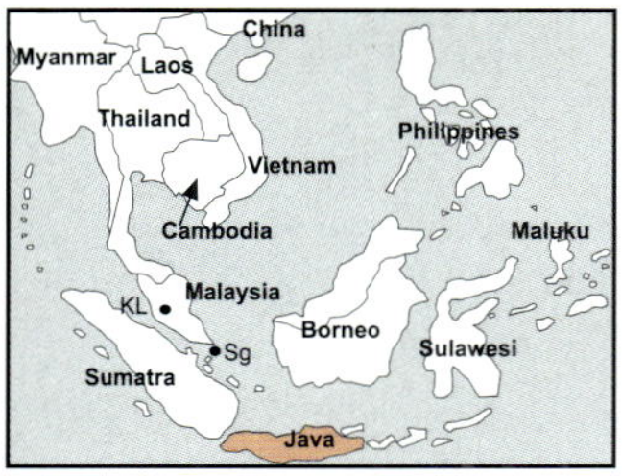


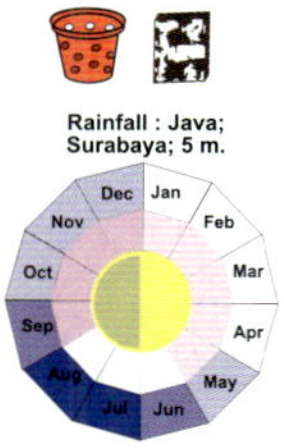


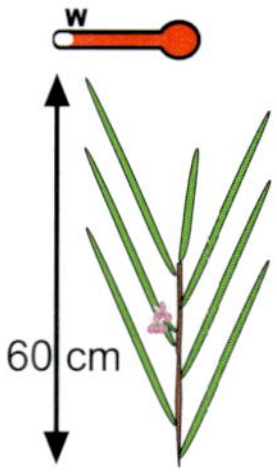

## *Luisia zollingeri* Rchb.f.

This widespread species can be found from sea-level to at least 800 metres altitude and is very common in parts of its range. I have seen it growing on mangroves in North Borneo, pastureland trees in Sumatra , and rainforest trees in Malaysia. It has an untidy growth habit, with long pendulous stems that branch frequently. It can be easily identified from the flowers, which have the petals and dorsal sepal held in the same plane.

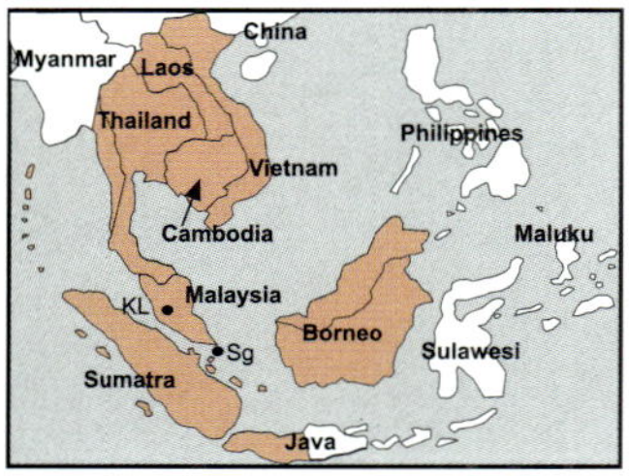


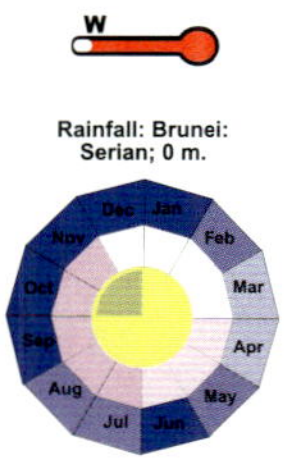


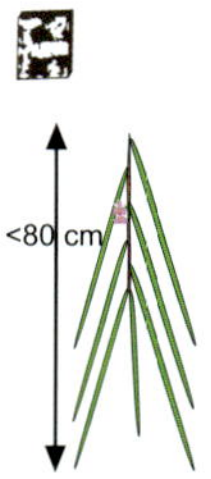

# M

## Macodes

This genus of about 10 small terrestrial species is distributed from Malaysia to the Pacific Islands. The plants have a creeping fleshy rhizome that sends up erect shoots at intervals. The shoot consists of a rosette of a few leaves and a terminal inflorescence. The leaves have brightly coloured veins; some species (especially *M. petola*) are known as "Jewel Orchids" and are cultivated as foliage plants. The inflorescence is hairy and has upside down flowers that are asymmetric because the lip and column are twisted.

### *Macodes petola* (Bl.) Lindl.

This species grows in humus and debris on the forest floor in hilly country from about 300-1500 metres altitude. It has a basal rosette of about 4-5 leaves and an erect 20 cm tall inflorescence that bears 10-40 flowers. The leaves are usually dark green with yellow veins; variety *argenteoreticulata* (illustrated) has silvery-white veins. The flowers usually have red-brown sepals, but a form with green sepals exists.

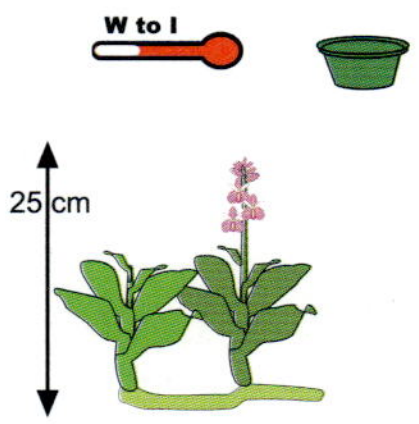


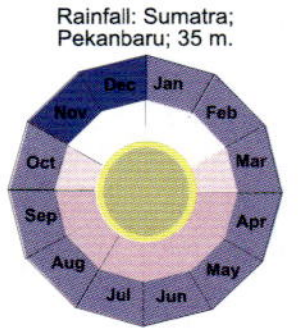


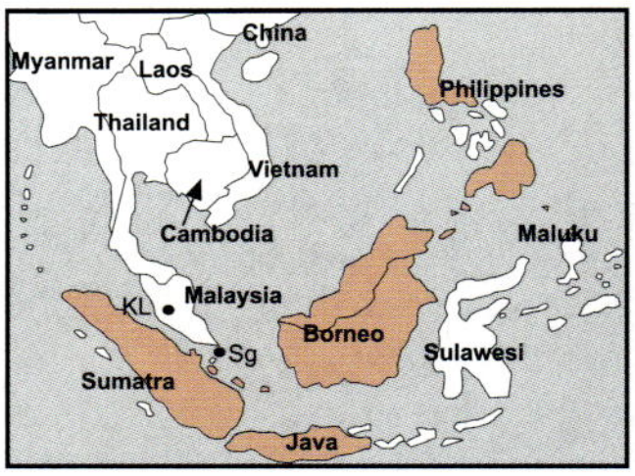

## Malaxis

This genus of about 300 species is found worldwide from the tropics to temperate regions. The majority of species occur in S. E. Asia and New Guinea. Most Asian species are terrestrial, though some are epiphytes or grow on rocks. Plant form is diverse; while many species have rather soft pseudobulbs, some have creeping stems. Some species have highly coloured leaves. All species have a terminal inflorescence of many small flowers that are held upside down with the lip uppermost. The lip is often deeply lacerated. *Malaxis* species are occasionally seen in collections, and a few species are quite well-known as cultivated plants.

### *Malaxis hoi* O'Byrne

This species was recently discovered in an orchid collection in Singapore; the owner had purchased the plants in a Malaysian nursery. It has a few leaves that are an iridescent bronze-green colour with a white central strip. The erect 15 cm tall inflorescence produces up to 100 flowers over a 3-4 month period.

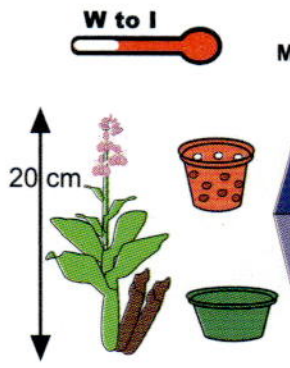


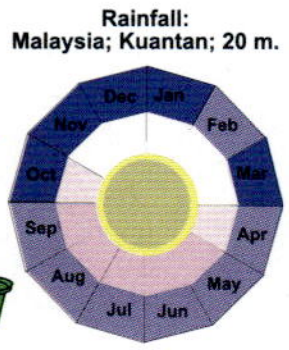


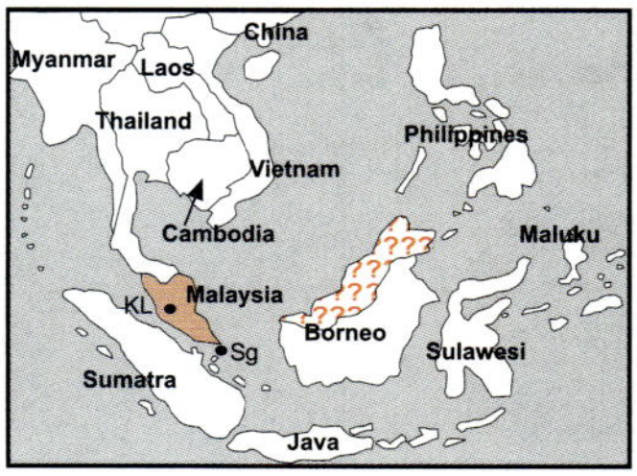

# M

## *Malaxis lowii* (E.Morren.) Ames

This small terrestrial is found growing in sandy soil on the forest floor in limestone areas from sea-level to 2000 metres altitude. It has 5-10 dark red-purple shiny leaves and a 10-15 cm long dark purple erect inflorescence that usually bears about 20-30 small flowers.

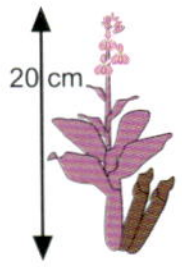


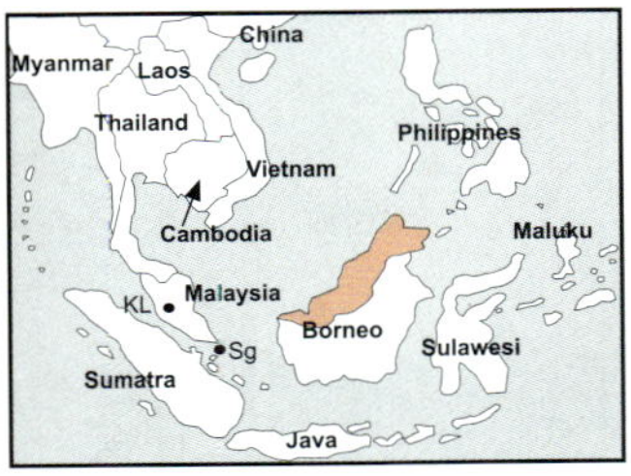


Rainfall: Sarawak: Kuching; 25 m.

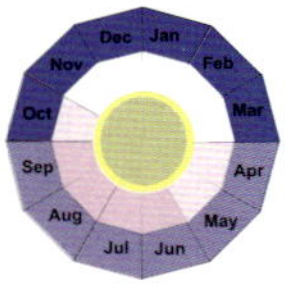


## *Malaxis metallica* (Rchb.f.) Kuntze

This species grows on the floor of the forest in hilly country from 400-1200 metres altitude. At higher altitudes it is often found on mossy rocks and boulders above the waterline in river beds. It is similar in habit to *M. lowii* but has different flowers. It has 4-6 shiny dark red-purple leaves and a 5-10 cm long dark purple inflorescence.

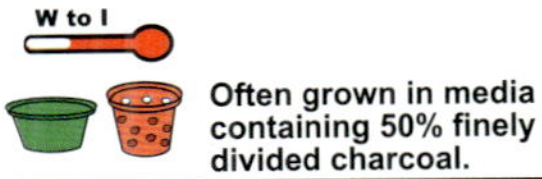


Rainfall: Sarawak: Kuching; 25 m.

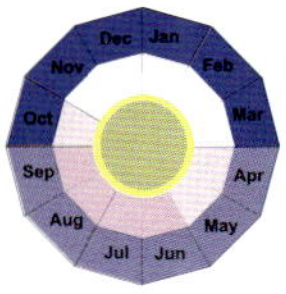


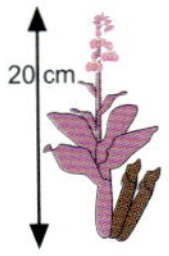


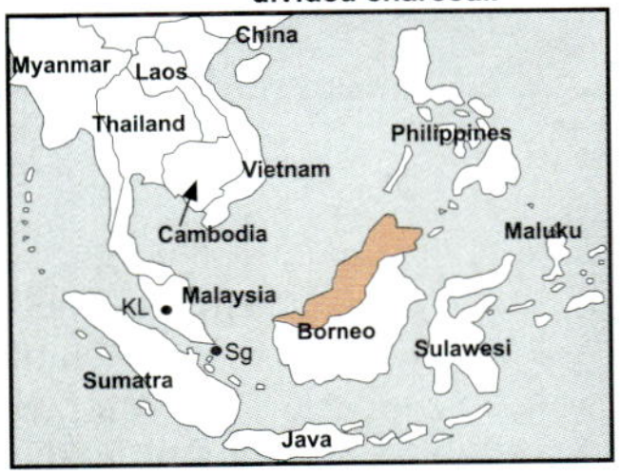

## Malleola

This genus of some 30 to 35 monopodial epiphytes is distributed from Thailand and Indochina to New Guinea and the Pacific Islands. Although the plants usually have short individual stems, old specimens can form quite large clumps. Many of the species have descending inflorescences with crowded racemes of long-spurred bright yellow flowers with red or orange markings. *Malleola* species are usually easy to grow, and although specimen-sized plants in full bloom are extremely attractive, they are seldom seen in orchid collections outside S. E. Asia.

# M

## *Malleola baliensis* J.J.Smith

Although this attractive species is restricted to Bali and eastern Java, it is still quite common in forests from sea-level to over 1000 metres altitude. It has fairly short stems with leaves that are purple-green in colour, and usually forms several 5-10 cm long inflorescences at a time. This species is easily identified by the lip, which has a large white callus ( rather like a rhino's horn) in the middle, and a downturned-hook at the tip.

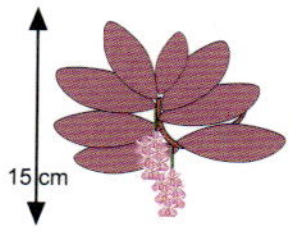


Rainfall:
Bali; 5 m.

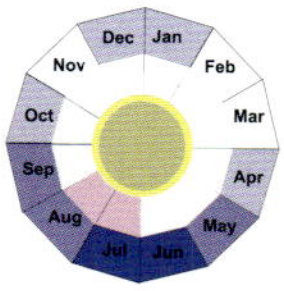


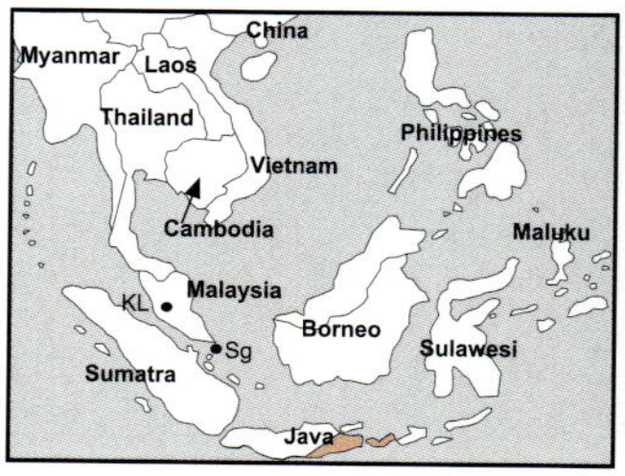

## *Malleola dentifera* J.J.Smith

This species is quite common in forests and plantations from sea-level to at least 1200 metres altitude. It is often seen growing as a 2-3 cm long twig epiphyte, but in the right situation it can develop into a 30 cm long bushy specimen. Each stem on the plant produces several 4-5 cm long inflorescences at a time. The brightly-coloured flowers have a lip with complex sidelobes which have a short tooth on the inside.

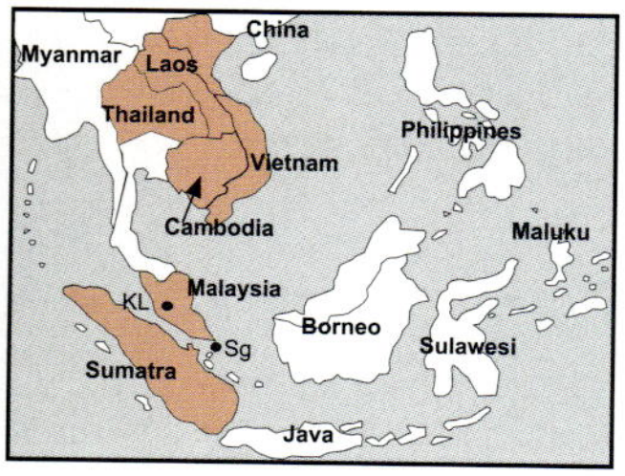


Rainfall and Flowering:
Singapore; 10 m.

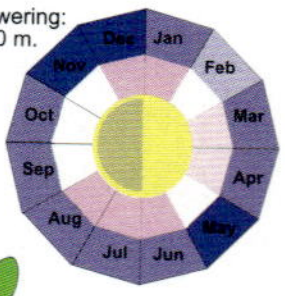


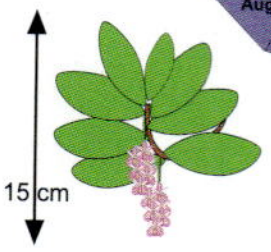

## *Malleola insectifera* J.J.Smith

This species grows in forests from sea-level to at least 1300 metres altitude. Each plant produces several inflorescences at a time; the flower colour is quite variable.

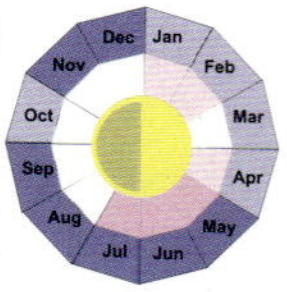


Rainfall: Sabah:
Tawau; 20 m.

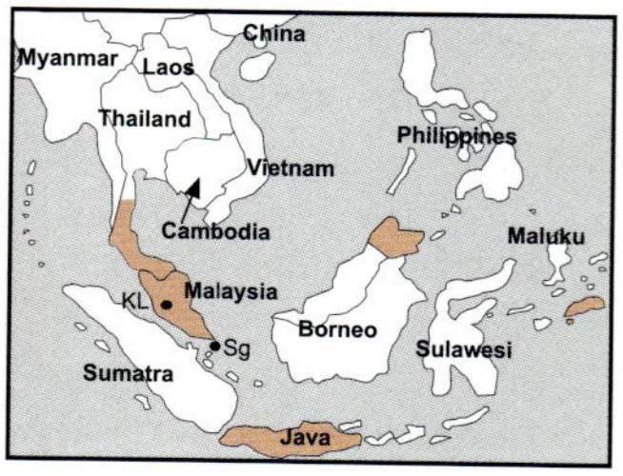


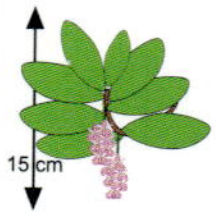

# M

## Mediocalcar

This genus of 15 species and five subspecies is related to *Eria* and is distributed from Sulawesi to the Pacific Islands. The majority of the "Cherry Orchids" occur in the mountains of New Guinea; only 2 species extend into our area. They are creeping plants with long rhizomes and seem happiest growing on mossy logs. They have pseudobulbs with 1-4 leaves at the apex; in most species the inflorescence is formed on the young shoot while the pseudobulbs and leaves are developing. Although they are generally easy to grow, few of these delightful plants are ever seen in cultivation.

### *Mediocalcar paradoxum* (Krzl.) Schltr. *ssp. robustum* (Schltr.) Schltr.

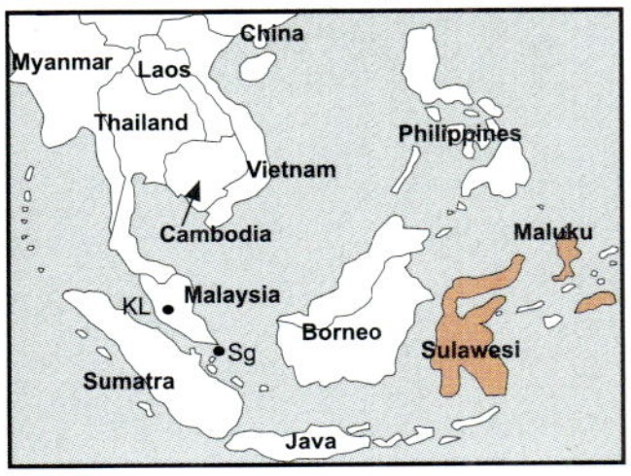


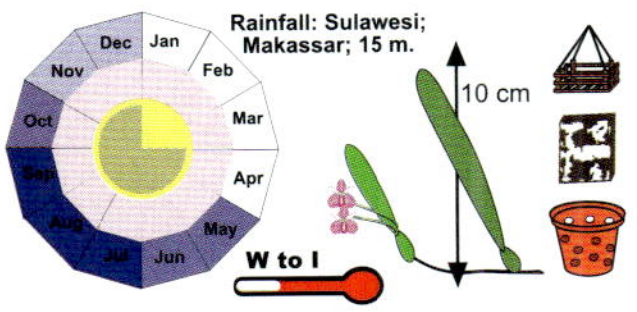

This species, which is large for the genus, grows from 400-2250 metres altitude. It produces 1-2 single-flowered inflorescences from each new shoot. The flowers last about a week.

## Micropera

This genus of about 15 monopodial species is distributed from India and Indochina to Australia and the Pacific Islands. All the species are climbing epiphytes and some are large, coarse plants. They usually have narrow leaves and an inflorescence that bends through a right angle so the rachis points either straight up or straight down; most species have cream to yellow flowers. *Micropera* species are sometimes seen in orchid collections, often labelled with the old genus name *Camarotis*.

### *Micropera callosa* (Bl.) Garay

This is a large coarse species that is often seen scrambling through tree branches overhanging rivers or climbing up the trunks of roadside trees. It has an upturned inflorescence that slowly extends to about 25 cm long with up to 5 flowers open at a time.

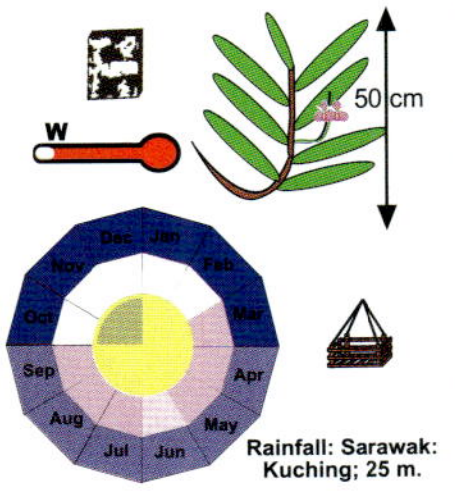


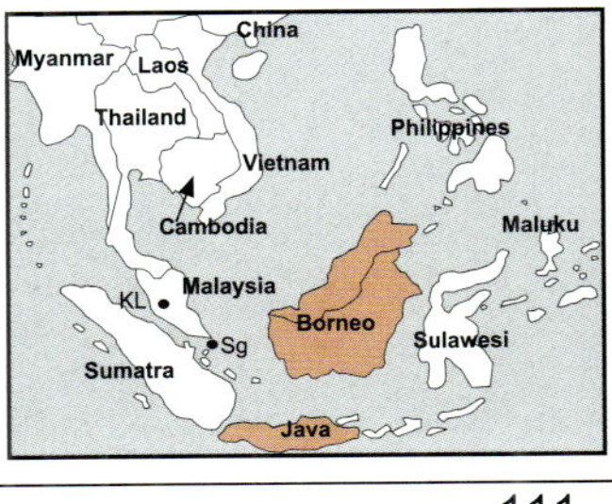

# M

## *Micropera fuscolutea* (Lindl.) Garay

This is a robust species that usually has 30-40 cm long stems. It grows in the lowland forests and is not as common as the other species mentioned here. It has an upturned inflorescence which grows slowly to about 20 cm long with 4-5 flowers open at a time. The flowers are the largest in the genus.

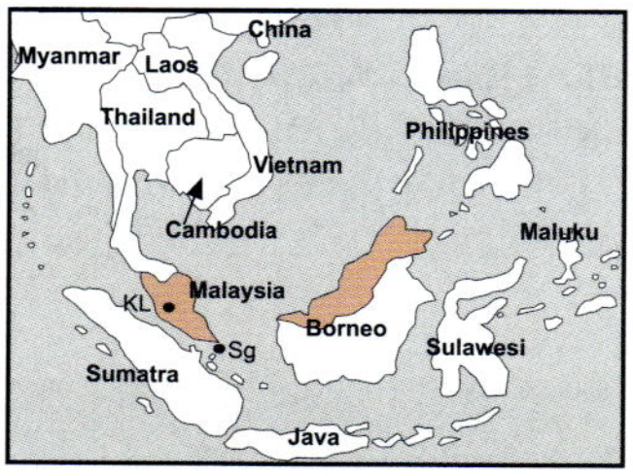


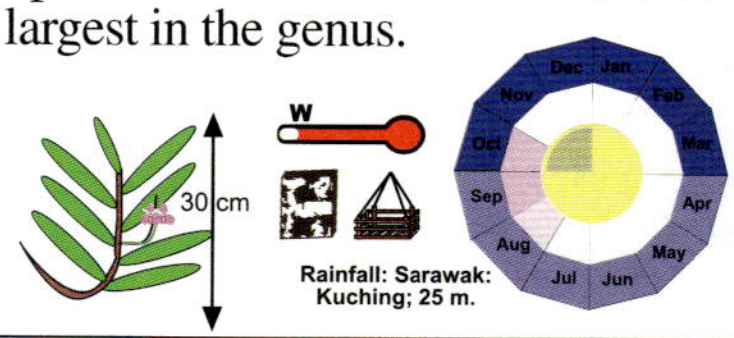

## *Micropera pallida* (Roxb.) Lindl.

This slender climbing species is found from sea-level to at least 1000 metres altitude and in some places it is very common. It produces several downturned inflorescences at a time; each one will flower continuously for several months.

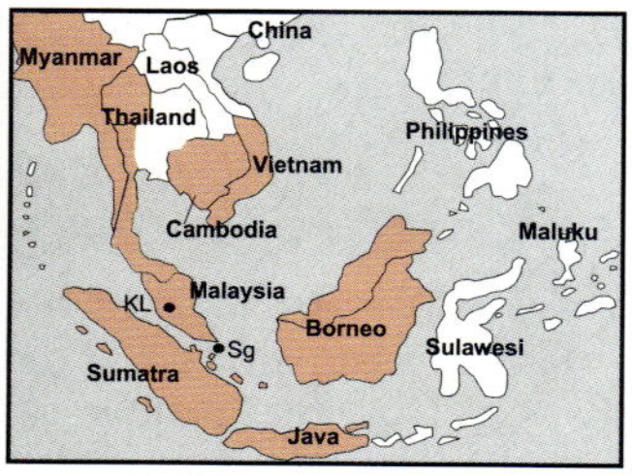


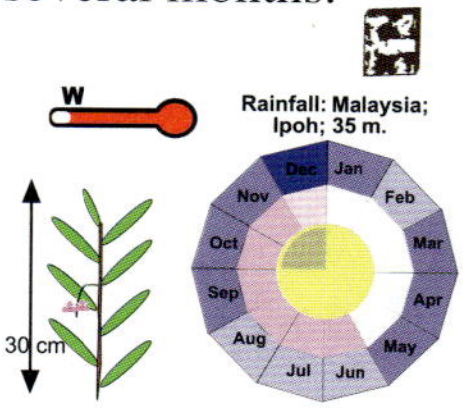


## Microtatorchis

This genus of about 50 monopodial species is distributed from Java and Borneo to the Pacific Islands; New Guinea has the most species. They are tiny epiphytic plants that usually grow twigs or small branches and are easily overlooked. Many of the species are leafless and use their green roots for photosynthesis. All species have an inflorescence that extends slowly, producing only a few tubular flowers at a time. *Microtatorchis* species are of no commercial interest and are almost never seen in collections.

## *Microtatorchis taenioides* O'Byrne

This very small epiphytic species is quite common at about 1000 metres altitude, growing on the branches and trunks of forest trees, and also on the branches of bushes in coffee plantations. It has a short stem with 2-3 small leaves, and a long green tapeworm-like pendulous inflorescence that remains active for several months, producing 1-2 tiny tubular flowers at a time.

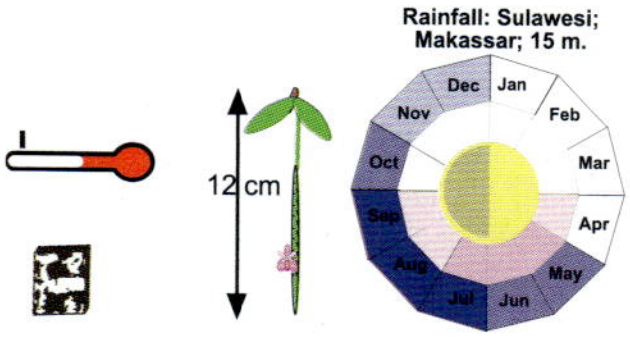


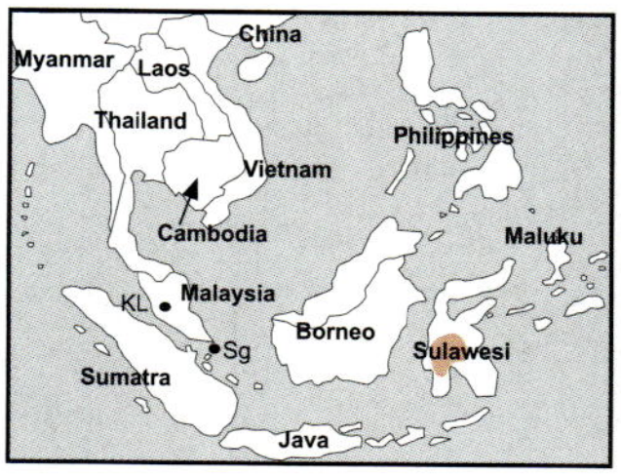

# N

## Nephelaphyllum

This genus of about 12 terrestrial species is distributed from China to Japan to Indonesia and the Philippines; Borneo, with 8 species, is the centre of distribution. The species have a short creeping rhizome and slender pseudobulbs that are hard to distinguish from leaf stalks. The heart-shaped leaves are quite fleshy and often marbled or otherwise coloured, especially when young. The flowers have a short spur. Only one species, *N. pulchrum*, is in cultivation and is quite often seen in collections.

### *Nephelaphyllum pulchrum* Bl.

This species is not uncommon throughout much of the region, growing in forests from 500-2000 metres altitude. It has a short rhizome that creeps slowly over the humus on the forest floor, rooting from the nodes. The pseudobulb is almost indistinguishable from the leaf base. The heart-shaped leaves are quite colourful, especially when young. The inflorescence has 10-15 crowded flowers at the tip; the sepals and petals point downwards while the lip points upwards.

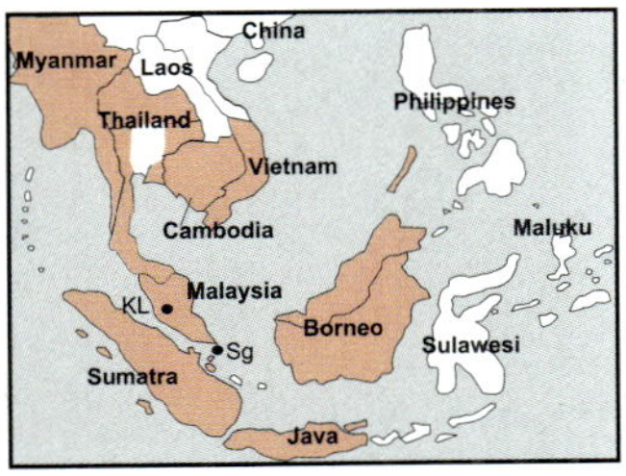


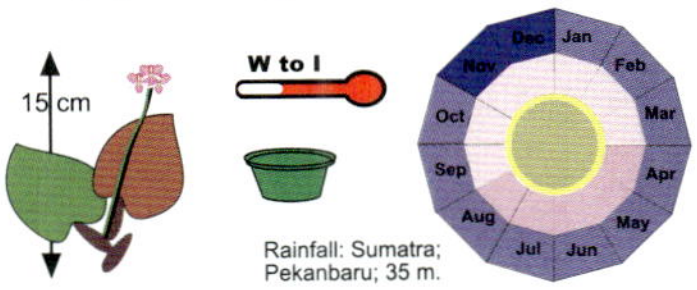


## Notheria

This genus, which was described as recently as 2000, has only one species and is endemic to Sulawesi. The epiphytic plant has the pseudobulbous stems of an *Eria* species, but *Ceratostylis*-like flowers that have a short spur behind the lip.

### *Notheria diaphana* O'Byrne & Vermeulen

This miniature species grows in montane forest at about 1300 metres altitude. It has small swollen pseudobulbs with 2 narrow leaves that each produce several 2-flowered inflorescences at a time. The very thin delicate flowers are very large and showy for such a small plant, but are not very long-lasting.

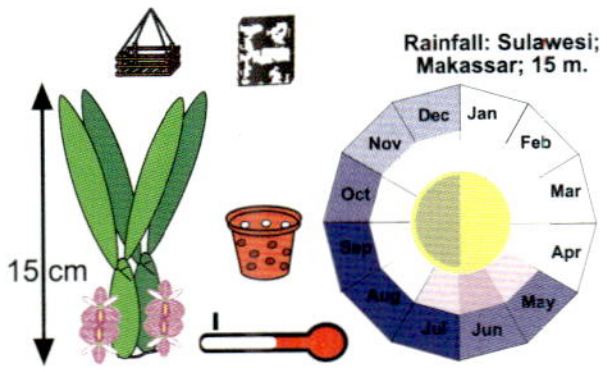


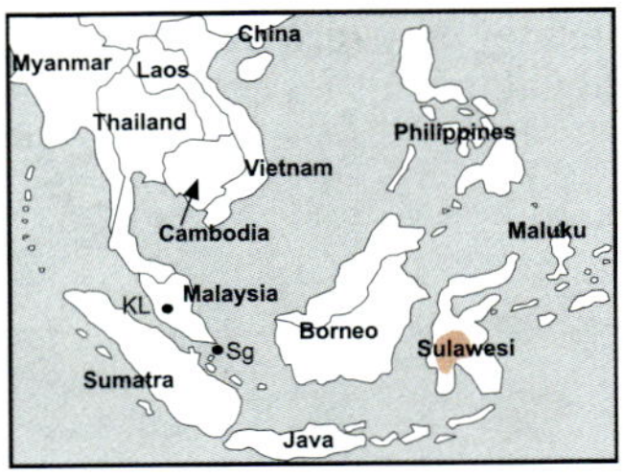

# O

## Oberonia

This genus of at least 350 epiphytic species is distributed from tropical Africa to the Pacific Islands. The majority of the species are found in S. E. Asia. Some species are virtually stemless and have leaves arranged in a fan, while others have quite long slender sinuous stems with 2 rows of leaves. The leaves are always bilaterally strongly flattened. When the shoot has matured, a single terminal is produced that has very many tiny flowers. *Oberonia* species have no commercial value and are seldom seen in collections.

### *Oberonia ciliolata* Hook.f.

This common and widespread species can be found from sea-level to at least 1000 metres altitude, growing on trees in forests, plantations, or village gardens, preferring well-lit situations. It has stems that are usually 10-12 cm long with 6-7 leaves on each side that spread outwards from the stem. The inflorescence is about 10-12 cm long and bears several hundred tiny greenish-yellow flowers. The sepals are externally hairy and the lip apex is irregularly lacerated.

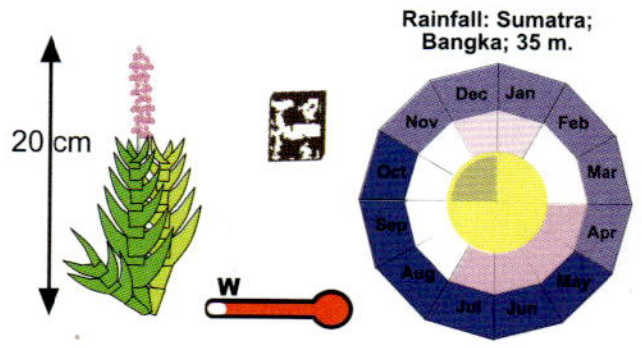


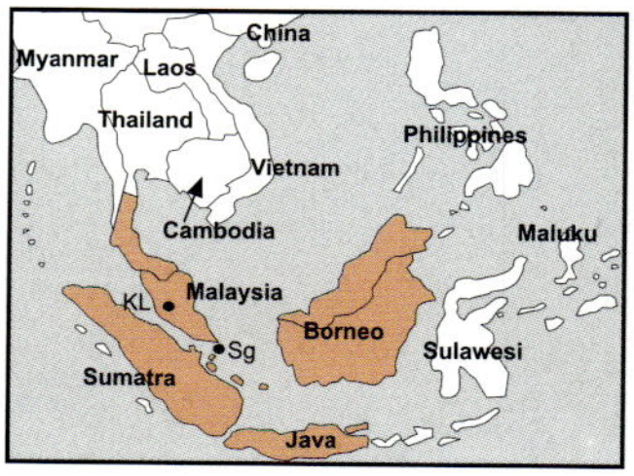

## Oeceoclades

This genus of about 20 terrestrial species is distributed across the tropical areas of the Old World, from Africa to the Pacific Islands. The species have slender pseudobulbs with 1-3 pleated leaves at the apex, and a lateral inflorescence that has many flowers and can be rather showy. The flowers have a short spur and a lip that is either 3-lobed or 4-lobed.

### *Oeceoclades pulchra* (Thou.) Cribb & Clem.

This terrestrial species is distributed from Africa across much of tropical Asia to New Guinea and Australia. It is still widely known by the old names *Eulophia pulchra* and *Eulophia macrostachya*. It has slender pseudobulbs that stand erect on stilt-like roots. The inflorescence may reach 75 cm tall, with many flowers on the upper half. The flowers are pale green with a white to yellow lip that sometimes has red veins inside; they turn yellow after a few days.

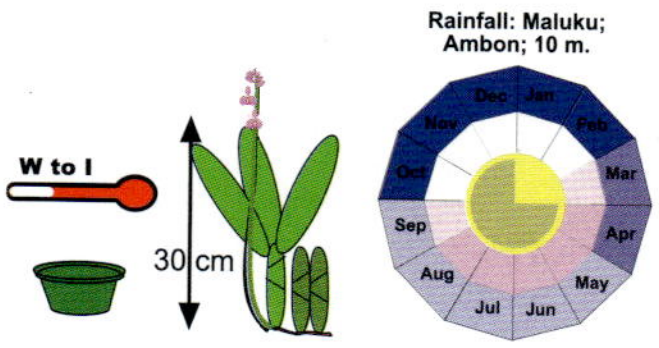


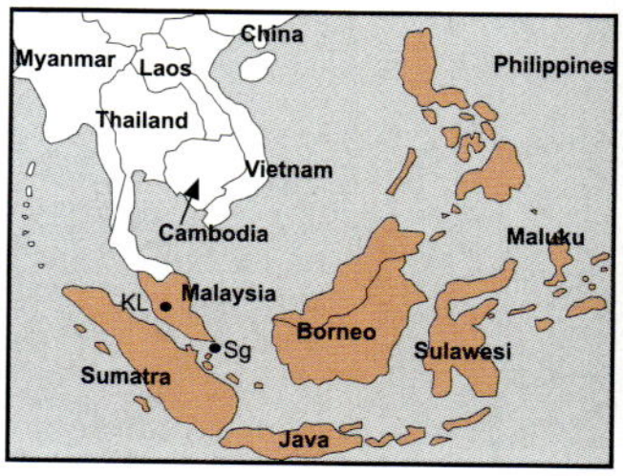

# P

## *Paphiopedilum*

This well-known genus of 'slipper orchids' containing about 70 species is distributed from India across tropical Asia to New Guinea and the Solomon Islands. Most species are terrestrial or lithophytic and are usually associated with a limestone substrate; some species are epiphytic. The plants are similar in habit, with each shoot forming 2 rows of rather soft leaves and a single inflorescence with 1 to a few large long-lasting flowers that have a slipper-shaped lip. *Paphiopedilum* species have been extensively hybridized; both the species and their hybrids are widely cultivated and highly sought after by orchid enthusiasts. Despite being placed on the CITES list of Endangered Species (which protects plants and animals by curtailing international trade) most *Paphiopedilum* species continue to be heavily collected and their wild populations are seriously endangered.

### *Paphiopedilum bellatulum* (Rchb.f.) Pfitz.

This large-flowered species is found on steep limestone slopes, rooted in a thin layer of humus in shady or open situations from 900-1500 metres altitude. It has 4-5 very dark green leaves that are mottled pale green on top and purple-spotted underneath. The inflorescence is only 2-5 cm long and carries a single 5-8 cm wide flower. Albino forms exist and are fairly common in cultivation.

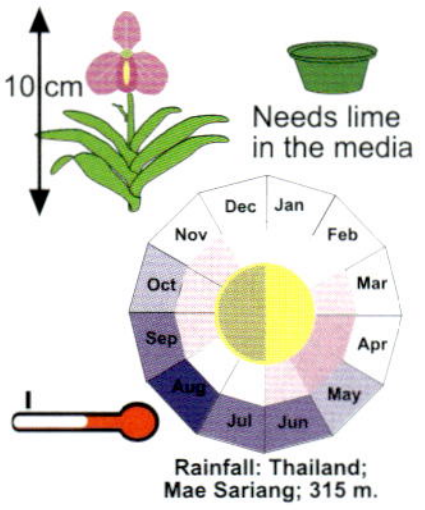


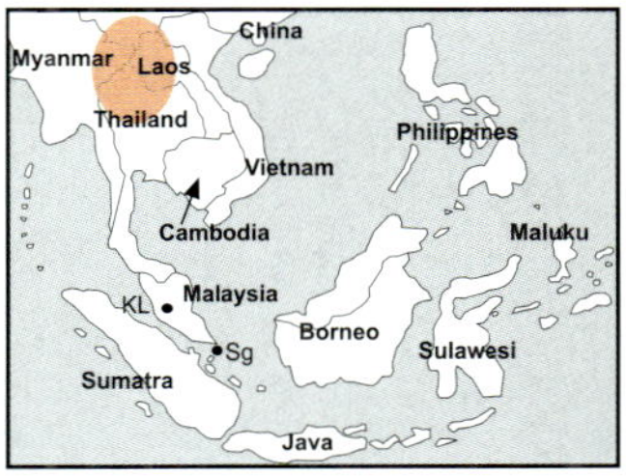

### *Paphiopedilum barbatum* (Lindl.) Pfitz.

This species is usually found growing in leaf litter on well-drained shady slopes from 600-1300 metres altitude, but it has also been recorded in the open and at lower altitudes. It has 5-8 leaves that are mottled pale and dark green on top. The 35 cm tall inflorescence has a single 8-10 cm wide flower. The colour is variable.

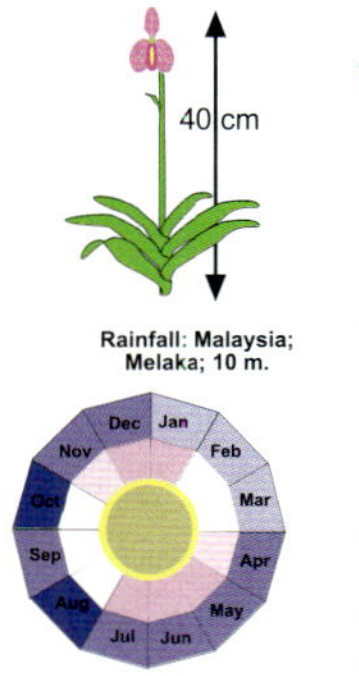


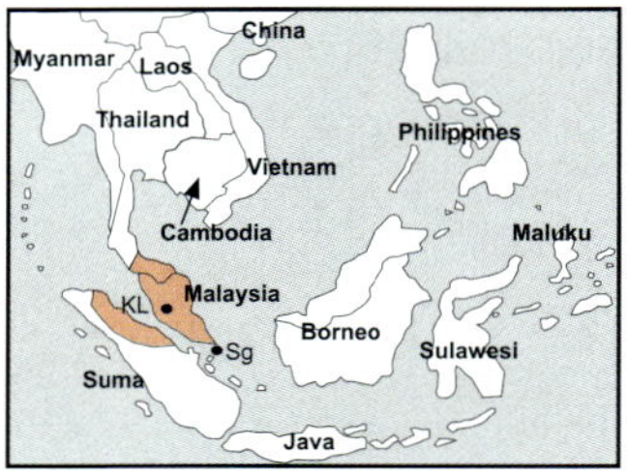

# P

## *Paphiopedilum callosum* (Rchb.f.) Stein **var. *sublaeve*** (Rchb.f.) Cribb.

This variety is found on the forest floor from 700-950 metres altitude, growing in sand, humus, or leaf litter, often in open and bright situations. It has 3-5 fairly short green leaves that are tesselated on top. The 20-40 cm tall inflorescence has a single 6-9 cm wide flower. *P. barbatum* var. *warnerianum* would be a more accurate name for this plant, which appears to be little more than a large-flowered variety of the widespread Malaysian species. The normal form of *P. callosum* is distributed across the north of Thailand and Indochina.

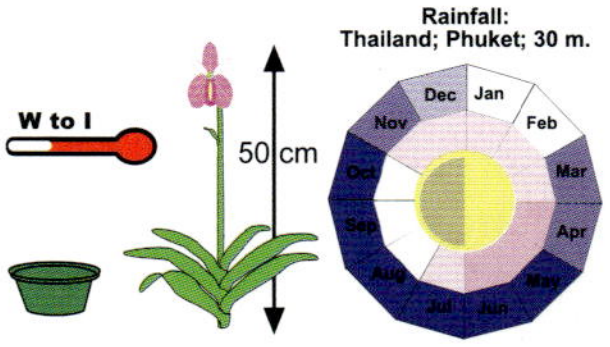


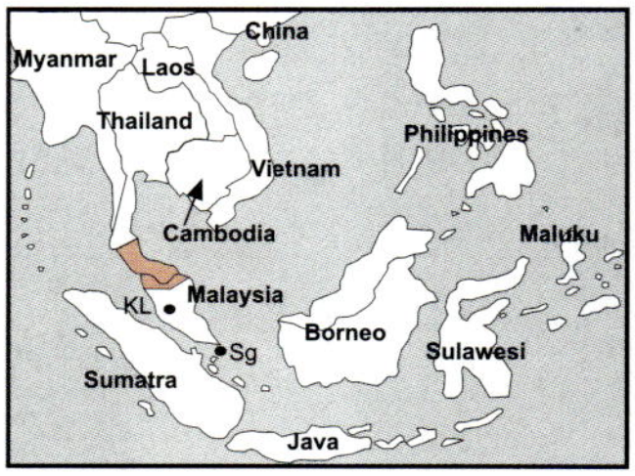

## *Paphiopedilum concolor* (Batem.) Pfitz.

This spectacular species is found growing on limestone hills from 90-1000 metres altitude. It has 4-6 short tessellated green leaves. There are usually two 5-7 cm diameter flowers; the colour and shape are rather variable.

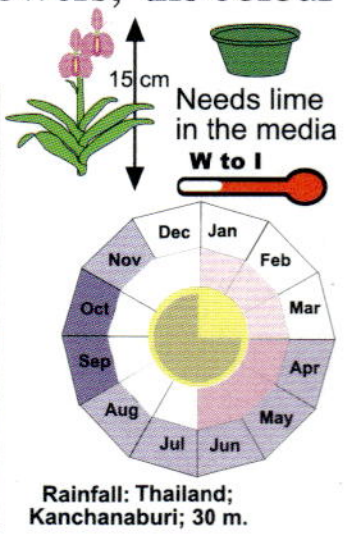


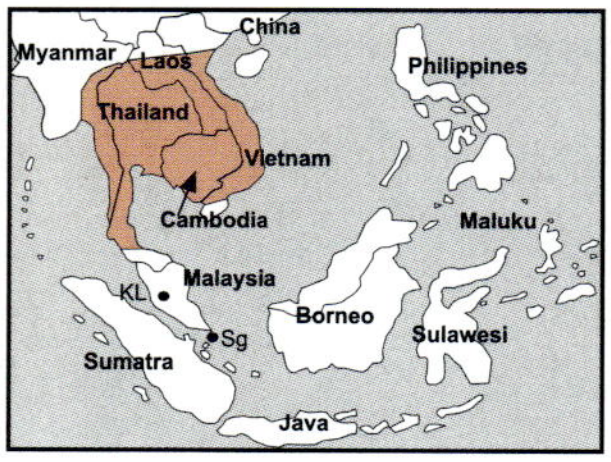


## *Paphiopedilum godefroyae* (Godefroy-Lebeuf) Stein

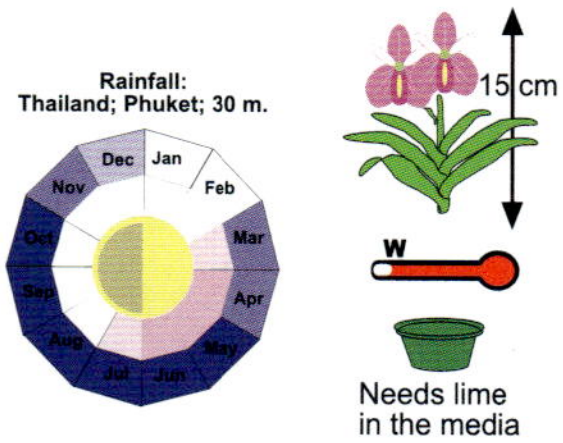


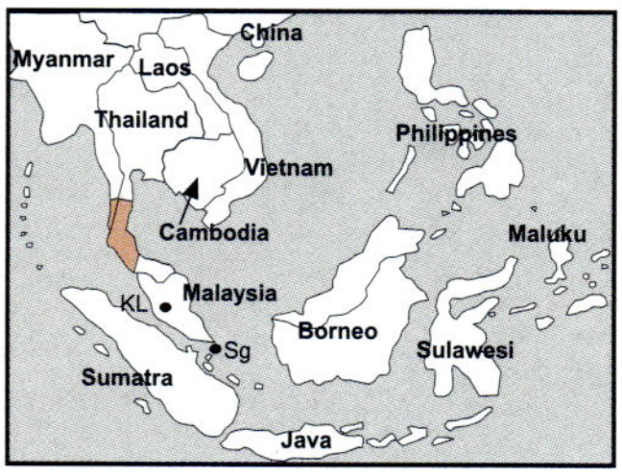


This attractive large-flowered species is found from 0-100 metres altitude, growing in humus-filled cracks on steep-sided limestone cliffs and islands. On some small islets it grows just a few metres above the waves. It has 4-6 fairly short green leaves that are tesselated on top and purple-spotted underneath. The inflorescence is only 4-8 cm long and carries one or two 5-9 cm wide flowers. The normal form of *P. godefroyae* is shown at the top. The bottom photo shows a colour form that lacks spots on the lip and is often called *P. godefroyae var. leucochilum.*

# P

## *Paphiopedilum helenae* Averyanov

This highly endangered miniature species was described as recently as 1996. It grows on lightly shaded *karst*-limestone hills from 850-900 metres altitude. It has 3-4 short green leaves that have yellow-white margins. A 4-7 cm long inflorescence bears a single 4-7 cm wide flower. The flower colour is rather variable; most plants are more yellow than the illustrated form.

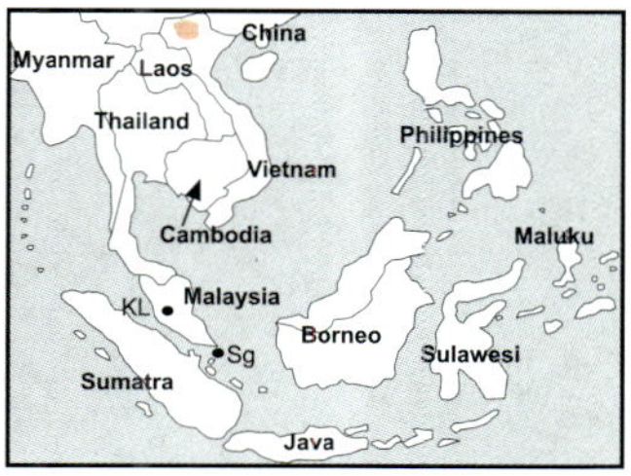


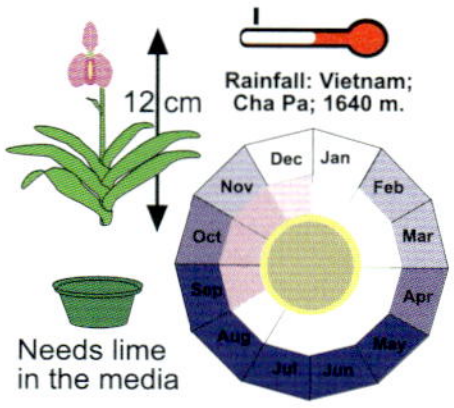


## *Paphiopedilum henryanum* Braem

This small-growing species is found on shaded limestone outcrops from 1000-1200 metres altitude. It has 3-4 green leaves and a 12-15 cm inflorescence. The single flower is about 5 cm wide. Wild populations of this species are highly endangered due to over-collection and habitat destruction.

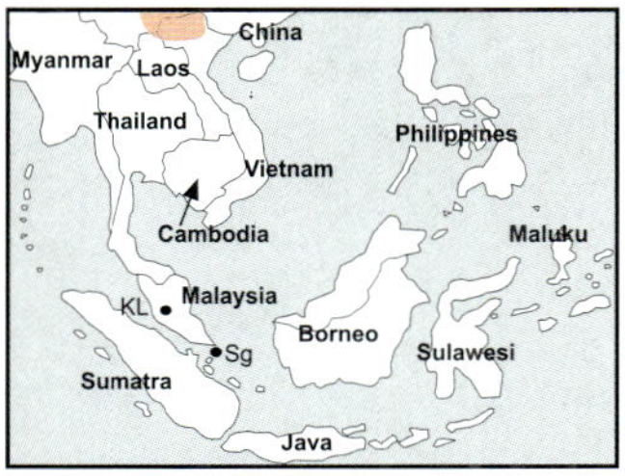


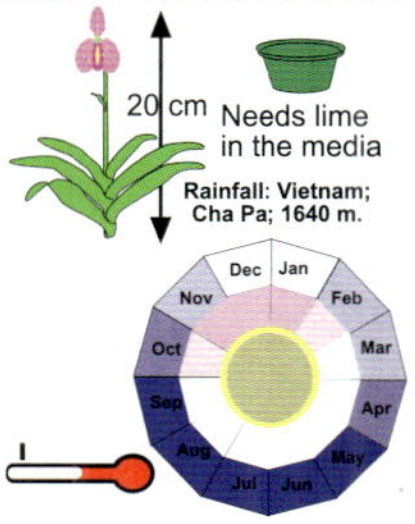

## *Paphiopedilum lowii* (Lindl.) Stein.

This spectacular epiphytic species is widespread across the region, growing on the branches of forest trees from 250-1600 metres altitude. It has 4-6 long leathery green leaves and a very long inflorescence of 3-7 flowers, all open at the same time. Each flower is 9-16 cm wide.

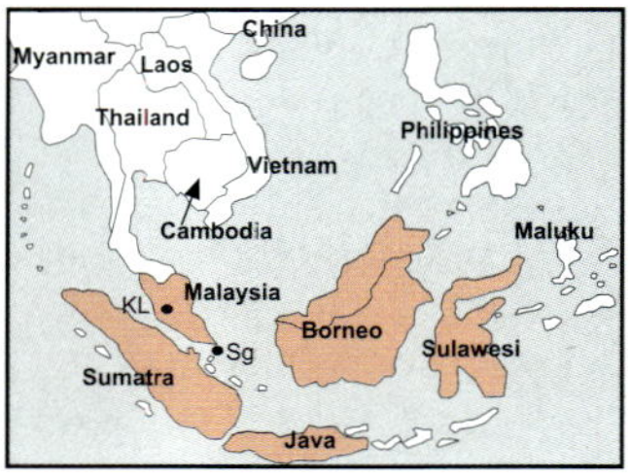


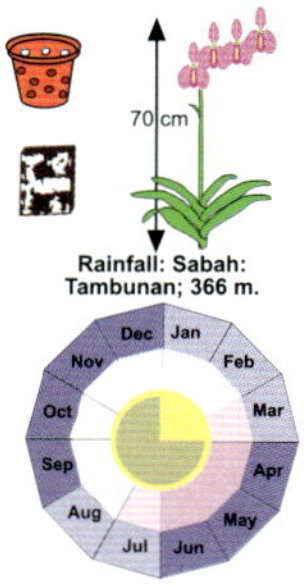

# P

## *Paphiopedilum malipoense* Chen S.C. & Tsi

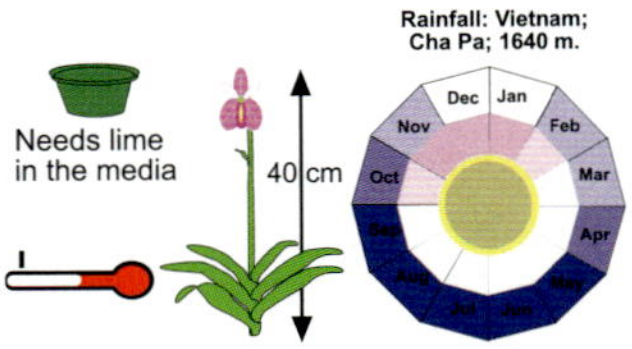


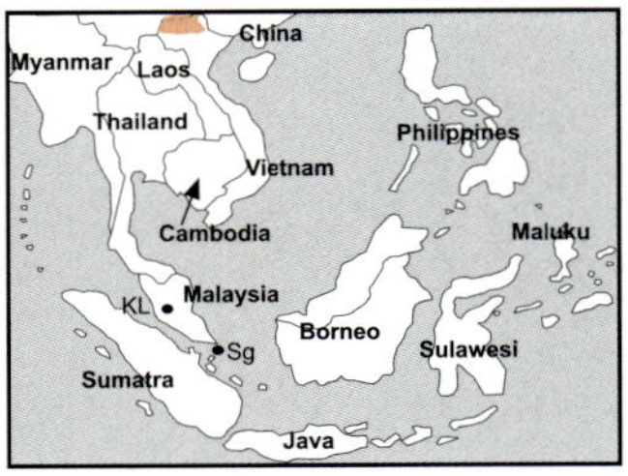


This highly endangered species grows in soil and leaf litter in montane forests on *karst*-limestone hills from 760-1300 metres altitude. It has 7-8 leathery green leaves with a chequered pattern on top. A single 8-12 cm wide rasberry-scented flower is borne on a 26-30 cm long stem. The flower illustrated to the right in the photo is an abnormal colour and lacks the usual maroon veins on the petals.

## *Paphiopedilum niveum* (Rchb.f.) Stein.

This seriously endangered terrestrial species grows from 0-200 metres altitude on small limestone islands and outcrops. It has 4-6 mottled green leaves and a 15-20 cm tall inflorescence that bears a single large white flower.

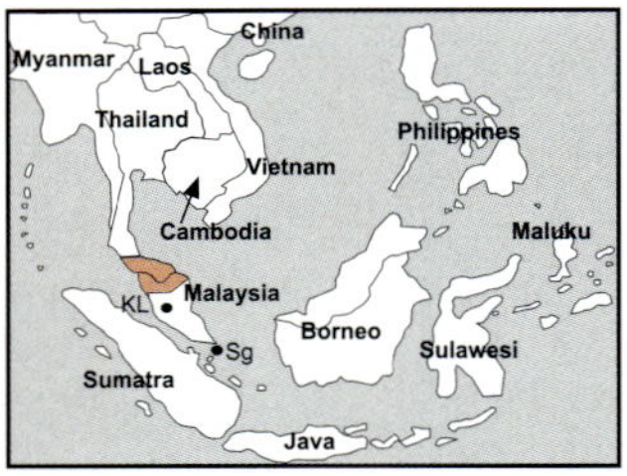


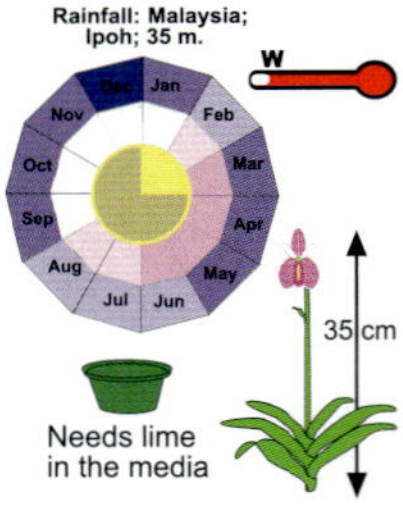


## *Paphiopedilum parishii* (Rchb.f.) Stein

This dramatic large-flowered species grows close to the ground on mossy tree branches or on shady moss-covered boulders in montane forest from 1200-2200 metres altitude. It has 5-8 very large leathery green leaves and a 30-50 cm tall inflorescence that has 2-9 flowers, all of which are open at the same time. Each flower is 9-10 cm across and may be as much as 15 cm tall. This species is easily distinguished by the long twisted petals which turn from an apple-green colour at the base to very dark maroon-black at the apex.

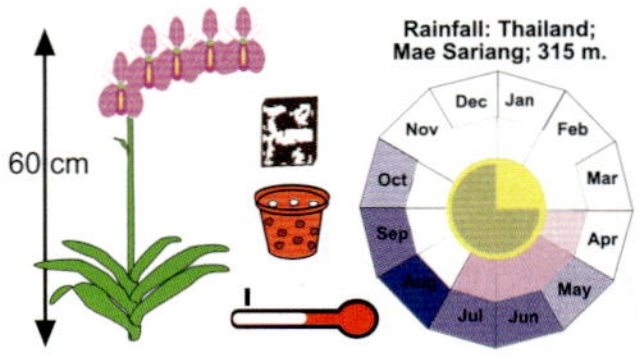


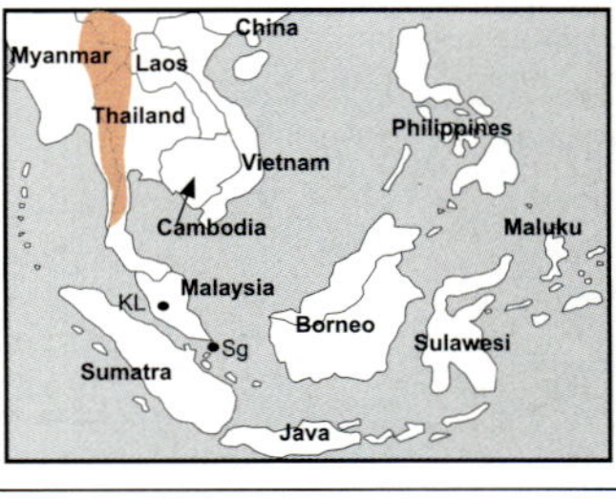

# P

## *Paphiopedilum philippinense* (Rchb.f.) Stein.

This is probably the most beautiful of the S. E. Asian *Paphiopedilum* species. It grows on limestone cliffs, hills and outcrops, usually rooted in limestone rubble in fairly open situations from 0-300 metres altitude. It has up to 9 very large thick, leathery green leaves. The inflorescence is up to 50 cm tall and has 2-5 flowers, all of which are open at once. Each flower is about 9 cm wide, and the petals can be as long as 18 cm.

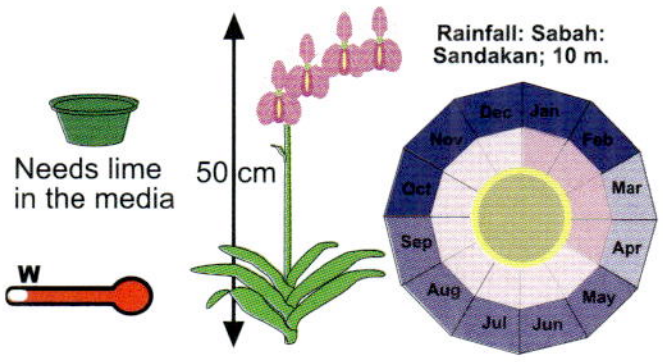


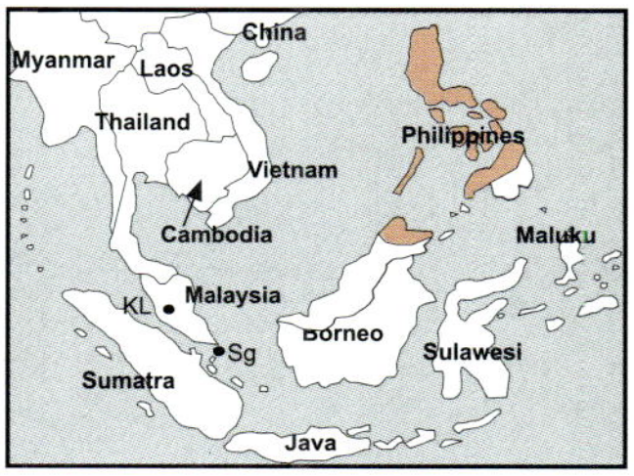

## *Paphiopedilum primulinum* M.Taylor & P.Taylor

This species grows as a lithophyte in humus on the top of limestone ridges from 400-500 metres altitude. It has 4-7 green leaves and an inflorescence that slowly extends to about 20-30 cm long, producing a large number of single flowers one after the other. A variety *purpurascens*, which is found on coralline rocks near the sea, has spots on the petals and an orange-pink flush on the lip.

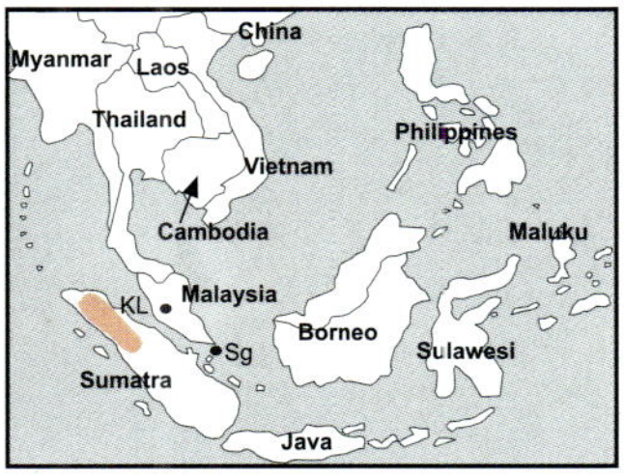


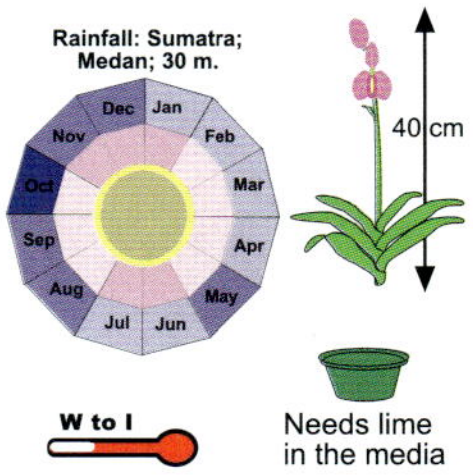


## *Paphiopedilum spicerianum* (Rchb.f.) Pfitz.

This distinctive species only just occurs in our region; most plants are known from N. E. India and N. W. Myanmar. It grows on limestone outcrops and cliffs from 300-1300 metres altitude in regions that receive frequent mists during the dry season. It has 4-5 long green leaves. A 13-35 cm tall inflorescence carries a single 5-7 cm wide flower. It is easily identified by the maroon stripe on the dorsal sepal.

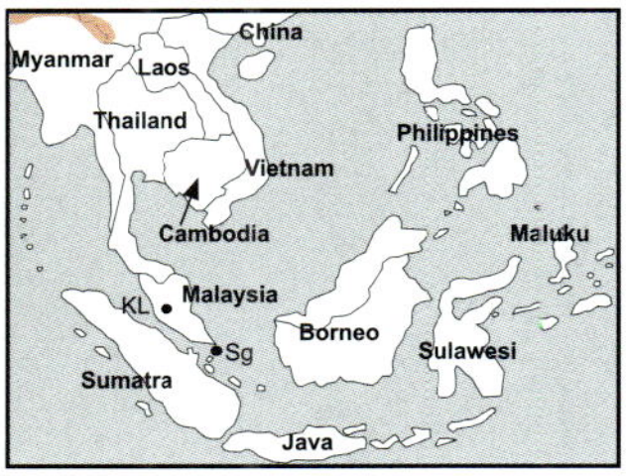


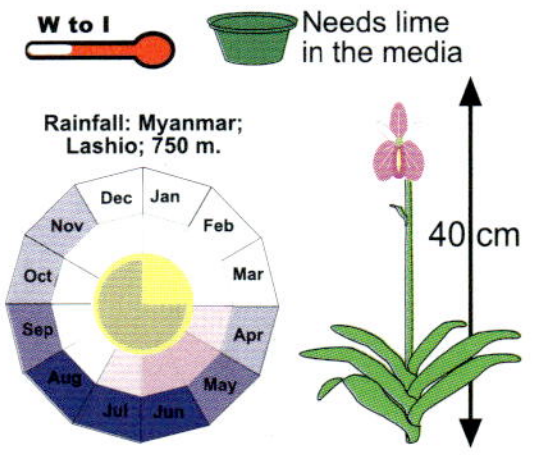

# P

## *Paphiopedilum stonei* (Hk.f.) Stein

This extremely rare species grows in deep shade on limestone cliffs and hills from 60-700 metres altitude. It has up 5-6 very long leathery green leaves. A 17-35 cm tall inflorescence carries 2-4 flowers, all of which are open at the same time. The flowers are up to 12 cm wide. The species is quite variable and several varieties are known.

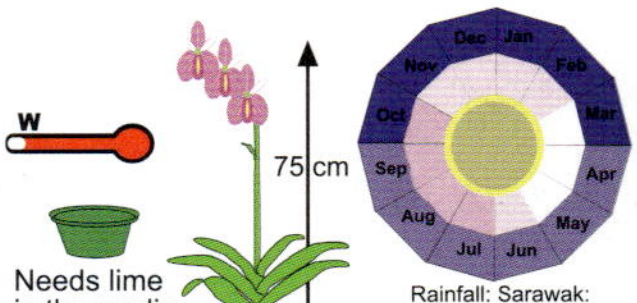

Rainfall: Sarawak: Kuching; 25 m.

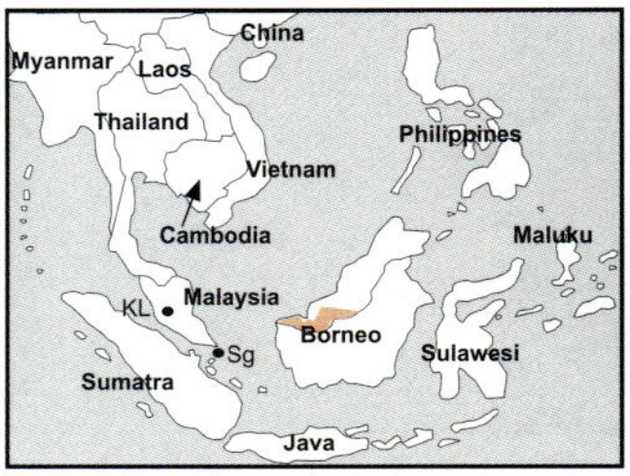

## *Paphiopedilum superbiens* (Rchb.f.) Stein.

This terrestrial species grows in humus and leaf litter on the floor of montane forest from 900-1700 metres altitude, often on steep slopes or ridgetops. It has 4-5 tesselated green leaves and a single 7-9 cm wide flower. The illustrated specimen is a variety known as *P. superbiens* var. *curtisii.*

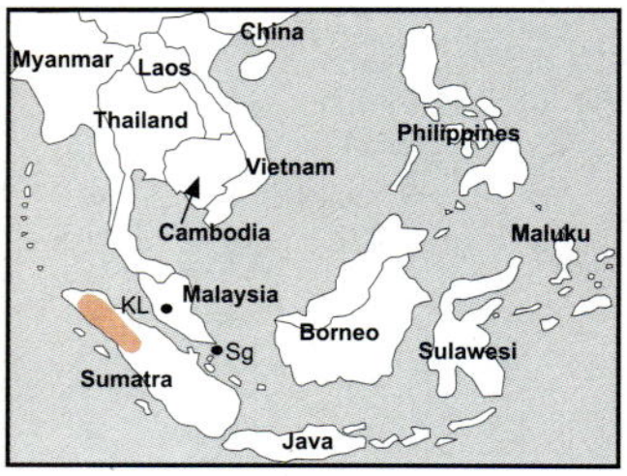


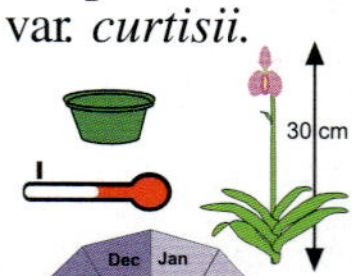


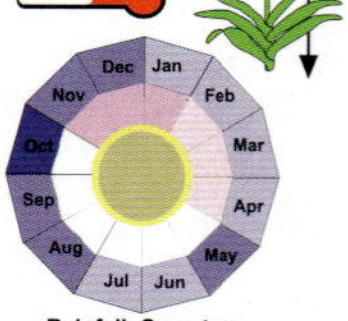

Rainfall: Sumatra; Medan; 30 m.

## *Paphiopedilum tonsum* Pfitz.

This species is not uncommon in deep humus on the floor of montane forest from 1000-1800 metres altitude. It is rather variable in flower size, petal shape, and the amount of spotting; several varieties are recognised in horticulture. It has 4-5 mottled green leaves and a single 12-14 cm wide flower on a 35 cm tall stem.

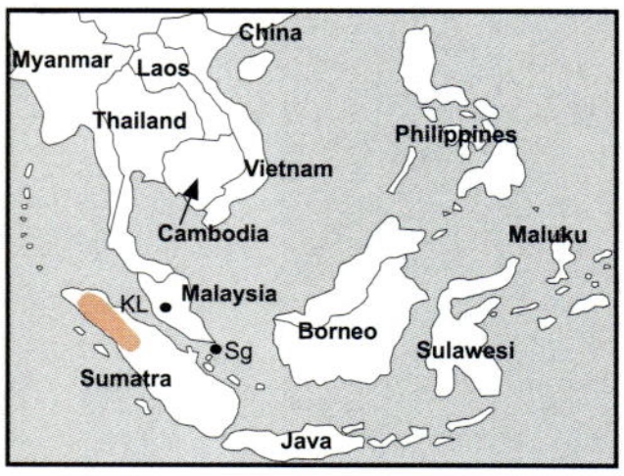


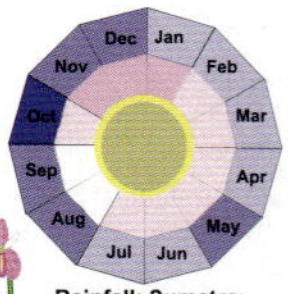

Rainfall: Sumatra; Medan; 30 m.

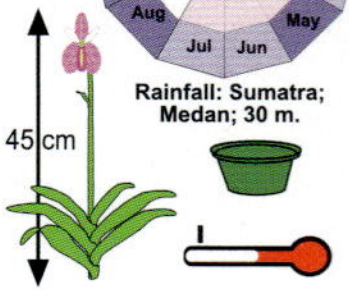

# P

## *Paphiopedilum victoria-regina* M. Wood

This attractive species is usually found growing on rocks or in leaf litter on the forest floor from 600-1300 metres altitude. It has 4-6 fairly large green leaves and a white-haired inflorescence that extends slowly, reaching up to 60 cm long. It may form as many as 32 flowers, with 1-2 open at a time. In cultivation it is often called *P. chamberlainianum* and is easy to identify because of the extensive dark purple-black markings on the dorsal sepal.

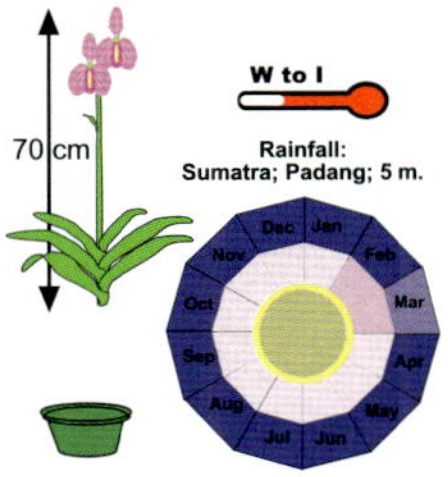


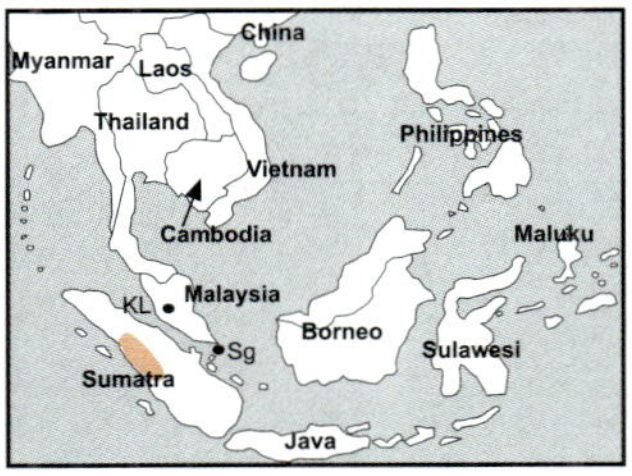

## *Papilionananthe*

This genus of 12 species is distributed from the Himalayas and Sri Lanka to Indochina and Indonesia. They are widely referred to as "terete-leaved *Vandas*" and are called *Vanda* in hybrid registration lists. Most of the species are tall climbing epiphytes with terete leaves and lateral inflorescences with large showy flowers. The best known species, *P. teres* and *P. hookeriana*, have been extensively interbred with various *Vanda* species and the resulting hybrids form a significant part of the S. E. Asian orchid industry. *Papilionanthe* species are seldom seen in orchid collections today; they have been replaced by hybrids which are easier to obtain and grow. The National Flower of Singapore, *Vanda* Miss Joaquim, is a natural hybrid between *P. teres* and *P. hookeriana*.

## *Papilionananthe tricuspidata* (J.J.Sm.) Garay

This rare species was originally found at 900 metres altitude in the Alor Islands, but more recent specimens are thought to have originated in North Maluku. It is a tall plant that climbs upwards through the canopy layer, flowering when the stems (which may be several metres long) reach sunlight. The inflorescences have up to 8 showy flowers that are about 6 cm in diameter.

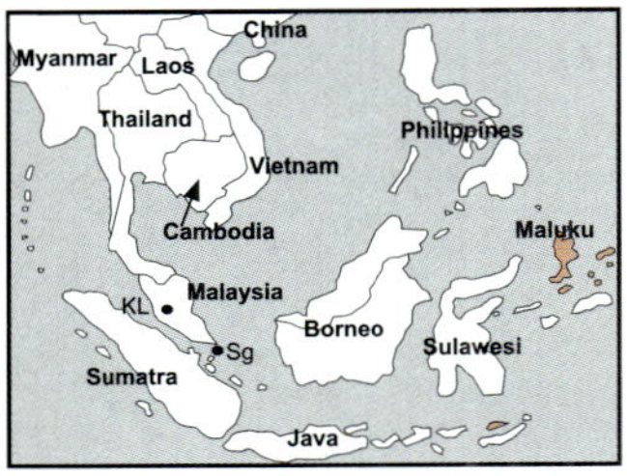


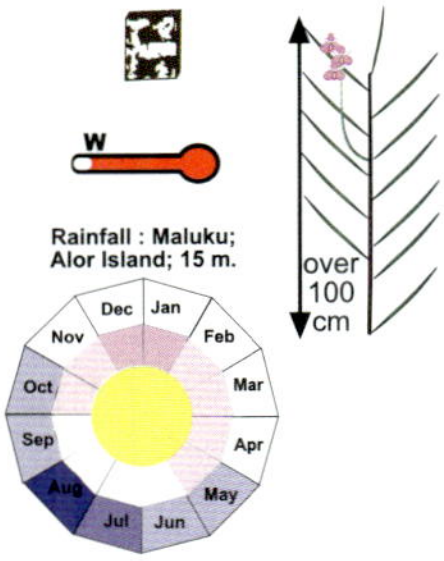

# P

## Paraphalaenopsis

This small genus has 4 species that are found only in Borneo. The plants were originally thought to belong to *Phalaenopsis*, and are still listed as *Phalaenopsis* for hybrid registration purposes. The species all have short stems with descending terete leaves and a short erect inflorescence of showy flowers. *Paraphalaenopsis* species are often seen in orchid collections and their hybrids (with related genera such as *Phalaenopsis, Vanda*, etc) are very popular.

### *Paraphalaenopsis denevei* (J.J.Sm.) Hawkes

This rare species is found in lowland rainforest, growing on riverside trees from sea-level to at about 300 metres altitude. It has leaves up to 70 cm long and a short inflorescence that is crowded with up to 13 flowers that are 4 cm across. The flowers are very sweetly scented. In cultivation the species flowers 3-4 times a year.

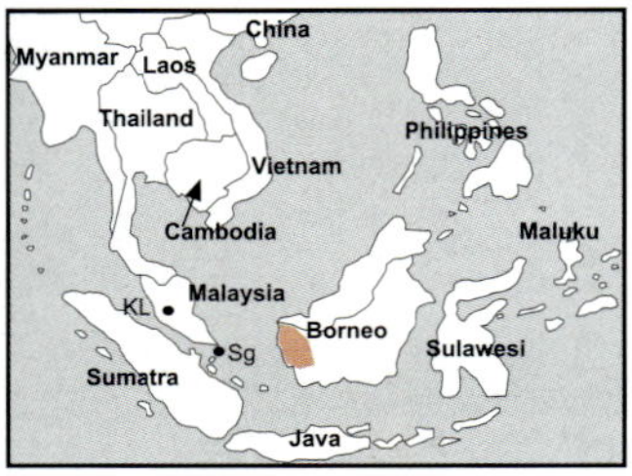


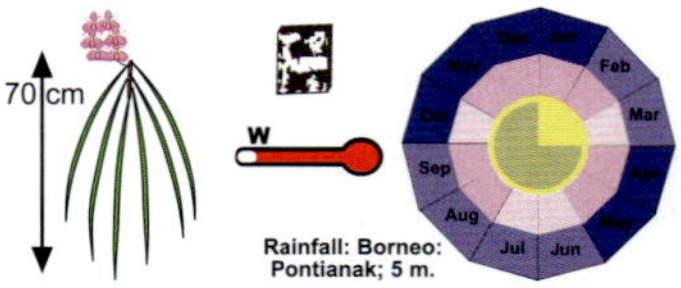


### *Paraphalaenopsis laycockii* (Henderson) Hawkes

This rare species is the second largest in the genus, with leaves up to 1 metre long and 1 cm in diameter. It is found in lowland rainforest from sea-level to at least 500 metres altitude. The inflorescence is crowded with up to 15 flowers that are 6-8 cm tall. The flowers have a sweet but musty scent.

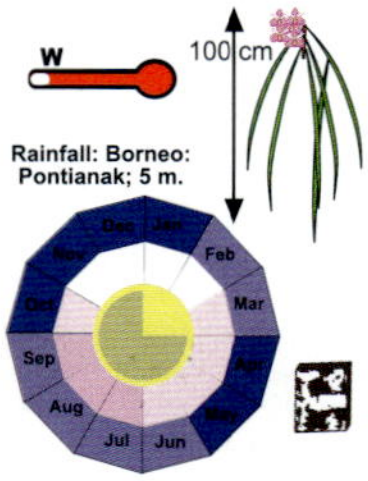


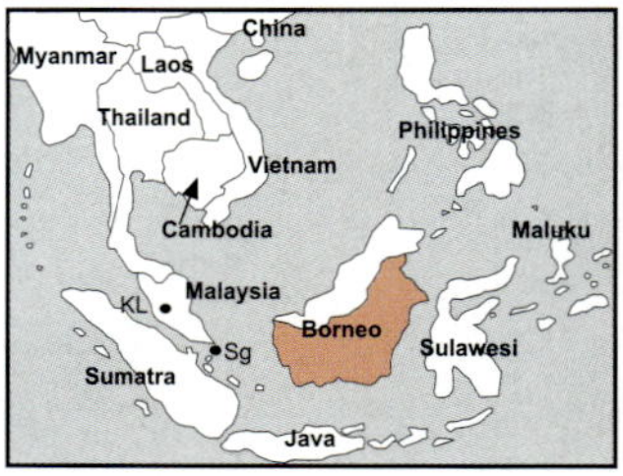

### *Paraphalaenopsis serpentilingua* J.J.Sm.

This rare species is the smallest in the genus. It is found in swampy forest and on moss-covered rocks from sea-level to 1000 metres altitude. It has leaves up to 30 cm long and a 35 cm long inflorescence of up to 7 sweetly-scented flowers that are about 2 cm across. It is easily identified by the forked midlobe of the lip. Two colour forms are shown in the illustration.

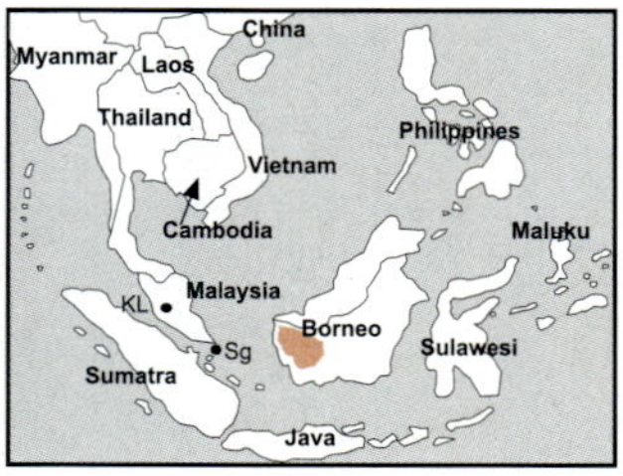


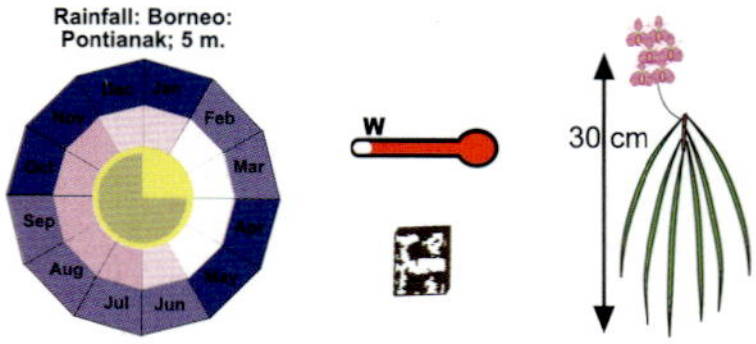

# P

## Parapteroceras

This small genus probably has 5 or 6 species that are found from Indochina to the Philippines and Indonesia. The species, which were originally placed in *Saccolabium* or *Tuberolabium*, are all epiphytes with fairly short stems and rather fleshy inflorescences that are often club-shaped. None of the species have any commercial importance, and are seldom seen in orchid collections because their flowers are relatively short-lived.

### ***Parapteroceras odoratissimum*** (J.J.Sm.) J.J.Wood

This sweet-scented species, which is often called *Tuberolabium odoratissimum*, deserves to be better known amongst orchid growers. For its size, it probably has the strongest fragrance of all Asian orchids. It is common in the hills and mountains, growing on trees in forest, plantations and villages from 500-1500 metres altitude. Numerous inflorescences are formed, each with up to 20 flowers that are over 1 cm in diameter and last for about 1 week. Flowers of Javanese plants are usually cream with purple marks.

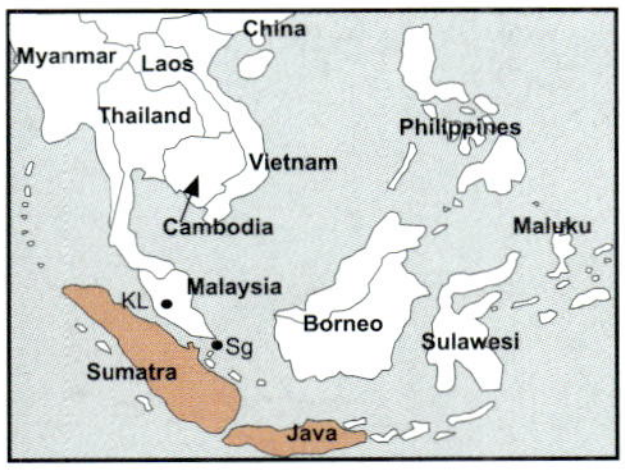


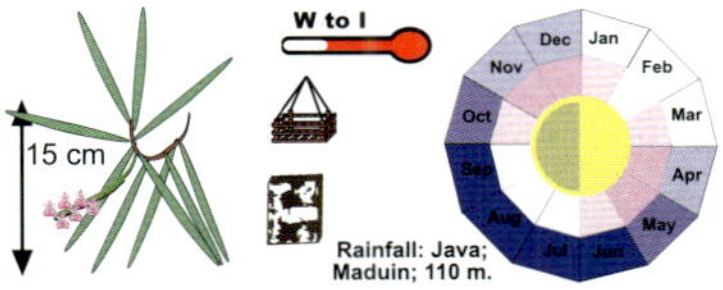


## Pecteilis

This small genus of 8 to 10 species is distributed from the Indian Himalayas and China to Maluku in Indonesia. The species, which are related to *Habenaria*, are all terrestrials with large underground tubers and large showy flowers that often have an elaborately fringed lip. At least 3 of the species are sought-after by orchid enthusiasts and can occasionally be seen in collections. In the case of *P. sagarikii* (below), over-collection to supply this trade means that the plant is now extremely rare in the wild.

### ***Pecteilis sagarikii*** Seidenfaden

This extremely rare deciduous terrestrial species has a rosette of 2-3 large leaves that are pressed against the soil. The single erect inflorescence is produced shortly before the leaves die away. It bears up to 12 long-lasting flowers that are about 3 cm wide and are fragrant after dusk. Two colour forms are illustrated, the only difference being the amount of yellow on the lip.

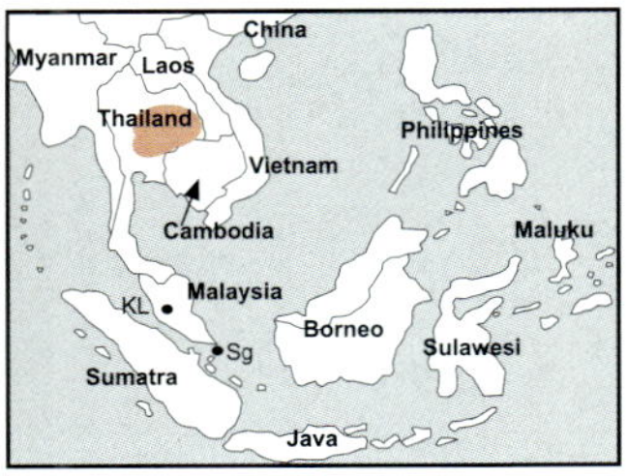


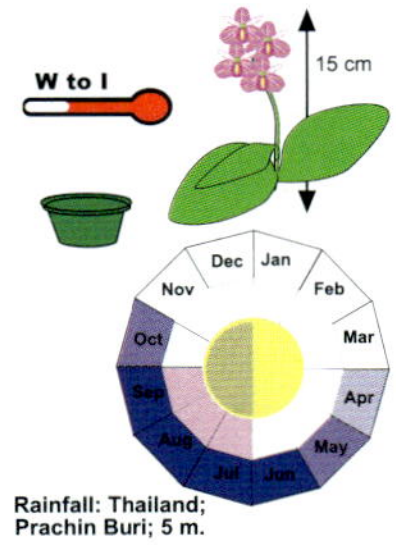

# P

## Pelatantheria

This genus of 5 species is distributed from Nepal and India to China, Malaysia and Sumatra. Four of these species can be found in Thailand. The plants are all long climbing epiphytes with rather short fleshy leaves and short inflorescences of a few small flowers that are not upside down. Although they may occasionally be seen in orchid collections, none have any commercial importance and they are not widely used to produce hybrids.

### *Pelatantheria insectivera* (Rchb.f.) Ridl.

This tall species climbs up through the canopy layer of trees in the forest from sea-level to at least 800 metres altitude. It has a fairly thick stem with rather short oval to linear leaves. It produces numerous short inflorescences that each bear up to five 10 mm long flowers and last for 7-10 days. *Pelatantheria rivesii*, which is found in Indochina and Yunnan Province of China, has almost identical flowers and can only be distinguished by examining the shape of the callus in the mouth of the spur; *P. insectivera* has a simple tongue-shaped callus.

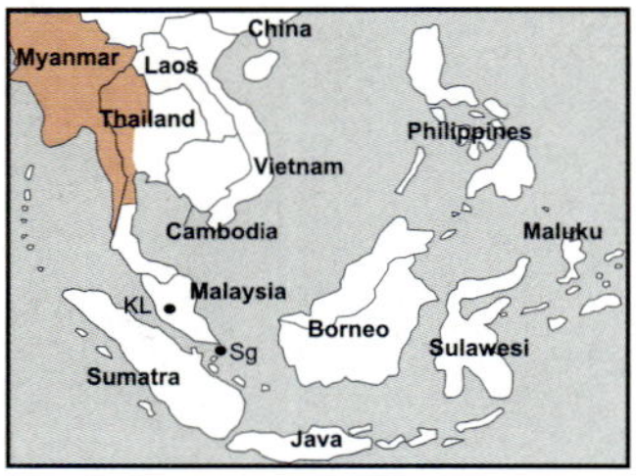


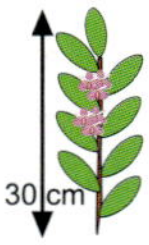

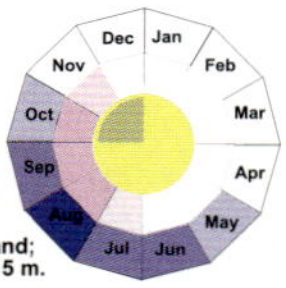


## Pennilabium

This genus of about 13 small epiphytic species is distributed from Assam through Thailand, Indochina and Malaysia to Indonesia and the Philippines. All the species appear to be uncommon. They have a short inflorescence ending in 2 rows of bracts that produce relatively large ephemeral flowers in succession, usually one at a time. The lips of most species have elongated scarf-like side lobes. They are seldom seen in orchid collections and have no commercial importance.

### *Pennilabium naja* O'Byrne

This unusual species was described as recently as 2000. It was found in Sulawesi at 1400 metres altitude, growing low down on hanging vines and the trunks of saplings and narrow trees in swampy forest. It has several inflorescences that remain active for at least a year, producing a succession of single 1 cm wide flowers at irregular intervals. The flowers are open for only one day.

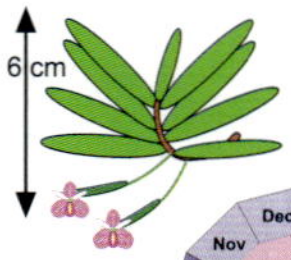

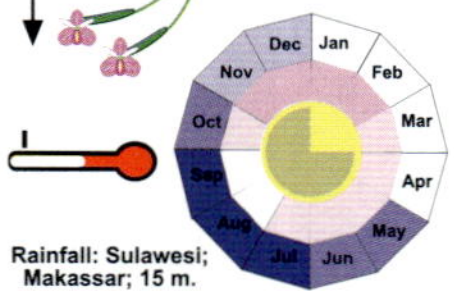


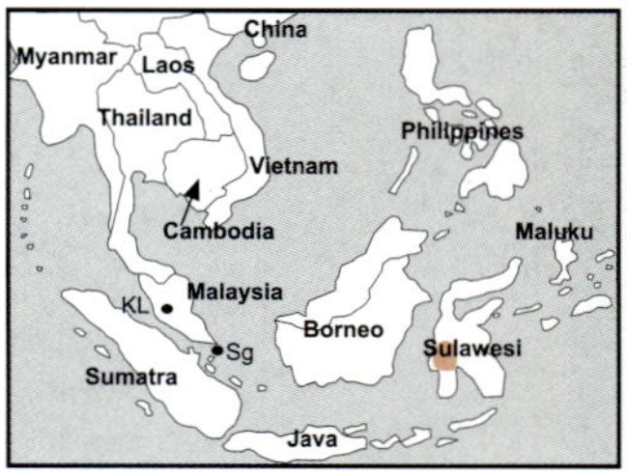

# P

## Peristylus

This genus is distributed from India across S. E. Asia to New Guinea and the Pacific Islands. The number of species is uncertain; it could be anywhere between 60 and 135. The species are closely related to *Habenaria*; they are terrestrials with underground tubers that produce an erect leafy shoot which terminates in a raceme of white, green or yellow flowers. The petals are joined to the dorsal sepal, forming a hood over the column. The lip has three lobes and a short spur. They are seldom seen in orchid collections and have no commercial importance.

### *Peristylus monticola* (Ridl.) Seidenf.

This easily-overlooked species is fairly common in mountain grasslands, roadside verges and alongside paths from 1000-1500 metres altitude. It prefers locations where the grass is thin, or where there are occasional fires. The flowers have a bilobed spur and thread-like sidelobes that are 5-15 mm long.

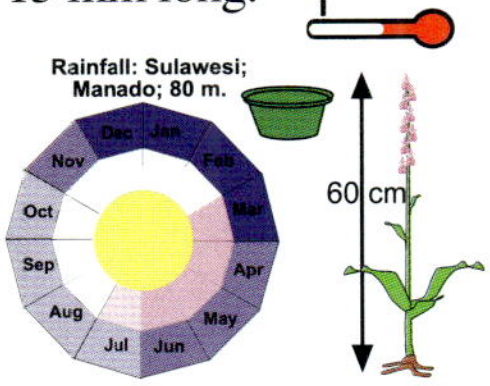


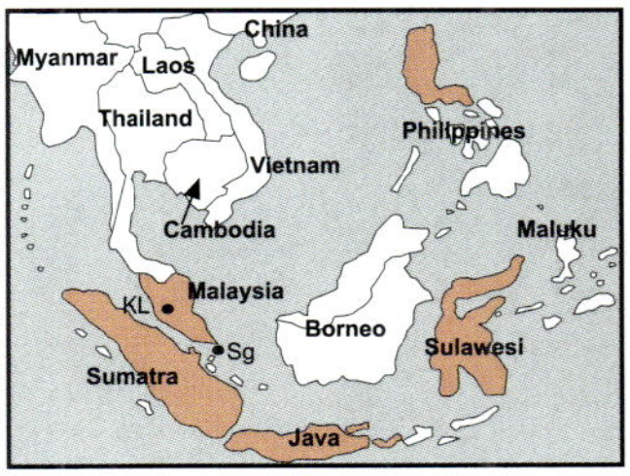

## Phaius

This genus of some 30-50 large terrestrial species that occur in forests and grasslands from Central Africa and the Malagasy Republic through Asia to China, Japan, Australia and the Pacific Islands. The plants are variable in form; some have pseudobulbs with large pleated leaves; others have slender leafy stems. They have large showy flowers with a lip that is partly joined to the column. They are popular garden plants in S. E. Asia, and are cultivated by orchid enthusiasts worldwide.

### *Phaius tankervilliae* (Banks ex L'Herit.) Bl.

This well-known species used to be common in grasslands, swampy areas, and forest margins from 300-1500 metres altitude, but has been heavily collected and is now seldom seen in the wild. It has small pseudobulbs and tall inflorescence that has up to 25 very large flowers which turn blue if bruised.

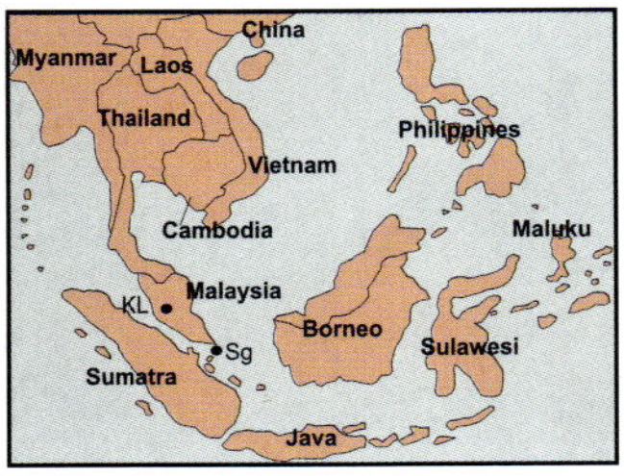

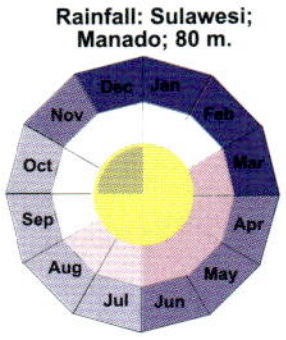

# P

## Phalaenopsis

This well-known genus of between 40 and 45 species is distributed from India across tropical S. E. Asia to New Guinea and Australia with the majority of species found in Indonesia and the Philippines. All the species are monopodial epiphytes with short stems and a few broad leaves. Inflorescences are usually long with large showy flowers; they either have all the flowers opening at once, or they extend slowly and the flowers open in succession. This is one of the most important commercial genera in the world. The production and sale of *Phalaenopsis* hybrid plants is a multi-million dollar industry in S.E.Asia (especially Taiwan) and elsewhere in the world. The species are sought after and cultivated worldwide. Most species are now very rare in the wild due to unscrupulous collecting and habitat destruction.

### *Phalaenopsis amabilis* (Lindl.) Bl.

This well-known species, the national flower of Indonesia, is still common in some parts of its range, growing on trees from 0-1200 metres altitude. It is a popular garden plant and can often be seen in villages and towns, sometimes growing in full sun.

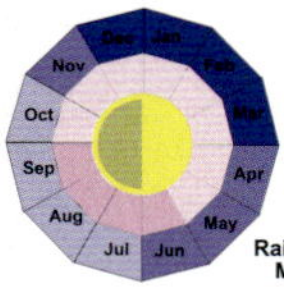

Rainfall: Sulawesi; Manado; 80 m.

### *Phalaenopsis aphrodite* Rchb.f.

Considered by some botanists to be a variety of *P. amabilis*, this species can be distinguished by its broad kite-shaped midlobe. It was once common but nowadays it is hard to see in the wild due to overcollection and habitat destruction. It occurs on shady trees in the hills and mountains from 500-1500 metres altitude. The flowers are rather variable in shape and size; there is also some variation in colour. Plants from Taiwan are sometimes called *P. formosana* in horticulture.

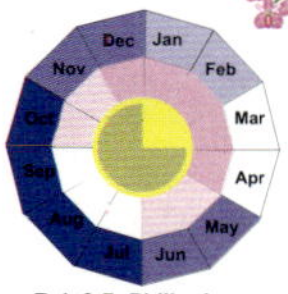

Rainfall: Philippines; Negros; 5 m.

# P

## *Phalaenopsis amboinensis* J.J.Sm.

This lowland species is rare in the wild but quite common in cultivation. It has inflorescences that extend slowly, producing a succession of 4-6 cm wide flowers. A large specimen may have 5 or 6 active inflorescences, each with 1-3 flowers open a time, creating a spectacular long-lasting display. Flower colour is extremely variable; the base colour can be white, yellow, gold or even pale green.

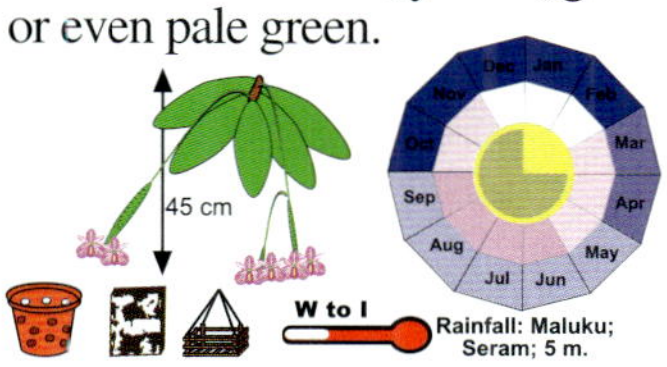


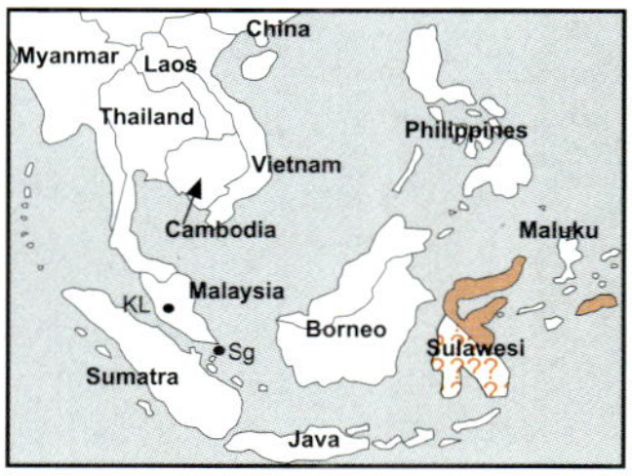

## *Phalaenopsis bellina* (Rchb.f.) E.A.Christenson

This species is widespread in cultivation under the name 'The Borneo form of *P. violacea*'. The state flower of Sarawak, *P. bellina* grows on trees and lianas overhanging rivers and swamps but is now almost extinct in the wild. Highly fragrant 4-6 cm flowers are produced in succession; colour is variable.

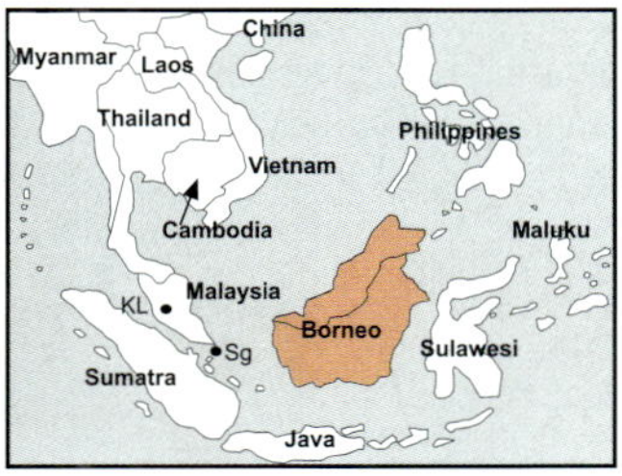


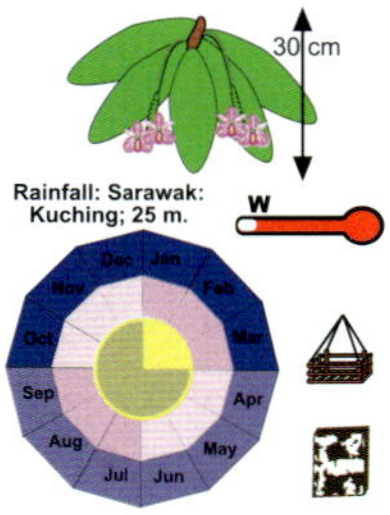


## *Phalaenopsis celebensis* Sweet

This small species, which was first described in 1980, probably grows on shady trees in the mountains of Sulawesi and appears to be extremely rare in the wild. It can be identified by its green leaves that are mottled silver-white. It has 30-40 cm long pendant inflorescences that bear up to 40 well-spaced 2 cm wide flowers that are all open at once but do not last very long.

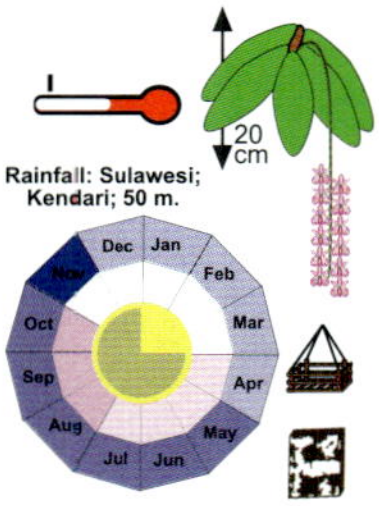


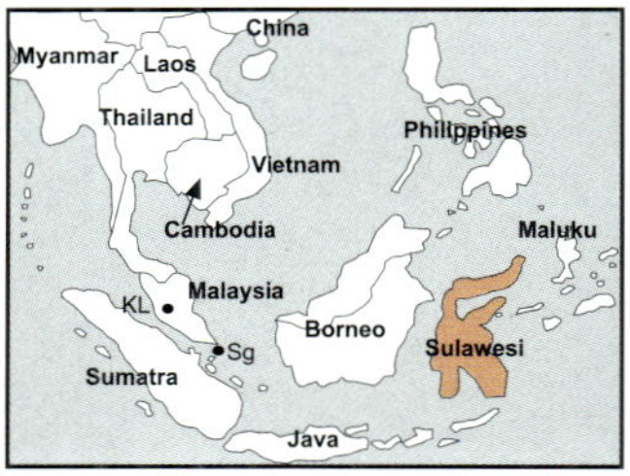

# P

## *Phalaenopsis cornucervi* (Breda) Bl. & Rchb.f.

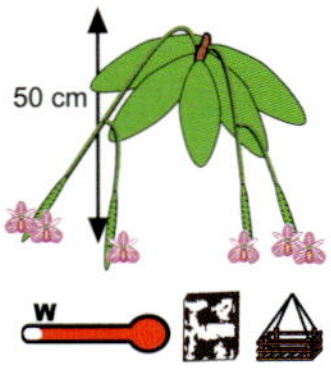


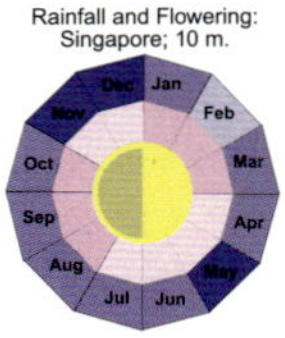


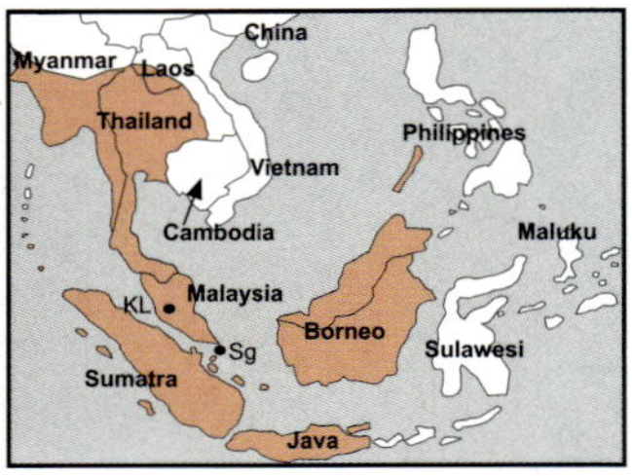


This large robust species, known in Borneo as the 'Crocodile Tail Orchid', grows from sea-level to 500 metres altitude and is usually found in bright situations in riverine forest and swamp forest. It has several active inflorescences that have a flattened rachis and extend slowly over several years, producing a succession of 4 cm wide long-lasting flowers, with 1-3 open at a time.

## *Phalaenopsis equestris* (Schauer) Rchb.f.

This lowland species is widespread throughout the Philippines, growing on trees and rocks from sea-level to 500 metres altitude. It can be distinguished from the closely-related *P. lindenii* by the absence of silver markings on the leaves. It has an arching branched inflorescence with many 25 mm wide flowers that are all open at once. Flowers are highly variable in colour; the base colour can be white to pink to dark purple, and many bicolour forms are known.

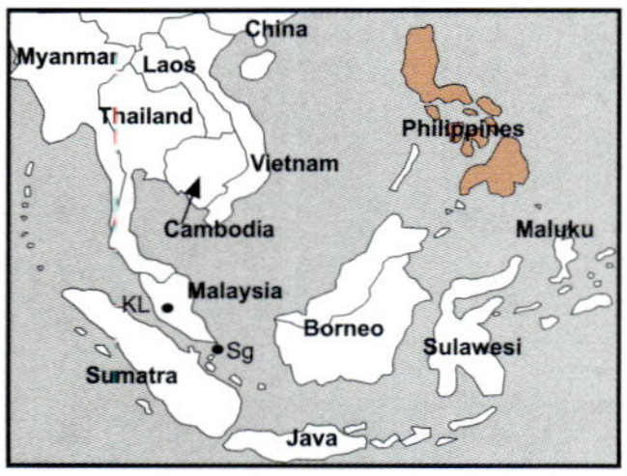


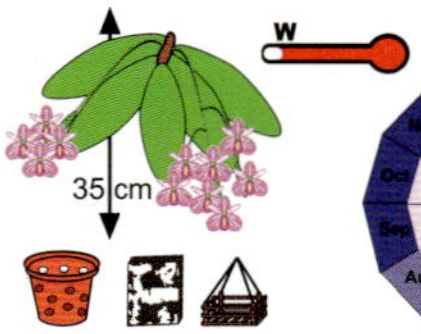


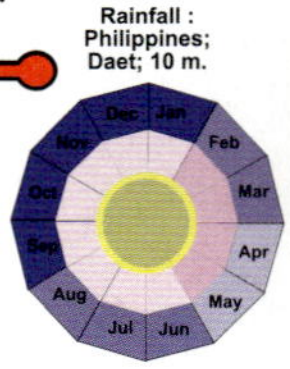


## *Phalaenopsis fasciata* Rchb.f.

This lowland species is widespread in the Philippines and is quite common in cultivation, but is frequently confused with other species. It has a long arching inflorescence that elongates slowly, flowering sporadically with 1-3 flowers open at a time. The rather fleshy 5 cm wide flowers are pale to dark yellow or orange-yellow, with heavy reddish-brown to cinnamon bars and blotches; the lip is bicoloured.

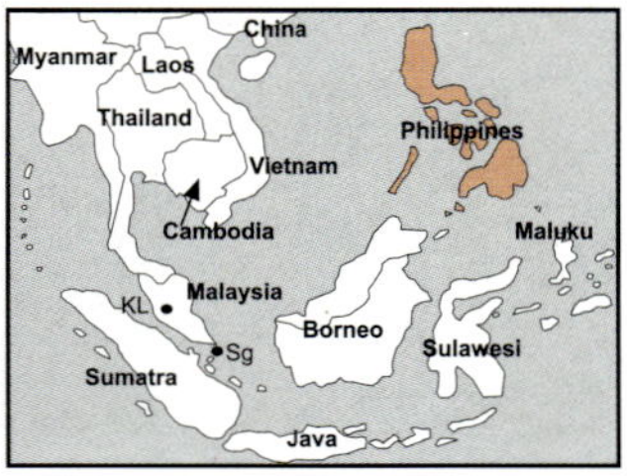


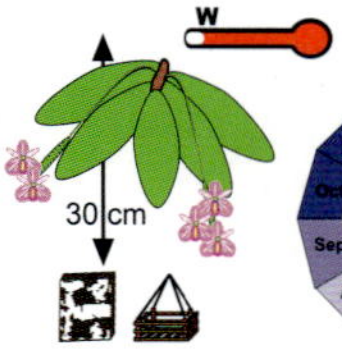


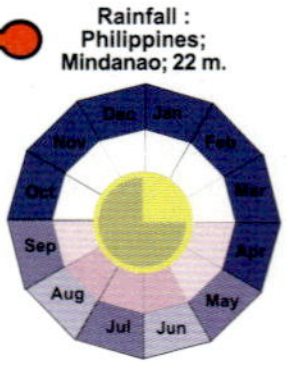

# P

## *Phalaenopsis fimbriata* J.J.Sm.

This beautiful free-flowering species has two known varieties; variety *sumatrana* (illustrated) grows on rocks and trees from 600-1400 metres altitude in Sumatra while the normal form, which has smaller flowers with magenta markings on the sepals and lip, grows on steep slopes in Java from 50-1000 metres altitude.

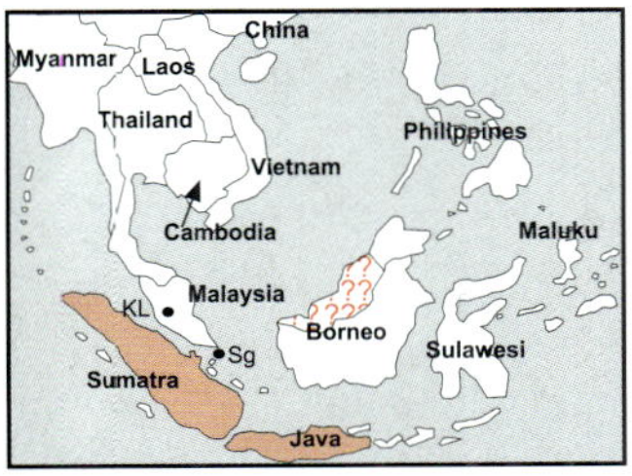


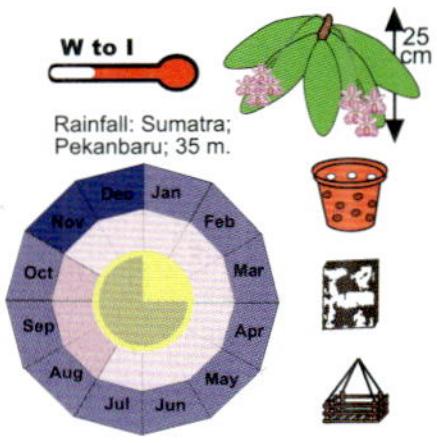

## *Phalaenopsis fuscata* Rchb.f.

This uncommon species is found growing on trees in shady locations near streams or rivers from sea-level to 1000 metres altitude. A large plant has several active inflorescences that elongate slowly, sporadically producing one to several flowers until a total of 10-12 have been formed. The fleshy 3 cm wide bicoloured flowers are easily confused with *P. kunstleri,* which has a shorter column and long keels at the lip base.

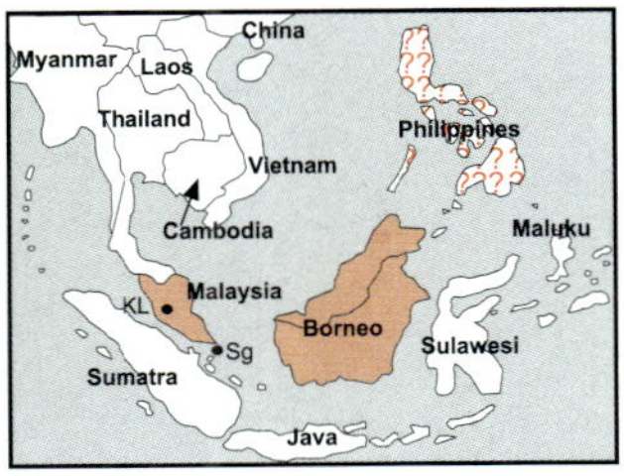


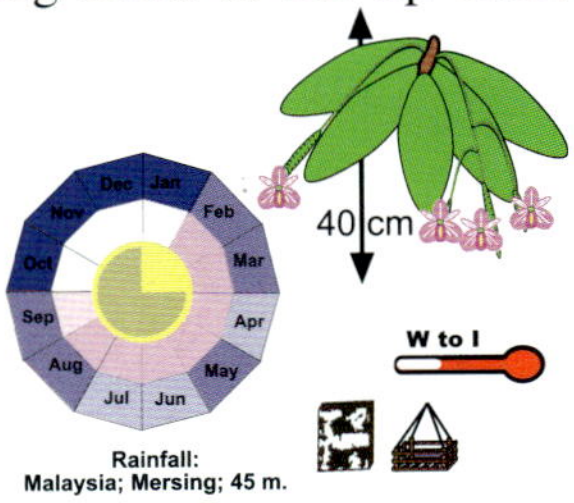


## *Phalaenopsis gigantea* J.J.Sm.

This giant lowland species is quite rare in the wild but increasingly common in cultivation. It grows on trees and shrubs in the forest from sea-level to 400 metres altitude. Plants with leaves 90 cm long have been recorded, but 50 cm is a more usual size. The species produces 40 cm long inflorescences with 15-30 crowded 5-6 cm wide scented flowers that all open at once; flower colour is extremely variable.

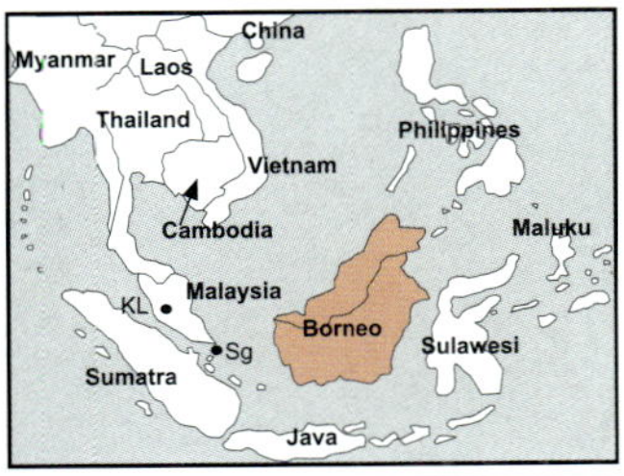


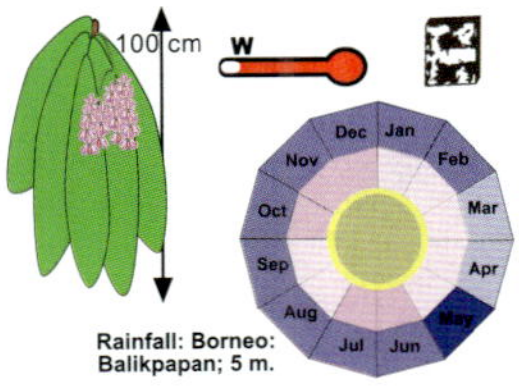

# P

## *Phalaenopsis javanica* J.J.Sm.

This species is probably extinct in the wild. The last known population was removed by an Indonesian orchid exporter in the 1970s. The inflorescences lengthen slowly with 1-2 flowers open at a time. The flower on the left in the photo is an *alba* form.

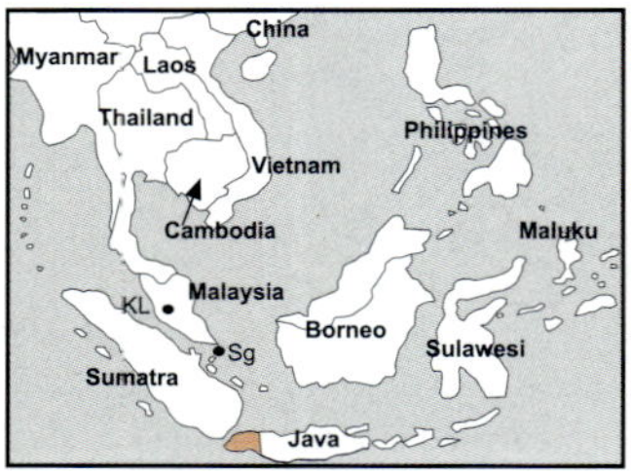


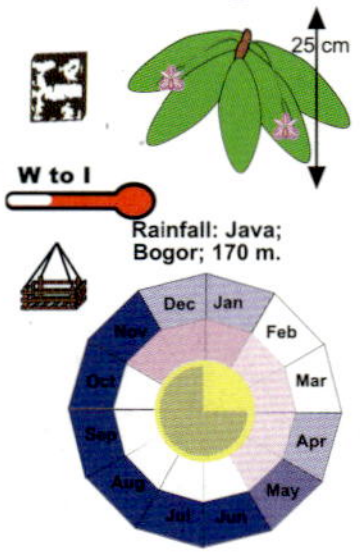

## *Phalaenopsis lindenii* Loher

This small species is found growing in shady locations on trees in the central mountains of north Luzon. It has silvery-white mottled green leaves and an erect to arching inflorescence with many 28 mm wide flowers that are all open at once. Most plants have white flowers with a rosy-pink lip; a few have rosy pink spots and stripes (illustration).

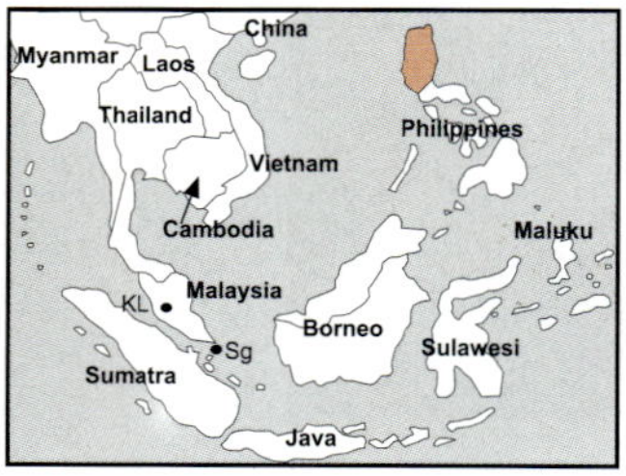


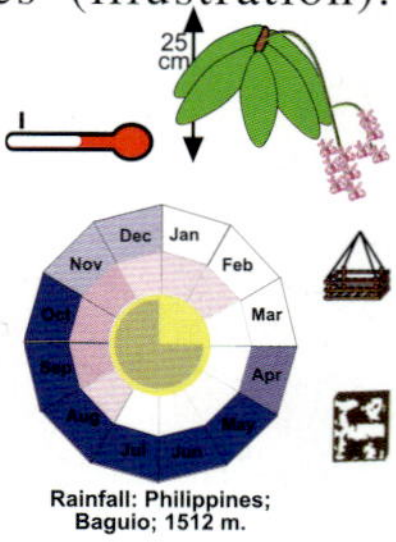


## *Phalaenopsis maculata* Rchb.f.

This small species grows in cool shady moist environments, near streams or rivers from sea-level to 1000 metres altitude. The inflorescence sporadically produces a 3 cm wide rather delicate flower that has a smooth red midlobe.

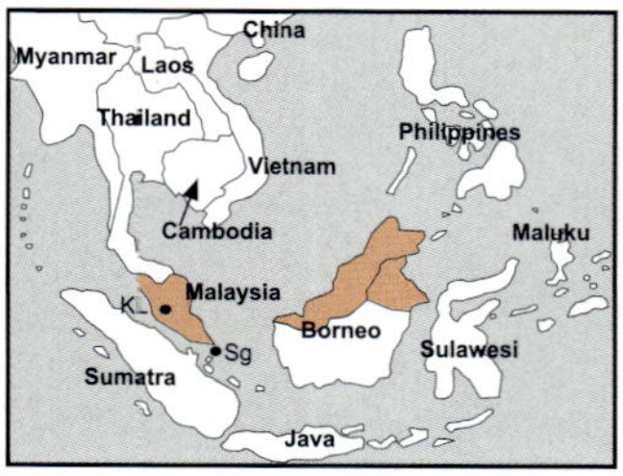


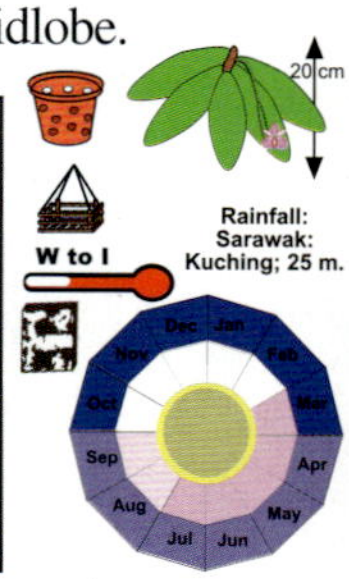

# P

## *Phalaenopsis micholitzii* Rolfe

This extremely rare small-growing species has a compact habit. It produces several inflorescences that are only 4-6 cm long and flower sporadically, producing 1-3 flowers at a time. The rather fleshy flowers, which are 5-6 cm across, do not open as widely as most other *Phalaenopsis* species.

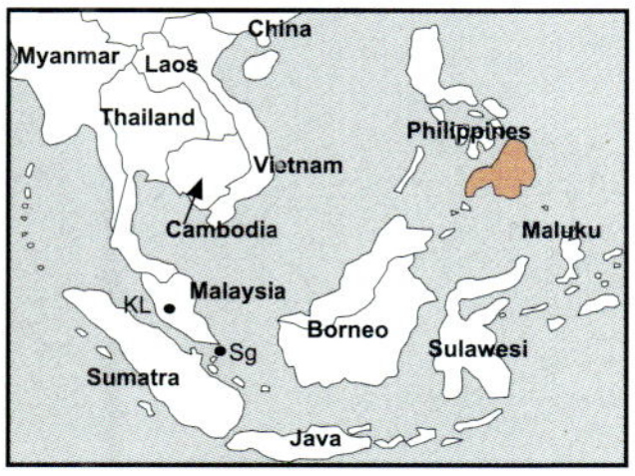


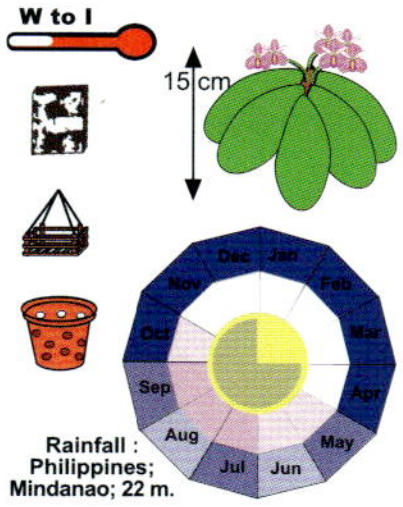


## *Phalaenopsis modesta* J.J.Sm.

This rare small-flowered species grows in deep shade, usually low down on tree trunks, exposed roots or lianas near rivers from 50 to 900 metres altitude. It has a rather short inflorescence that sporadically produces a few 3 cm wide delicate flowers. The flowers, which are strongly scented in the morning, are white with red or purple markings and turn greenish with age.

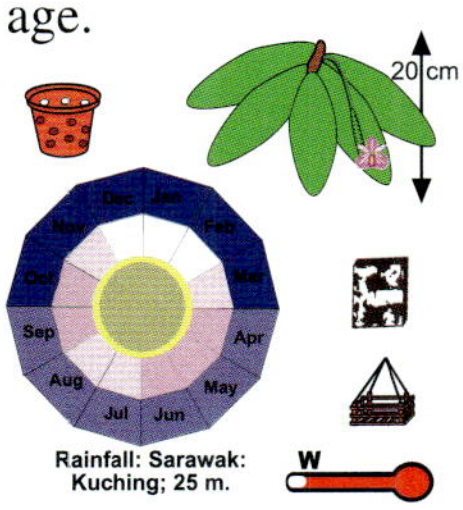


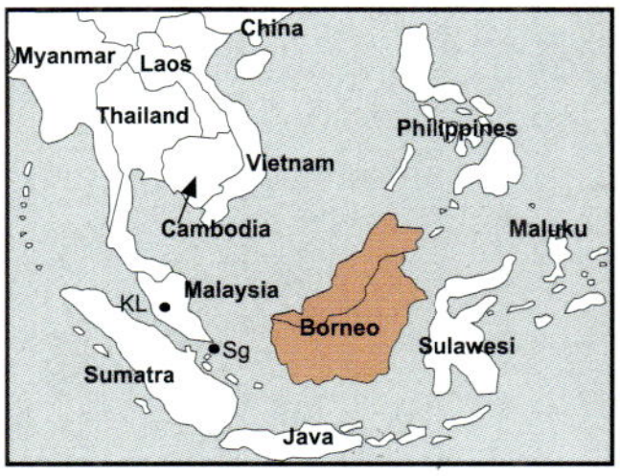

## *Phalaenopsis pallens* (Lindl.) Rchb.f.

This lowland species is quite common in cultivation, but is frequently confused with other species. The inflorescence elongates slowly, flowering sporadically with 1-2 flowers open at a time. The 4 cm wide flowers have a lip with toothed margins.

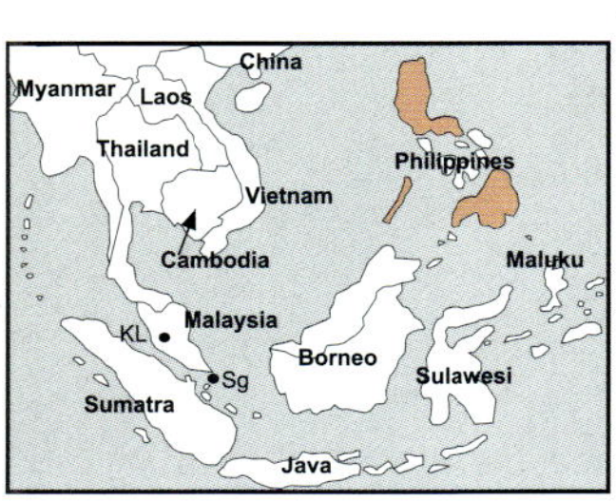


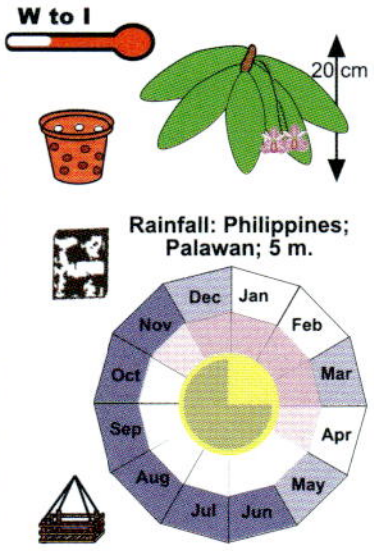

# P

## *Phalaenopsis pantherina* Rchb.f.

This species can be distinguished from *P. cornucervi* (which also has a flattened rachis) by the large anchor-shaped midlobe. It grows on the lightly shaded upper branches of tall forest trees from 0 - 800 metres altitude. The inflorescences extend slowly over 2-3 months, producing a succession of 5 cm wide fragrant flowers, with 1-2 open at a time.

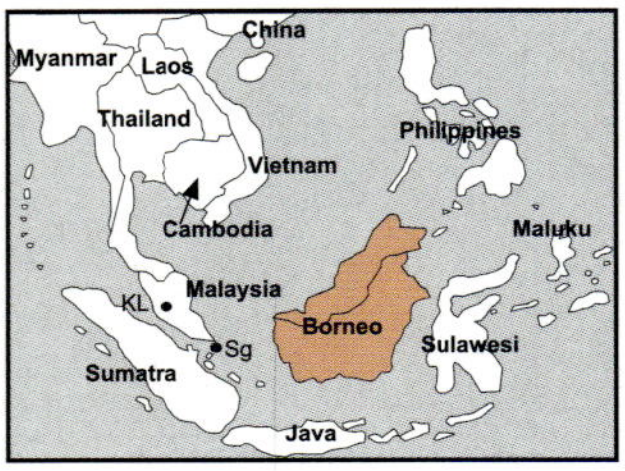


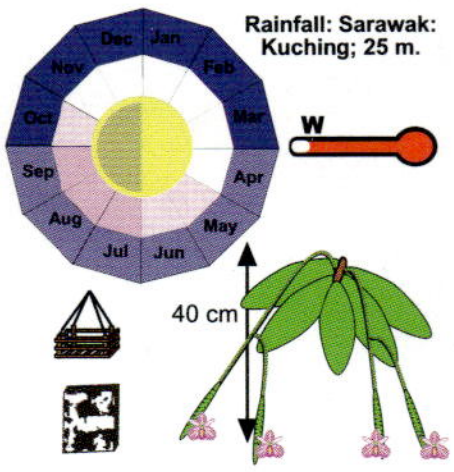


## *Phalaenopsis schilleriana* Rchb.f.

This well-known lowland species has silvery-grey mottled leaves and a branched inflorescence that can be up to 1 m long and carry over 200 flowers, all of which are open at once. The 7-8 cm wide flowers vary from white to pale pink, mauve or rosy purple, often with white margins.

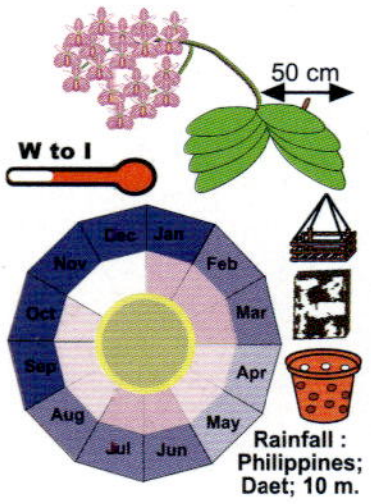


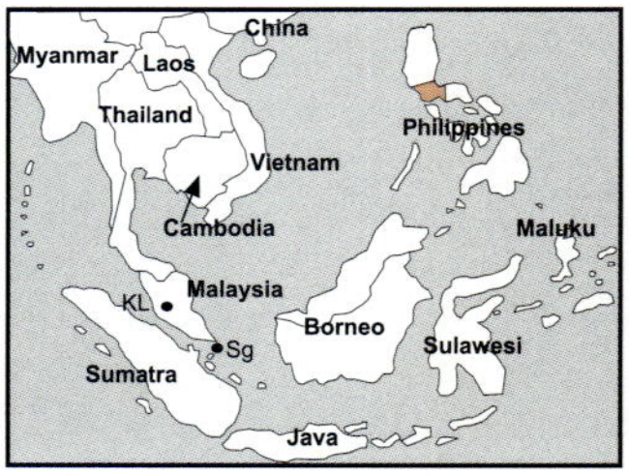

## *Phalaenopsis speciosa* Rchb.f.

This extremely rare lowland species was thought to be extinct until recently. It grows in deep shade on trunks and low branches of trees in freshwater swamp forest. The sepals and petals vary from white to rosy pink or purple (or combinations of them), often with rose, pink or purple markings. The all white form (illustrated) is often called *P. tetraspis*.

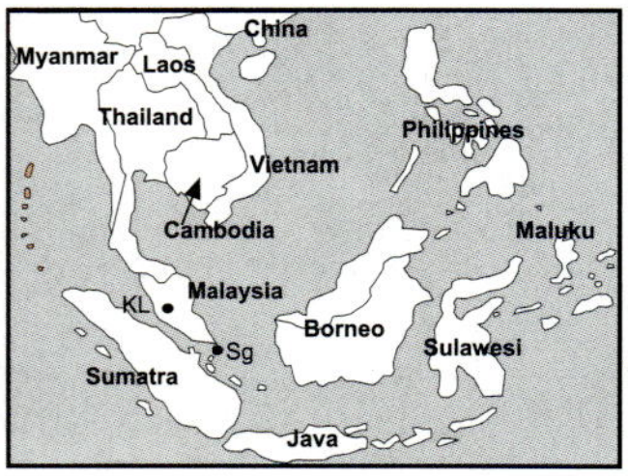


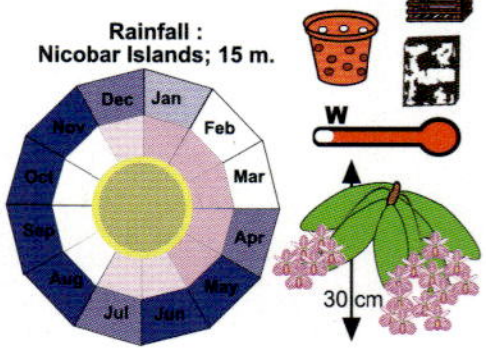

# P

## *Phalaenopsis stuartiana* Rchb.f.

This well-known species grows on trees in the valleys of northern Mindanao. It has silvery-grey mottled leaves and an arching branched inflorescence that can be up to 75 cm long and carry over 60 lightly fragrant flowers, all of which are open at once. The 6-7 cm wide flowers are fairly constant in colour, although one variety has spotted sepals and petals while a second has red marks on the sidelobes.

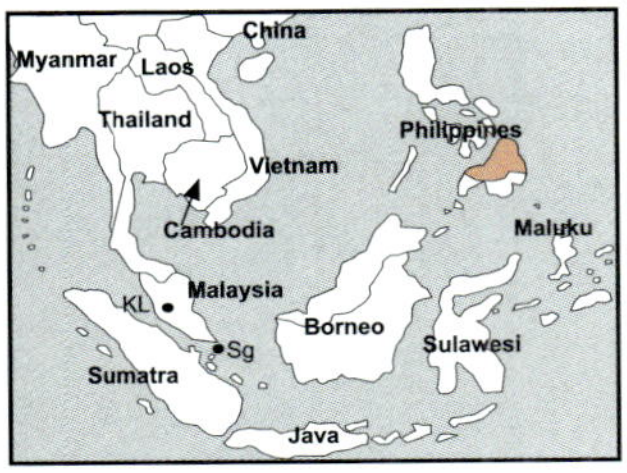


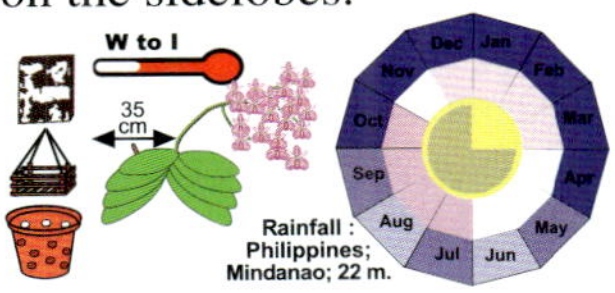


## *Phalaenopsis sumatrana* Korth. & Rchb.f.

This species grows low down on tree trunks in deep shade from sea-level to 800 metres altitude. It is now very rare in Borneo but can still be seen in some parts of Sumatra. A large plant will have several active inflorescences that flower sporadically, producing few to several 4-5 cm wide flowers on each occasion. Several different varieties (colour forms) are known in horticulture.

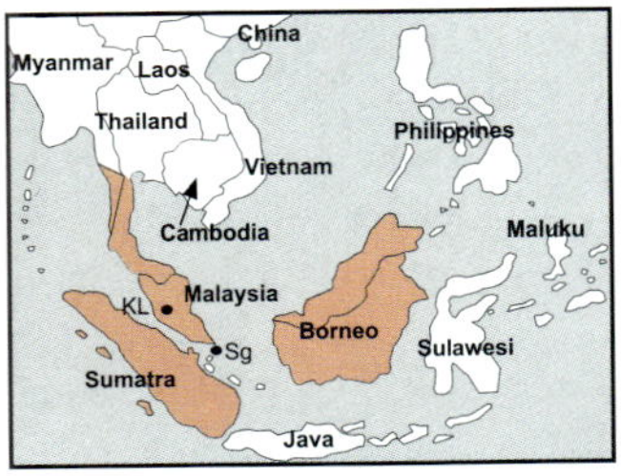


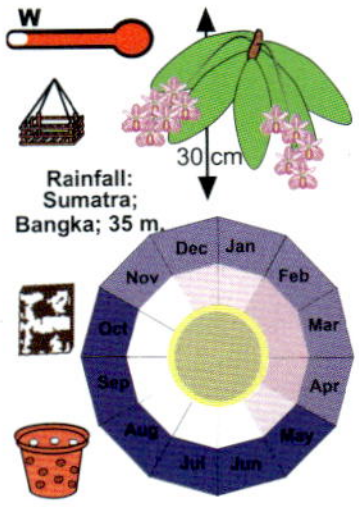


## *Phalaenopsis venosa* Shim & Fowlie

This little-known species grows on tree trunks in the understorey of lower montane forest from 700-1500 metres altitude. The arching 20-70 cm inflorescence extends slowly, sporadically producing 1-2 flowers at a time. The 5 cm wide flowers are yellow to yellow-green, suffused and barred in shades of orange, red or brown.

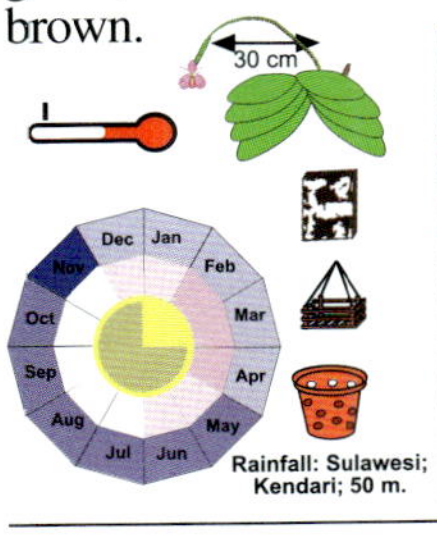


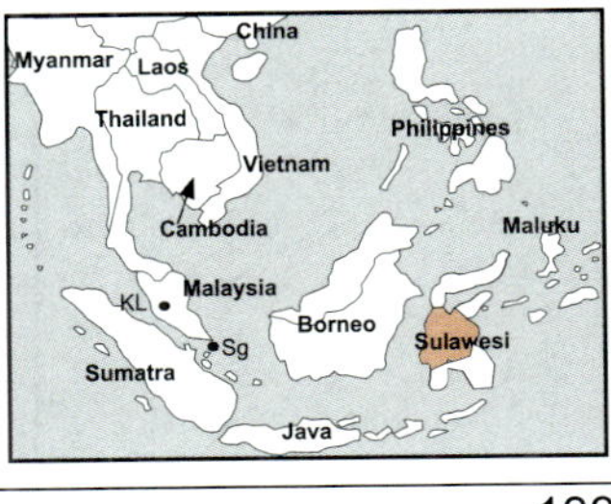

# P

## Pholidota

This genus of 28 epiphytic or lithophytic species is found across S. E. Asia to Australia and the Pacific Islands. Recent DNA analysis indicates that the species in this genus actually belong in *Coelogyne*, but the new name combinations have not yet been published. All the species have single-noded pseudobulbs with one or two leaves. The zig-zagging inflorescence is usually formed on the young shoot before the pseudobulb and leaves have fully developed, and has many small white flowers. Some species are common in orchid collections in the region.

### ***Pholidota articulata*** Lindl.

This peculiar species grows on isolated trees and bushes and in scrub and rocks. It is usually found from 700-1000 metres altitude, although it is occasionally seen much lower or much higher. Each 8-10 cm long pseudobulb grows from the tip of the predecessor, resulting in a chain with a pair of leaves at the top. The 10-15 cm long inflorescence is formed on a young shoot, and has up to 15 flowers that are about 15 mm wide. The lip has a constriction in the middle, behind which is a saccate region with 5 yellow ribs.

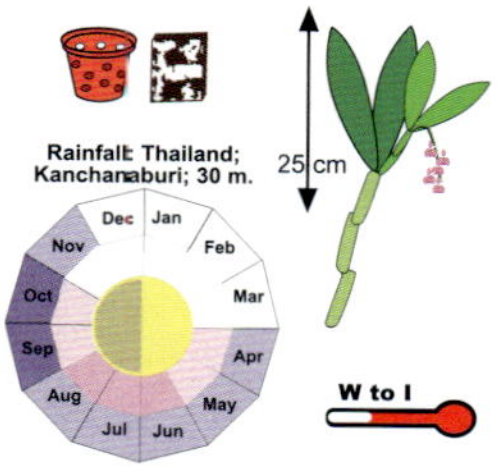

### ***Pholidota camelostalix*** Rchb.f.

This species occurs in the mountains from 1000-2000 metres altitude. There are two common varieties, both found in Sumatra and Java. Variety *vaginata* (illustrated) likes high light levels and grows on isolated trees or in the canopy layer of montane forests. Variety *camelostalix* usually grows in deep shade, low down on forest trees. The inflorescence emerges from the tip of a fully-matured pseudobulb and may reach 60 cm in length. Var. *vaginata* has smaller flowers than var. *camelostalix* and has a constriction near the tip of the petals.

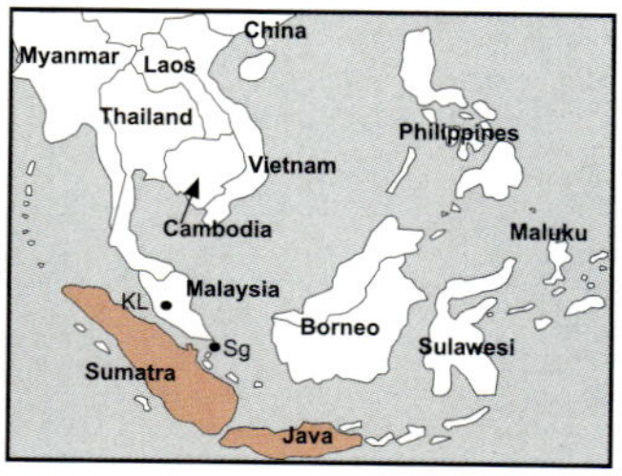


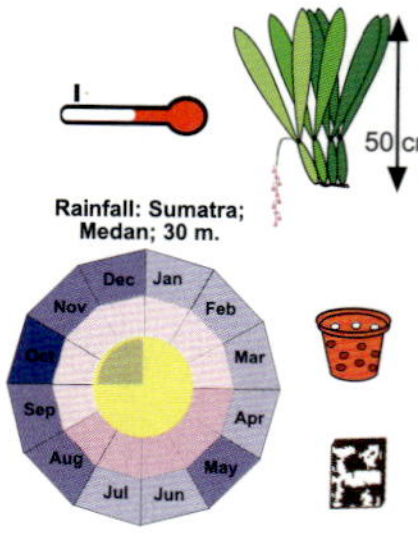

# P

## *Pholidota carnea* (Bl.) Lind.

This widespread and variable species is very common, creeping along or hanging off mossy tree trunks and branches throughout the region from 1000-2000 metres. It can also be found at lower altitudes. Three varieties are recognised; var. *carnea* (illustrated) occurs from Thailand to New Guinea, while the other two varieties are confined to Malaysia. The inflorescence emerges from a very young shoot and grows up to 25 cm long. The flowers are about 6 mm wide.

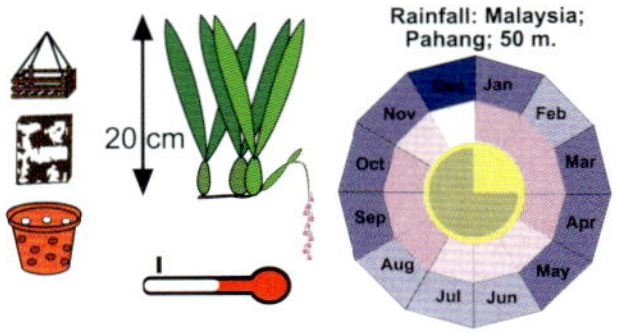


## *Pholidota imbricata* Lind.

This well-known species can be found throughout the region from sea-level to 1200 metres altitude. It is usually epiphytic, but is also common on limestone *karst* formations. It has 4-angled pseudobulbs and large leathery grey-green leaves. The 40-50 cm long inflorescence emerges from the young shoot and has flowers arranged in 2 spiral rows. The flowers are about 10 mm high.

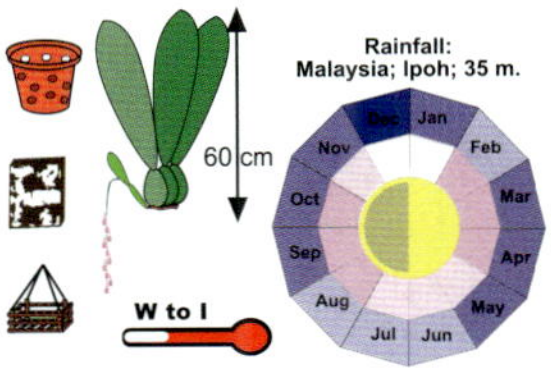

## *Pholidota gibbosa* (Bl.) de Vriese

This species is very common in the mountains at around 1000 metres altitude, where it forms large masses on mossy tree branches and can also be found on shaded rocks or even on road cuttings. The 20-50 cm long inflorescence emerges from the tip of a partly-developed pseudobulb. The 1 cm long flowers have 2 high rounded keels on the lip.

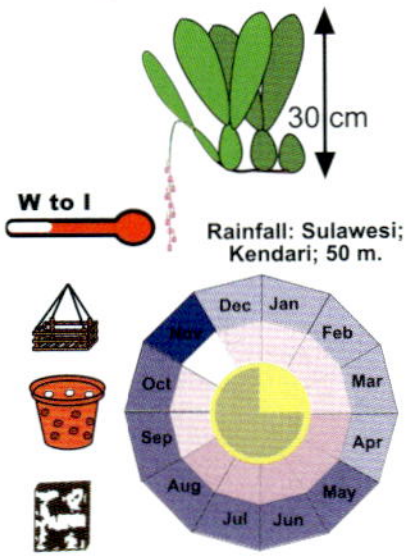

# P

## *Phreatia*

This genus of about 150 species is found across S. E. Asia to Australia and the Pacific Islands. The species are rather diverse; some have pseudobulbs, others a short stem, and others a long stem. The inflorescences usually emerge in the axils of the lower leaves and bear a large number of tiny white or green flowers that have the lip uppermost. Although they are common and widespread in the region, especially in the mountains, *Phreatia* species are seldom seen in orchid collections and have no commercial significance.

### ***Phreatia secunda*** (Bl.) Lind.

This unobtrusive little species is common in many parts of the region, although it is easily overlooked. It grows in exposed situations on the high branches of trees in swamp-forest, rainforest or plantations, from sea-level to about 2000 metres altitude. The plant has a short stem with up to 8 thick, fleshy leaves and usually produces 2 inflorescences at a time. The flowers are about 3 mm wide.

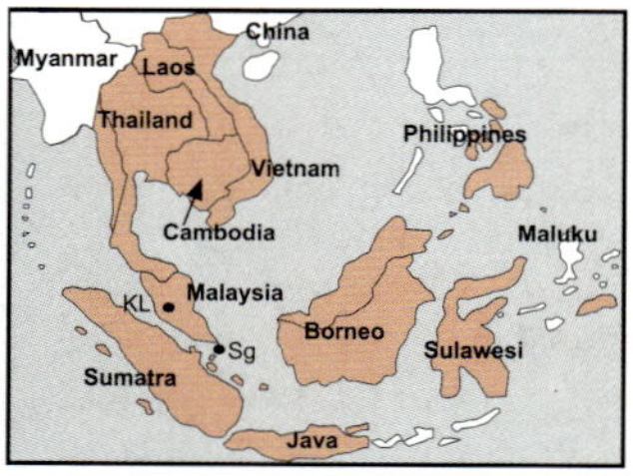


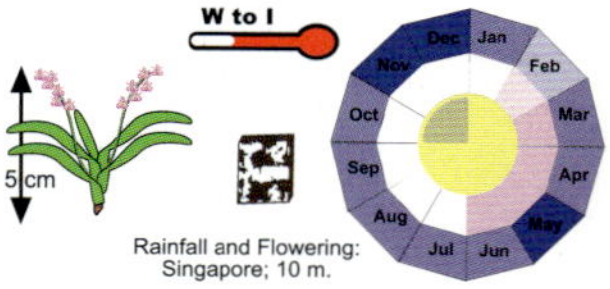


## *Plocoglottis*

This genus of about 40-45 species is distributed from Indochina through Malaysia and Indonesia to New Guinea. Most species are forest-floor terrestrials, with slender pseudobulbs that bear a single leaf. The tall inflorescence emerges from the base of the pseudobulb and bears a succession of flowers that open a few at a time. The lip has a peculiar elastic joint at the base that springs closed when touched. Despite having attractive and unusual flowers, *Plocoglottis* species are seldom seen in orchid collections and have no commercial significance.

### ***Plocoglottis acuminata*** Bl.

This species grows in moist soil on the forest floor from sea-level to about 1200 metres altitude. It has large leaves that are sometimes blotched with yellow. The inflorescence may reach 50 cm long and forms up to 15 flowers, with 1-3 open at a time. The flowers are about 3 cm wide and have a lip that closes after contact with pollinating insects.

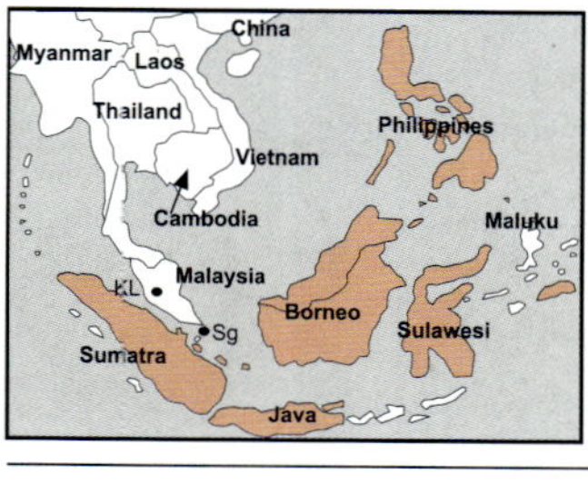


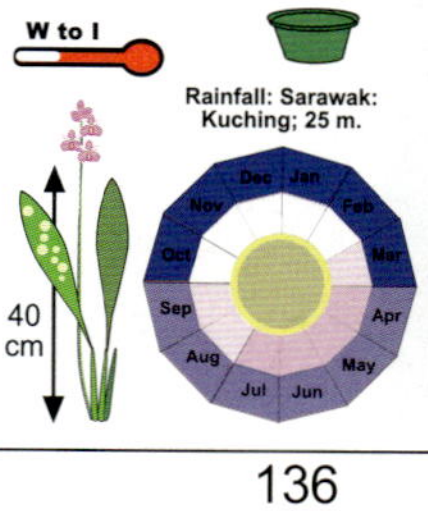

# P

## *Podochilus*

This genus of about 60 species is distributed from India and China through S. E. Asia to New Guinea and the Pacific Islands. Most of these species have small stems and leaves, and form dense mats that completely cover boulders, logs or tree branches. The overall effect is rather moss-like, and although these species are often common, especially in the hills and mountains, they are easily overlooked. Since the inflorescences are very small and the flowers are tiny, *Podochilus* species have little appeal to orchid enthusiasts and are seldom seen in collections.

### ***Podochilus microphyllus*** Lindl.

This small mossy species can be found from 200-1500 metres altitude and commonly forms dense mats on rocks and trees in moist shady spots in the mountains. It has short branching stems with small sharp-pointed leaves. The tiny flowers are usually all-white; some plants have pink-purple streaks on the petals and sepals.

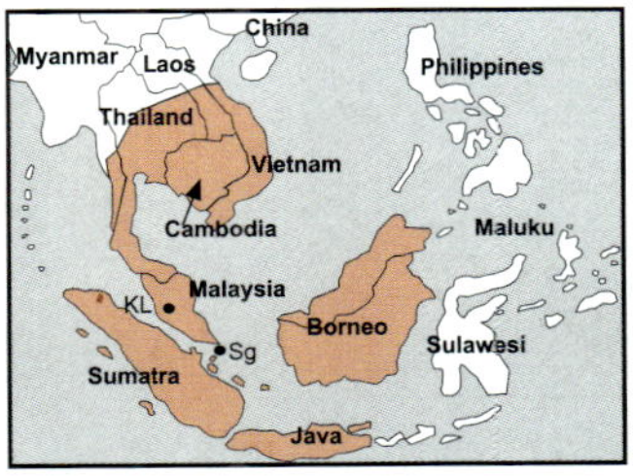


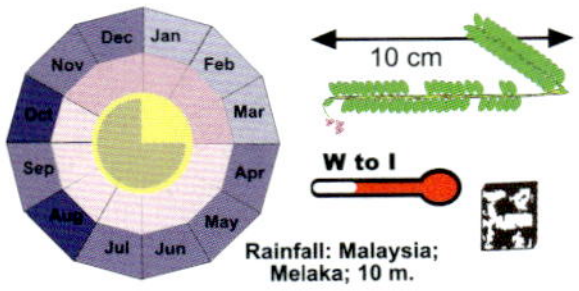


## *Polystachya*

This genus of about 200 species occurs principally in Africa; there are two species in tropical America and one species in Asia from Sri Lanka to the Philippines. These species are epiphytes or lithophytes; most have pseudobulbs or leafy stems and inflorescences with many small flowers. Some of the more beautiful African species have brightly coloured flowers and are often seen in orchid collections in Europe and USA; this cannot be said about the only Asian representative of this genus.

### ***Polystachya concreta*** (Jacq.) Garay & Sweet

This species has a worldwide tropical distribution. In Asia it grows on trees and occasionally on mossy rocks, from sea-level to about 1500 metres altitude. It is common in some areas. The 10-30 cm long inflorescence usually has several short branches and carries up to 30 rather variable flowers. Many Asian specimens are self-pollinating and have flowers that hardly open.

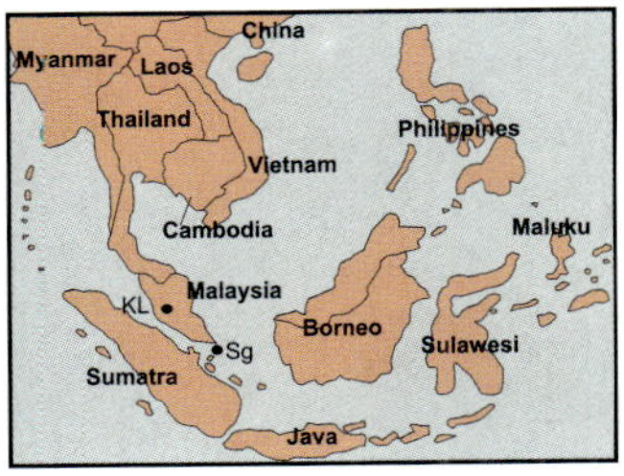


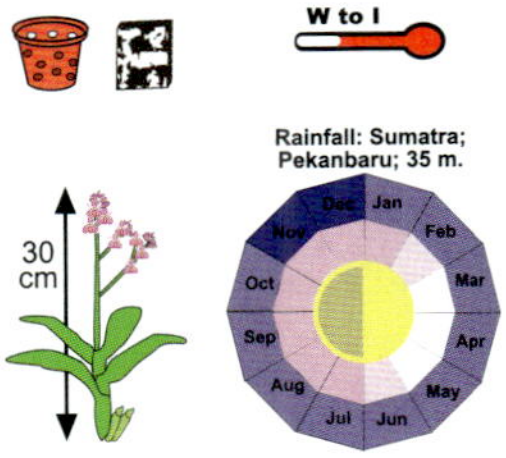

# P

## Pomatocalpa

This genus of about 30-40 species is distributed from Myanmar through S. E. Asia to Taiwan, New Guinea and the Pacific Islands. The species are usually epiphytes with either long climbing stems or short clinging stems and flat leaves. Inflorescences are often branched and are usually crowded with many rather small flowers that have a well-developed sac or spur at the back of the lip. Some *Pomatocalpa* species have attractive flowers and are occasionally seen in orchid collections in the region.

### *Pomatocalpa kunstleri* (Hk.f.) J.J.Sm.

Although widespread, this graceful species is not very common. It grows on the branches of forest trees from sea-level to about 600 metres altitude. The 20-40 cm white-haired inflorescences usually have 3-5 branches and may eventually produce 50-70 flowers. The 7 mm long flowers last for about 5-7 days.

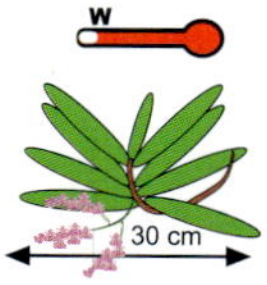


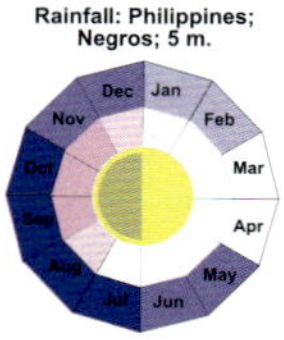


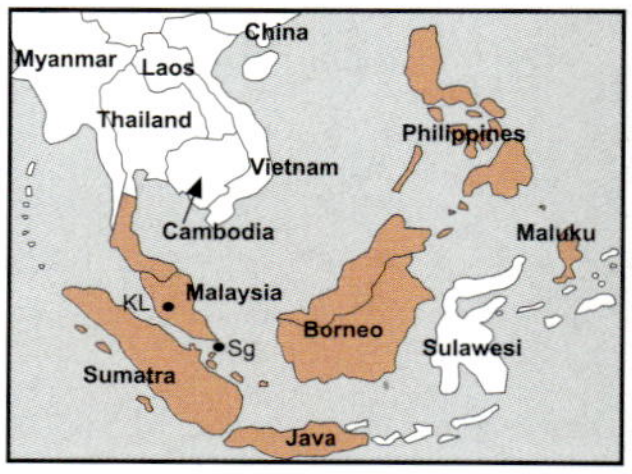


### *Pomatocalpa naevata* J.J.Sm.

This tall climbing species grows from sea-level to about 600 metres altitude, favouring the trunks and upper branches of exposed trees, and also grows on exposed rocks. 1-3 inflorescences are formed at a time, each slowly elongating to about 50 cm long with about 10 flowers open at a time. The 1 cm long flowers last for about 1 week.

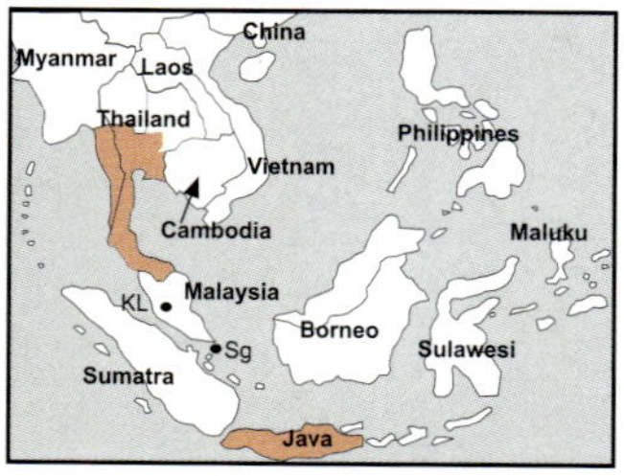


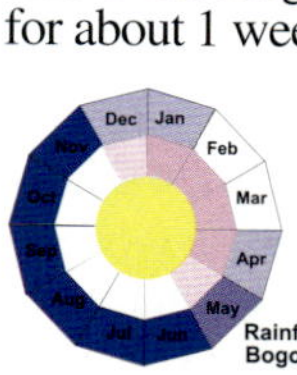

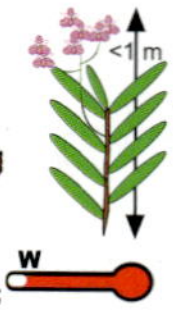


### *Pomatocalpa spicata* Breda

This widespread epiphytic species occurs from sea-level to at least 700 metres altitude. Plants without red markings are often called *P. setulensis*.

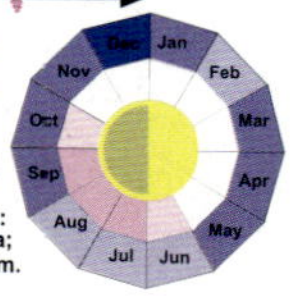

# P

## Porphyrodesme

This genus probably has only 4 species; one is endemic to Borneo and another to Maluku; a third is found in New Guinea and Borneo, and the fourth is listed below. The plant habits vary considerably, but all possess short branching inflorescences of small bright red flowers. None of the species is particularly common in the wild and little is known about them. I have never seen any of them in orchid collections.

### *Porphyrodesme sarcanthoides* J.J.Sm.

This seldom-seen species appears to be very rare throughout most of its range, probably occurring from 160-250 metres altitude in colonies that are widely scattered. It grows on the high branches of emergent trees in the forest where it can hang unimpeded. It has stems up to 250 cm long that are rooted only at the base and have few branches. The narrow linear leaves are 15 cm long. It produces several stiffly-descending 15 cm long branched inflorescences that bear a large number of tiny flowers only a few being open at once.

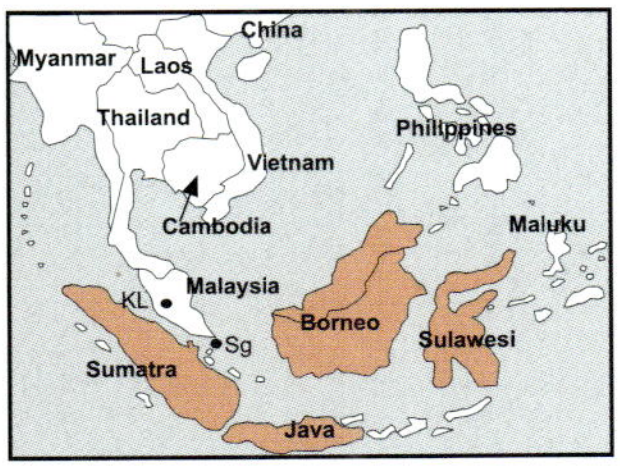


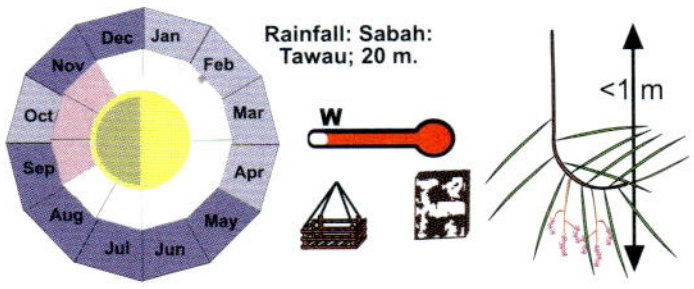


## Porrorhachis

This genus has only 2 species; one is endemic to Sulawesi and the other is listed below. Both species are epiphytic monopodials with short to medium length stems and fleshy leaves. The stiff descending inflorescences are usually longer than the leaves and have up to 25 small green to yellow flowers that are dominated by a conical spur. Neither species is in cultivation.

### *Porrorhachis galbina* (J.J.Sm.) Garay

This small epiphytic species grows on the trunks and branches of forest trees between 1000- 1500 metres altitude. Although it is very rare in Java and Borneo, in the mountains of central Sulawesi it is quite common on mature bushes in coffee plantations. It has short thick pointed leaves and a 5-8 cm long inflorescence that bears up to 25 well-spaced fleshy flowers, 3-5 of which are open at once. The flowers are about 5 mm high and last for about 5 days.

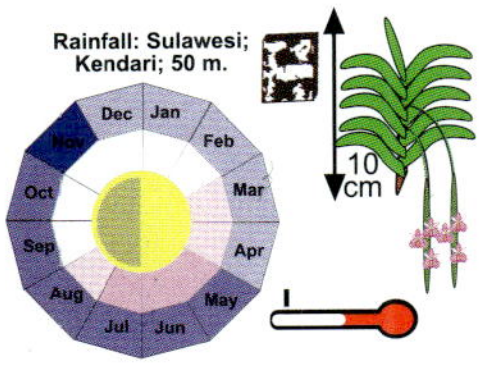


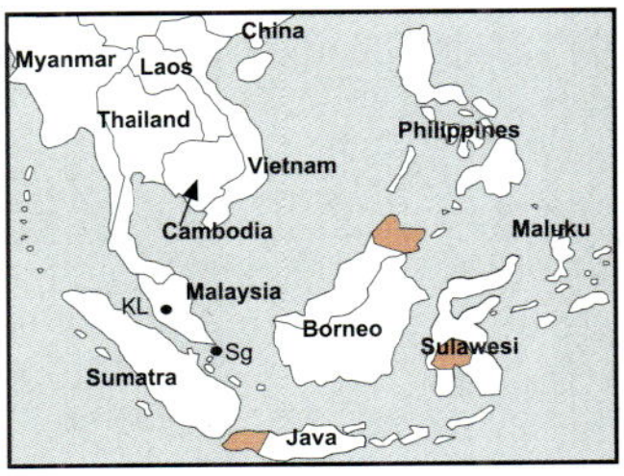

# P

## Pteroceras

This genus of 19 species is distributed from India to the Philippines and the Lesser Sunda Islands. They are small to medium-sized epiphytes or lithophytes with short leafy stems and inflorescences that (in most species) extend slowly over a long period of time, sporadically producing a few short-lived flowers. Some of the species are very closely related and hard to identify accurately. Although many *Pteroceras* species have attractive sweet-scented flowers, they are seldom seen in orchid collections.

### *Pteroceras cladostachyum* (Hk.f.) H.Ae.Peders.

This common species is found from sea-level to at least 1200 metres altitude, growing low down on forest trees, usually in deep shade. It has a short inflorescence that is active for 1-2 years, sporadically producing 1-3 flowers at a time. The flowers are 30-35 mm across and last for 2 days.

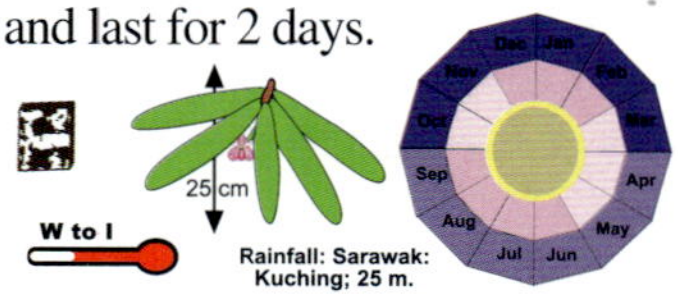

### *Pteroceras fraternum* (J.J.Sm.) Bakh.f.

This mountain species is found from 750-1600 metres altitude, growing on forest trees in deep shade and on isolated trees in quite bright situations. It has a flattened inflorescence with raised ridges between the flower nodes. One to three flowers at a time are produced sporadically for 1-2 years; the flowers are 15 mm across and last for 1 day. It can be distinguished from the closely related *P. compressum* by the shorter inflorescence and yellow background colour of the sepals.

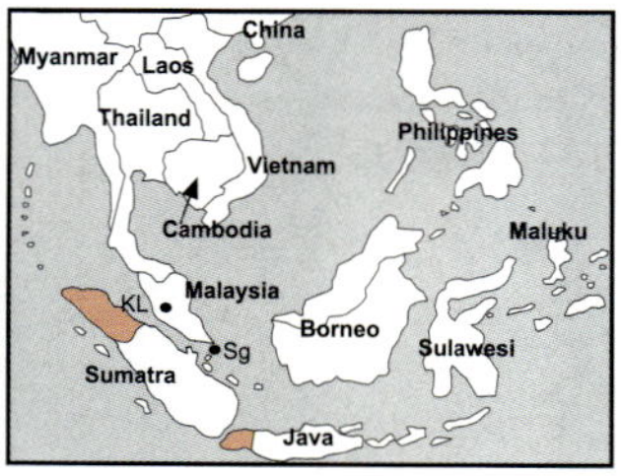


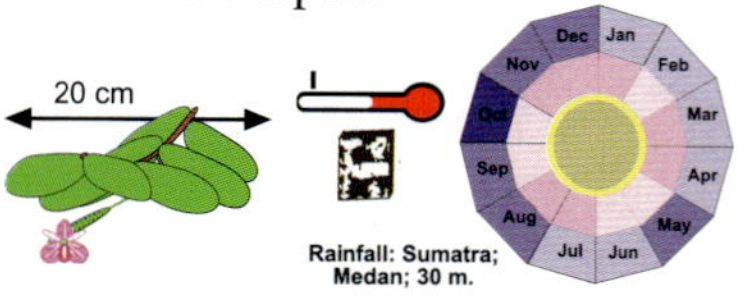


### *Pteroceras pallidum* (Bl.) Holtt.

This species is found from sea-level to about 600 metres altitude, growing low down on forest trees and lianas in deep shade. It can be distinguished from *P. cladostachyum* by it's smaller flowers and all-red sidelobes. The flowers are 20-30 mm across, last for 1 day, and often have a musty odour.

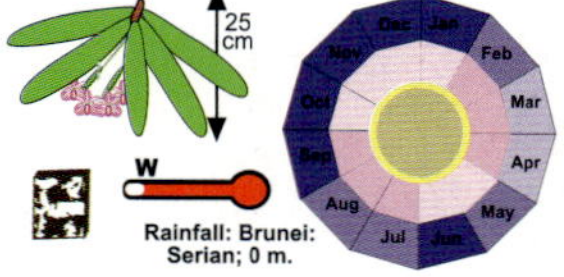


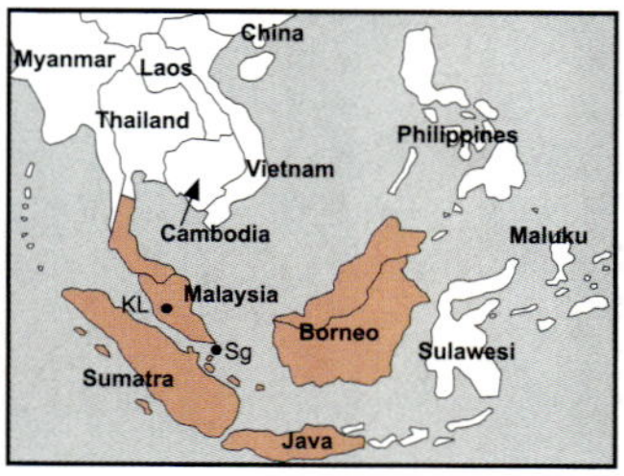

# R

## Renanthera

This commercially important genus of about 15 species is distributed from India and Nepal to China, New Guinea and the Solomon Islands. *Renanthera* species and their hybrids are tall sun-loving plants that are grown in collections around the world. The species are climbing plants with long stems and short leathery leaves. The long branched inflorescences usually bear a spray of many long-lasting small red or orange flowers.

### *Renanthera angustifolia* Hook.f.

This species is still widely known by its former name *R. matutina* (Bl.) Lindl. In some parts of its range it is a mountain plant, growing at about 1000-1400 metres altitude, while in other areas it grows from 100-600 metres altitude. It prefers growing in exposed places and may be found hanging from a high branch by only one or two root-tips. The inflorescence is about 30 cm long and has 1-2 branches. In Java the flowers are orange-red with red spots; elsewhere the base colour is yellow.

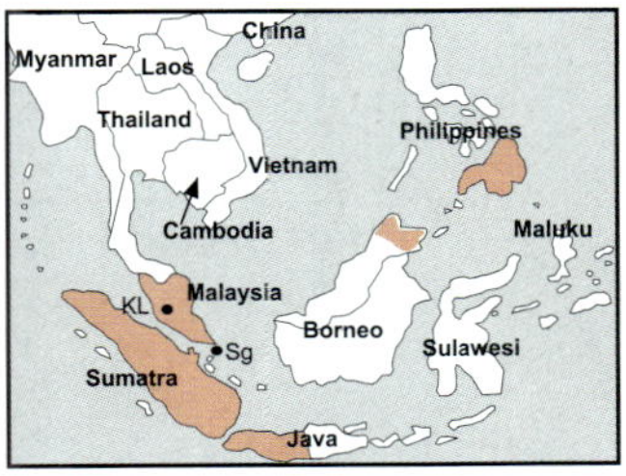


Rainfall: Malaysia; Pahang; 50 m.

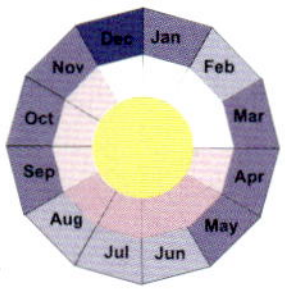


### *Renanthera bella* J.J.Wood

This rare species is confined to a few locations in the hill-country of Sabah, growing from 800-1100 metres altitude. It prefers to grow in slightly shaded places, often low down near the ground. The stems are 20-50 cm long, and the inflorescence sometimes has 1-2 branches. This species is considered highly endangered because of over-collecting, and only seedlings in flasks are allowed to be traded internationally.

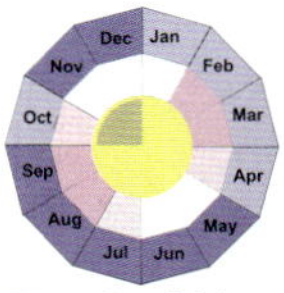


Data: Sabah: Danum Valley; 20 m.

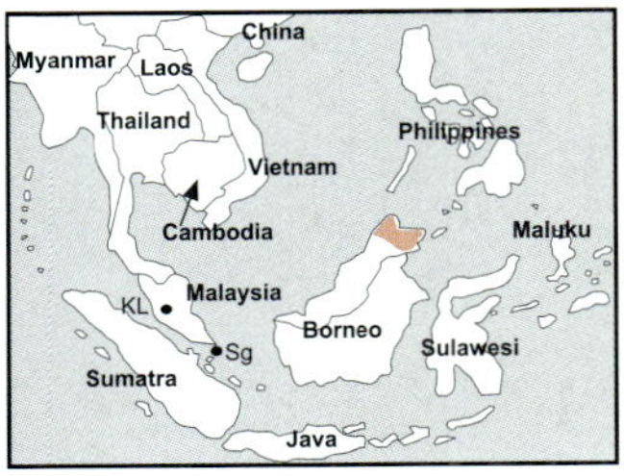

# R

## *Renanthera elongata* (Hk.f.) Lindl.

This widespread species has been found from sea-level to at least 1000 metres altitude, climbing up tree trunks and branches in swamp-forest and both primary and secondary forest. It also grows as scrambler over limestone outcrops, and has been found as a terrestrial in grassland. The stems are often longer than 1-2 metres. The 20 cm branched flower spike carries 40-70 small flowers.

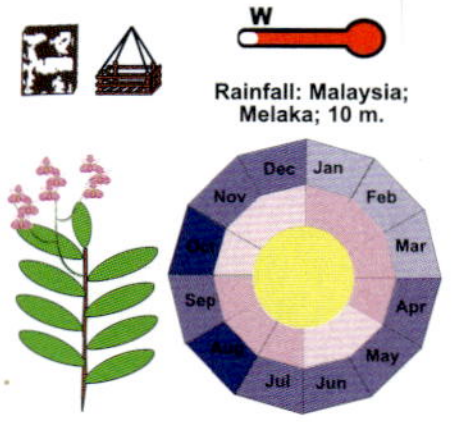

## *Renanthera monachica* Ames

This species grows on the trunks and branches of isolated trees in grassland at about 100 metres altitude. Although it is rare in the wild, it is commonly cultivated in the Philippines. It has short stems and narrow strap-shaped leaves. The 20-30 cm inflorescence is sometimes branched, and has 5-10 loosely arranged large flowers.

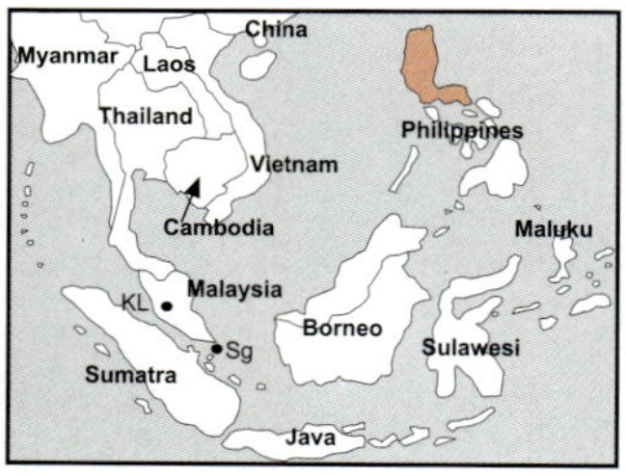


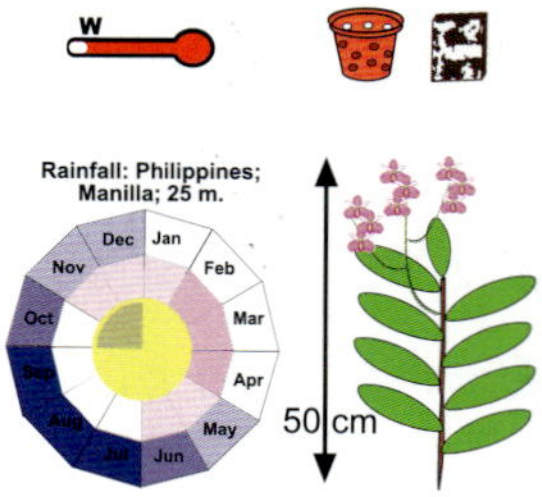


## *Renanthera philippinensis* Ames & Quisumbing

This dramatically beautiful species has a very limited distribution in Quezon Province and is rare in the wild, where it favours growing on trees in mangrove swamps. It has fairly short erect or climbing stems with short oblong leaves. The 30-40 cm inflorescence has several branches and carries up to 50 lightly scented flowers.

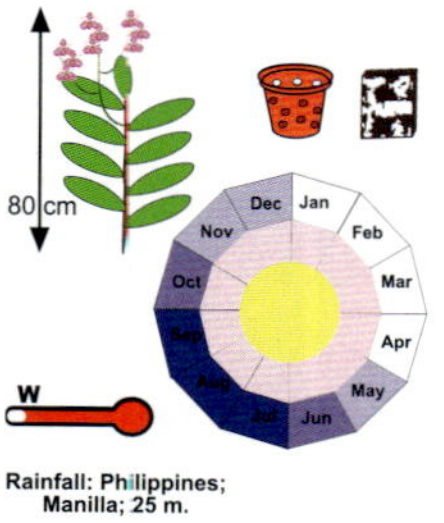


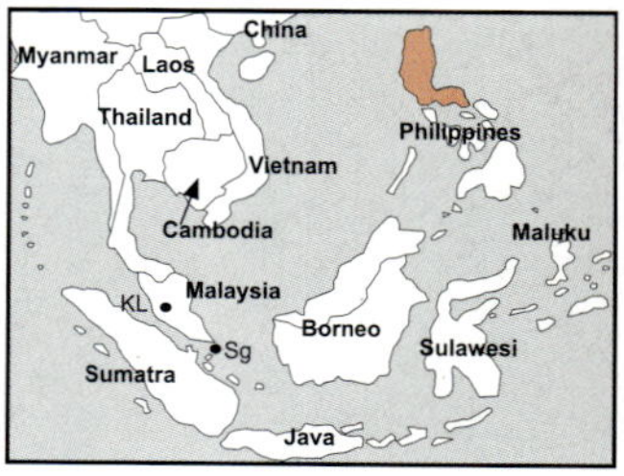

# R

## Renantherella

This genus contains only one species, *R. histrionica*. Some authors prefer to call this plant by its original name *Renanthera histrionica*. The species differs from *Renanthera* by not having a high-climbing habit and not having long branching inflorescences. The column on the upside-down flowers is long and slender. A subspecies, *R. histrionica ssp. auyongii* (Christenson) Senghas, is found in Sarawak.

### ***Renantherella histrionica*** (Rchb.f.) Ridl.

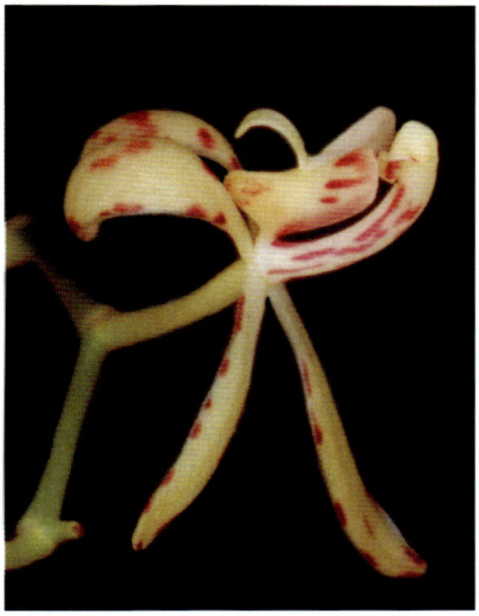

This distinctive species prefers to grow on mangroves and other trees near the coast, although it is sometimes found in hilly country, growing at up to 200 metres altitude on trees and rocks. The climbing or hanging stems are usually about 20 cm long, (occasionally much longer), and have thick sharp-pointed 10 cm leaves The inflorescence slowly elongates to 10 cm, with 1 or 2 flowers open at a time.

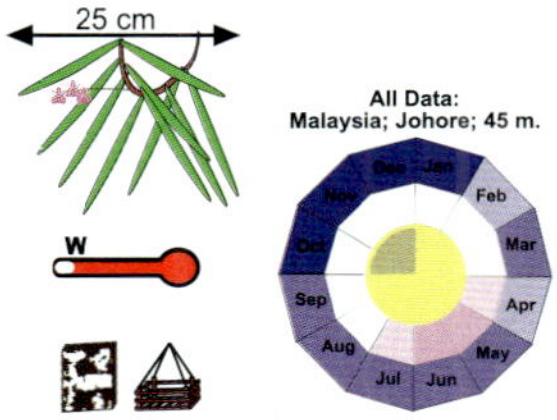


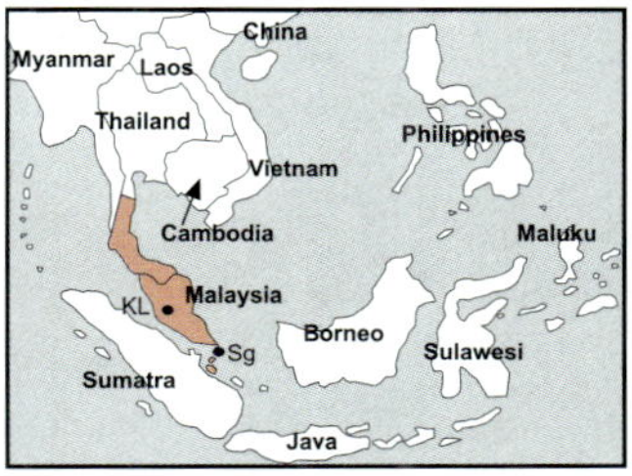


## Rhynchostylis

Although this epiphytic monopodial genus has only 4 well-known species, it is of great importance as the plants are all extremely beautiful and are interfertile with a wide range of other monopodial genera and have been used extensively to introduce various desirable characters to hybrids. The species remain very popular with orchid growers, and a variety of "improved" strains have been created by line-breeding programs. The 4th species, *R. violacea*, is found in the Philippines.

### ***Rhynchostylis coelestis*** Rchb.f.

This much sought-after species grows in deciduous forests in hilly country. It has been cultivated for many years, and different colour forms are becoming available, such as the "dark blue" form illustrated. A good plant will flower 2-3 times a year, producing several inflorescences each with up to 50 small but fragrant flowers.

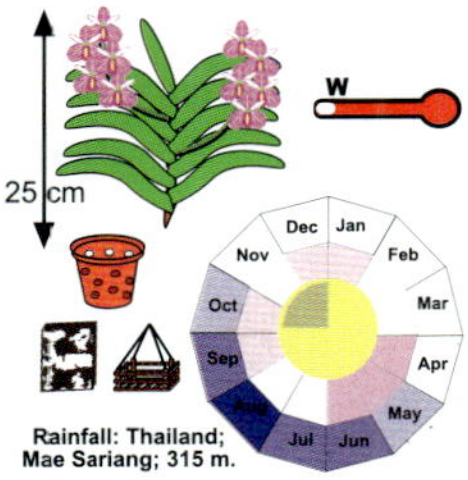


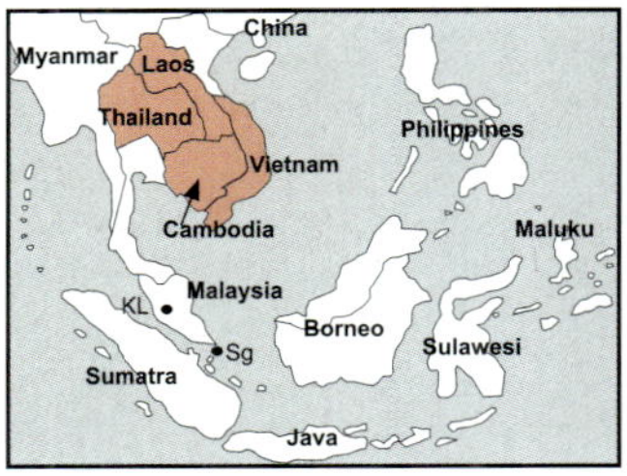

# R

*Rhynchostylis* species are prized for their ability to produce spectacular floral displays. A well-grown plant will produce several inflorescences that have all their flowers open at the same time. They are usually very reliable, flowering at the same time each year.

Left to right: *Rhynchostylis coelestis, R. coelestis* 'alba', *R. gigantea.*

## ***Rhynchostylis gigantea*** (Lindl.) Ridl.

This highly fragrant species is often grown in village and town gardens. It has thick leathery leaves and usually produces several descending inflorescences at a time. The 3 most common colour forms are dark violet, white with dark purple spots, and white.

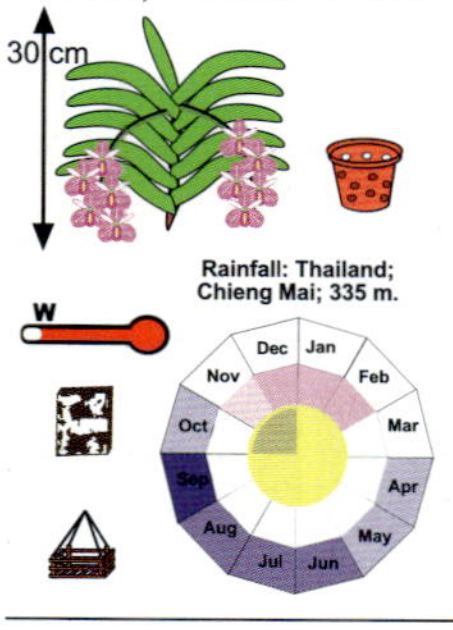


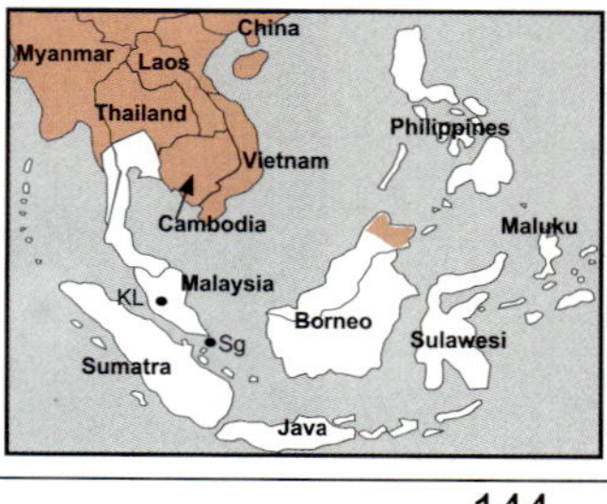

# R

## *Rhynchostylis retusa* Bl.

The "Foxtail Orchid" grows in forests and plantations and is quite common in some areas where the dry season is not too long. It is a popular garden plant throughout the region. Although it can grow to about 50 cm tall, it is usually much smaller. At flowering time each stem will produce a few spectacular 40-50 cm long pendulous inflorescences. A rare pure white form exists.

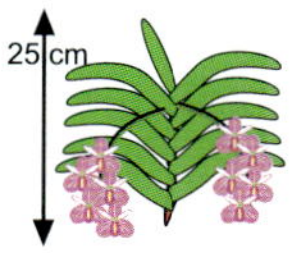


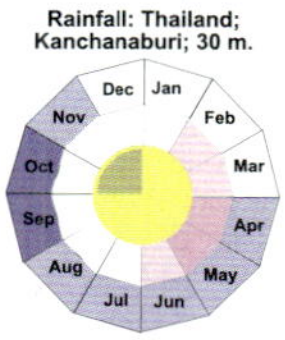

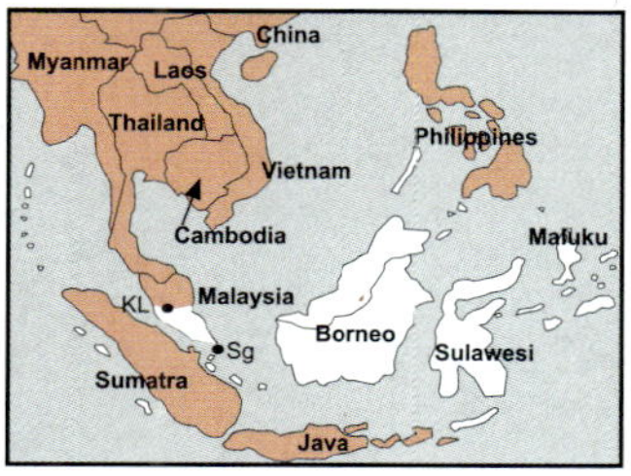

## Robiquetia

There are about 40 monopodial epiphytic species in this genus, which extends from Sikkim to China, Australia and the Pacific Islands. The majority of species are found in Indonesia, although little is known about many of them. The plant size is variable; some are medium-small while others are large and robust. The majority have inflorescences that descend stiffly from the stem, with a raceme of many rather small flowers. The flowers are usually yellow to yellow-brown with red to brown spots. Few of the species are seen in orchid collections, although some of the New Guinea species are extremely beautiful and deserve to be better known.

## *Robiquetia succisa* (Lindl.) Garay & Seidenfaden

This slender fast-growing species has rather short oblong leathery leaves. It produces inflorescences that are up to 25 cm long and usually have 3 branches, bearing up to 100 small flowers, many of which are open at the same time. The most common colour form has golden-yellow flowers that are suffused red on the outside. In Thailand this species is still often referred to by the old name *R. paniculata*.

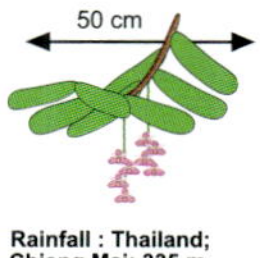


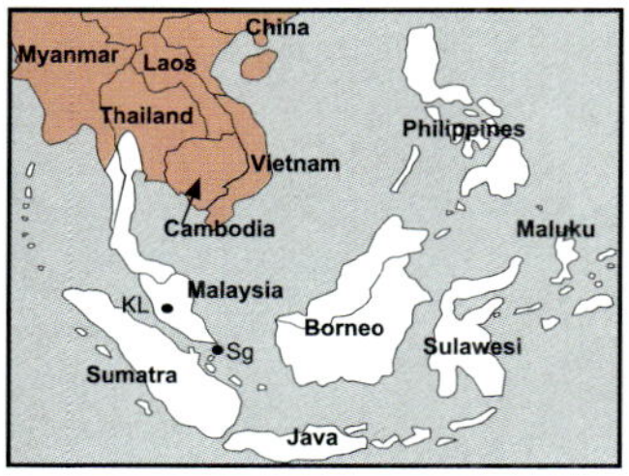


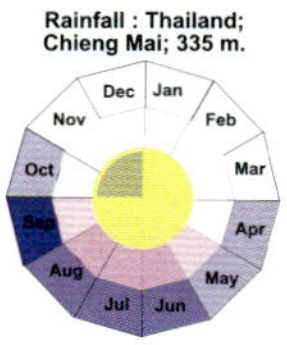

# S

## *Sarcoglyphis*

The 10 or 11 epiphytic monopodial species in this genus are found from Myanmar and China to Sumatra, Java and Borneo. They are small plants with short stems and strap-shaped leaves. The inflorescences are branched with many small flowers. They can be distinguished from the closely-related *Cleisostoma* by the arrangement of the anther and pollinia. They are occasionally seen in orchid collections in S. E. Asia.

### ***Sarcoglyphis comberi*** (J.J.Wood) J.J.Wood.

This rare species is found from 200-600 metres altitude in hilly country in Borneo, and at 1000 metres altitude in one location in Java. It has a short stem with strap-shaped leaves that have a deep notch in the tip. It produces several slender branched inflorescences at a time that are much longer than the leaves, sometimes as much as 40 cm long. Each inflorescence has up to 40 beautiful 12 mm flowers, most of which are open at the same time.

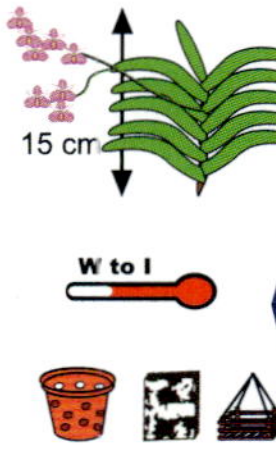


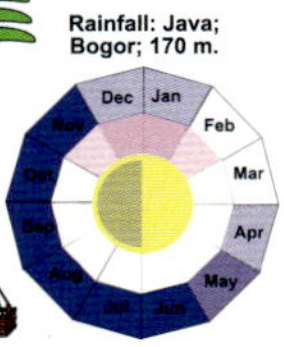


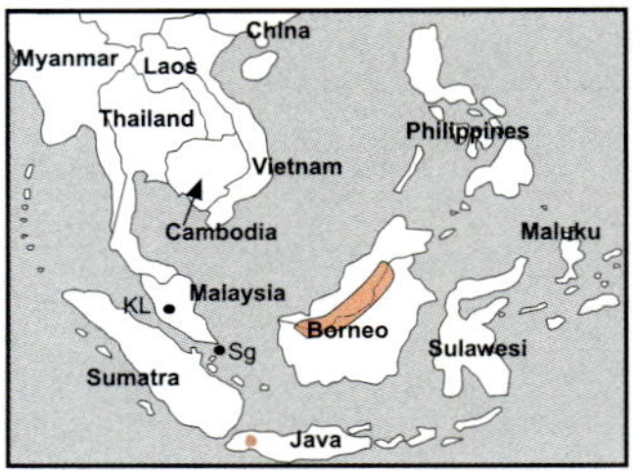


### ***Sarcoglyphis pomatophila*** (Schltr.) Garay & Kitt.

This rare species is found at about 150 metres altitude on the branches of rainforest trees. It has 10 cm stems that branch from the base, forming a small clump. The leaves have a deep notch in the tip. It usually flowers twice a year, forming 1-3 inflorescences that are longer than the leaves and each bear up to 25 attractive 13 mm flowers, most of which are open at the same time.

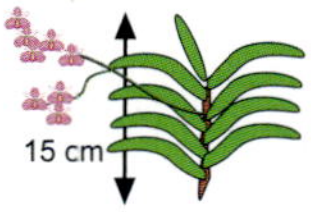

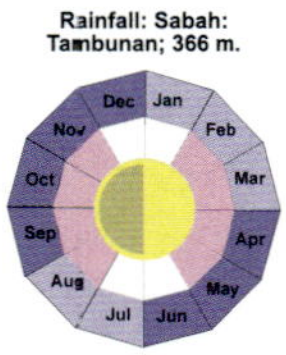


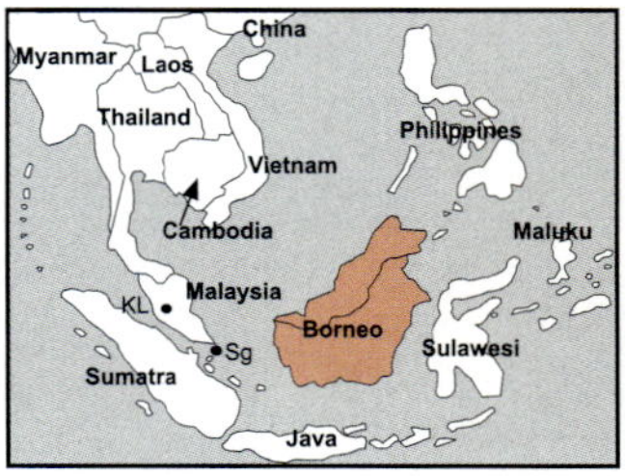

# S

## Schoenorchis

This genus of 15 to 24 epiphytic monopodial species is distributed from north India and the Himalayas to China, Australia and the Pacific islands. Some plants are extremely small while others are medium sized; some are erect while others are pendulous; most have fleshy or terete leaves. Even though the sepals are not fused together, the flowers are rather tubular and not wide-opening. *S. micrantha*, *S. seidenfadenii* and the closely-related *S. fragrans* have become quite popular amongst collectors in recent years.

### *Schoenorchis secundiflora* (Ridl.) J.J.Sm.

This showy species is quite common in many lowland localities in Malaysia, growing on trees in rainforest mangroves and plantations. It has also been found in Kalimantan, where it appears to be rare. It has an upright 10-15 cm stem with 2 rows of fleshy strap-shaped leaves that have a notch in the tip. It produces several slender descending 10-12 cm inflorescences that have 20-30 flowers arranged so that they all face one direction. The 10 mm long flowers are sweetly scented.

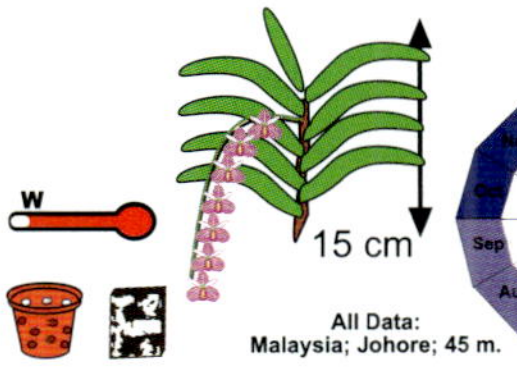


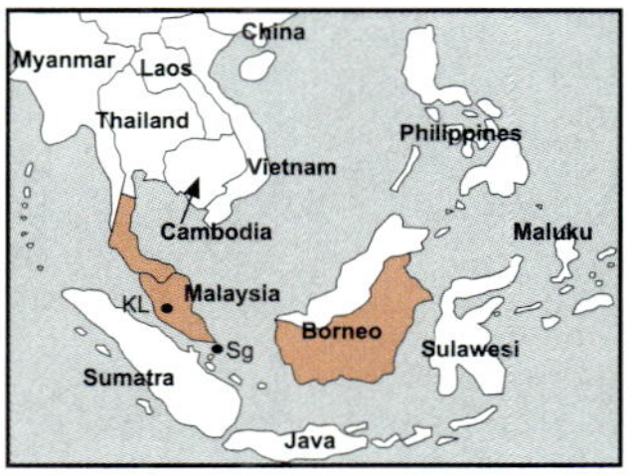


### *Schoenorchis seidenfadenii* Pradhan

This miniature species has two rows of thick succulent-like leaves to enable it to survive the long dry season. It produces so many inflorescences that the leaves can be totally hidden behind the tiny bright pink flowers.

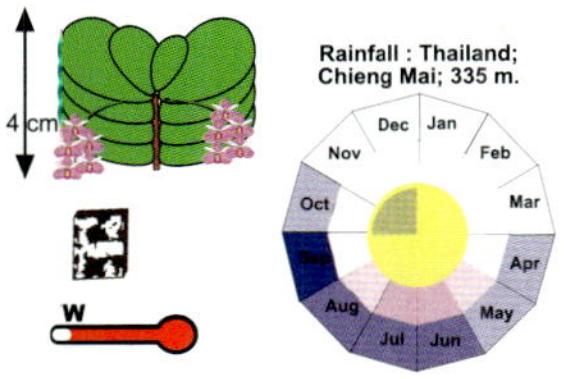


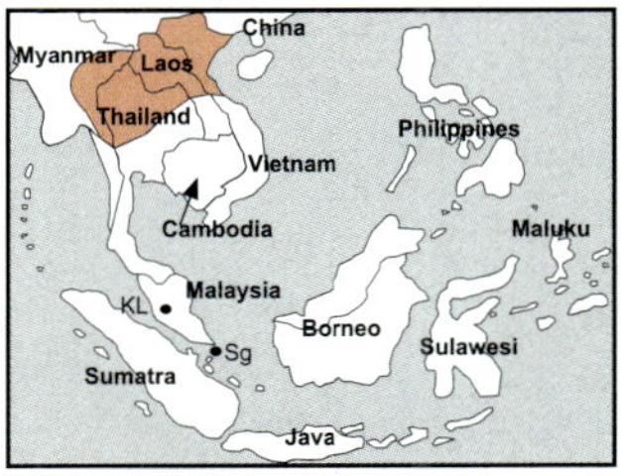


### *Schoenorchis sumatrana* J.J.Sm.

This uncommon mountain species forms 1-4 inflorescences that have many 3-4 mm long flowers, only a few of which are open at the same time.

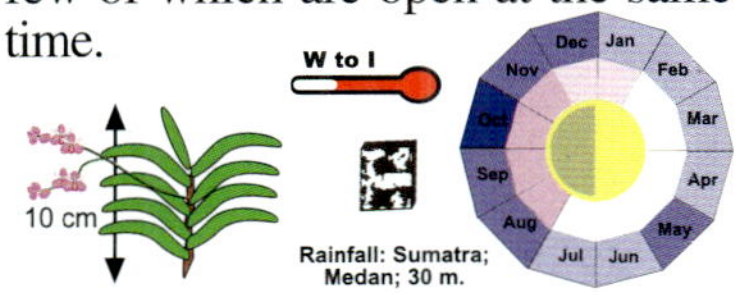


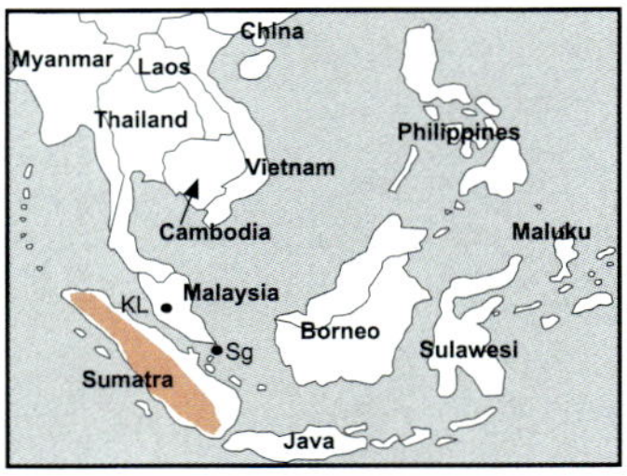

# S

## Seidenfadenia

This genus contains only one species. Until recently it was called *Aerides mitrata*, and it still appears in some nursery catalogues under that name.

### ***Seidenfadenia mitrata*** (Rchb.f.) Garay

This showy species grows in deciduous forest in hilly areas, on trees that lose their leaves during the long dry season. It has an extremely short stem with a long root system and several long pendulous terete leaves that can resist full exposure to the sun for the long dry period. Towards the end of the dry season it produces several upright 15-25 cm inflorescences that have a crowded raceme of lilac and white flowers .

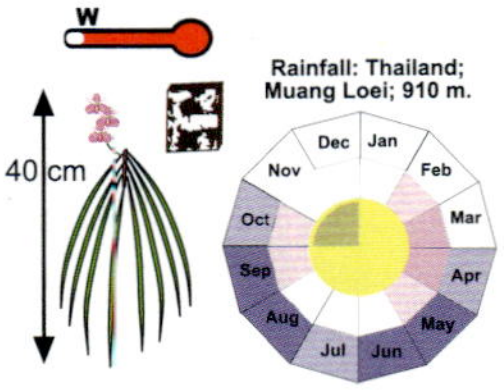


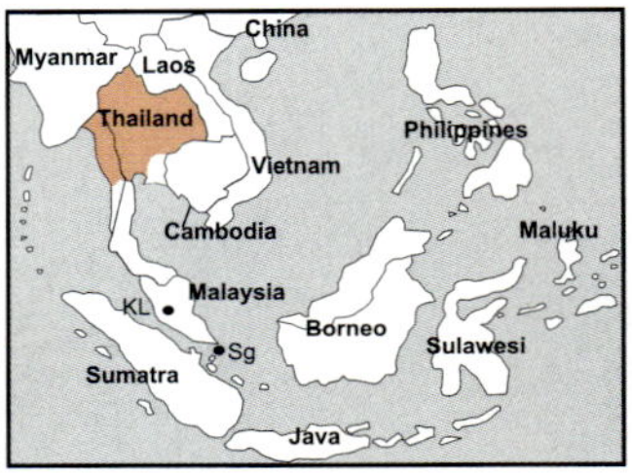


## Smitinandia

This genus contains three epiphytic monopodial species. *S. helferi* is found from Myanmar across Thailand to Indochina. *S. micrantha* is distributed from the northwest Himalayas to Indochina, Malaysia and Borneo. All species form clumps with fairly short stems, short thick strap-shaped leaves, and racemes of small fleshy flowers that have a small spur with a high callus across the opening.

### ***Smitinandia selebensis*** (J.J.Sm.) Garay

This little-known species grows on trees in lowland rainforest and swamp forests, and is occasionally found on mangroves. It has stems that are usually about 10 cm long but branch freely, eventually forming an extensive clump. Many 4-5 cm long inflorescences are formed that arch outwards and downwards and carry up 40 small long-lasting flowers, with about 8-10 open at a time. The cream-coloured flowers have a faint pink flush on the middle of the lip and red marks on the anther and column.

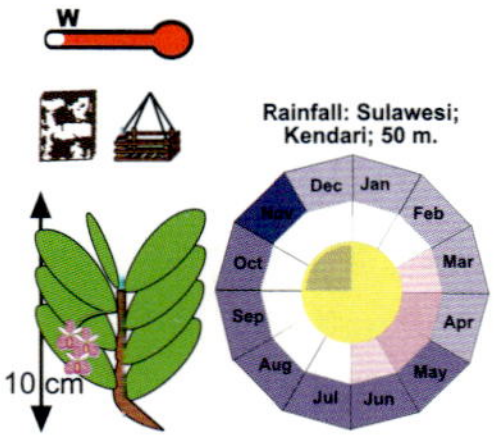


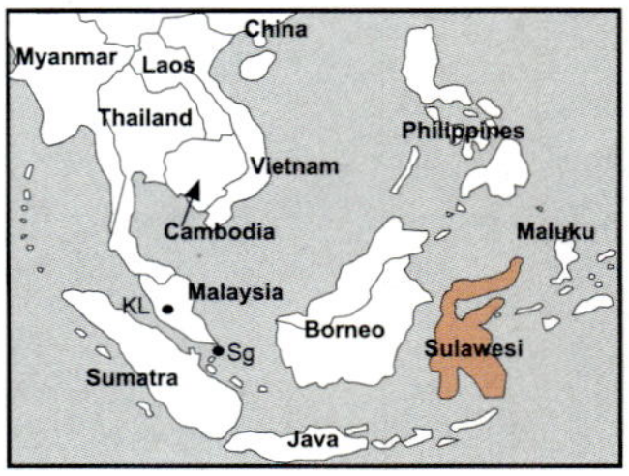

# S

## Spathoglottis

This genus of about 55 terrestrial species is distributed from the Himalayas across Asia to the Pacific Islands. These sun-loving plants have distinct pseudobulbs with one to several pleated leaves and tall inflorescences that remain in bloom for several months, producing a succession of showy, colourful flowers. They are popular with orchid enthusiasts and gardeners and several species are commercially propagated. *Spathoglottis* has been the subject of much hybridization and a variety of colour forms are available.

### ***Spathoglottis affinis*** de Vriese

This deciduous species grows in rocky areas of seasonally-dry grassland from 600-1000 metres altitude and has large subterranean pseudobulbs that can survive dry-season fires. The flowers are well-spaced along the inflorescence; up to 6 at a time are open. Plants from Myanmar (illustrated) and Thailand are often known by the old name *S. lobbii*.

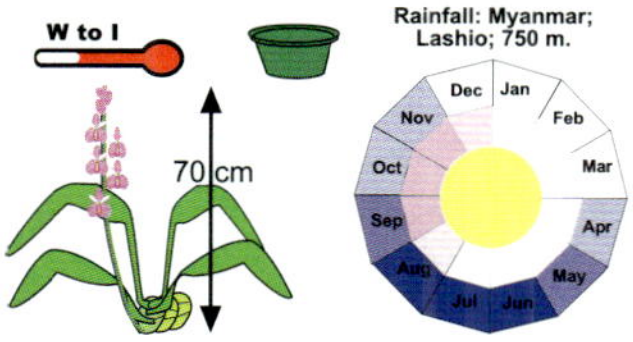


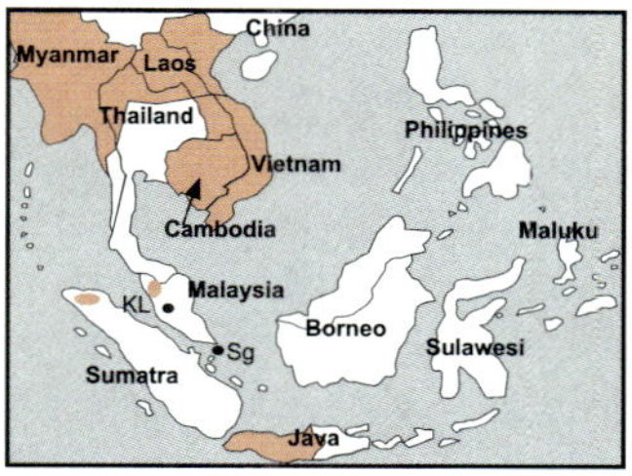

### ***Spathoglottis kimballiana*** Hort. **var. *angustiflora*** Ames

Although *S. kimballiana* is found only in Borneo, this narrow-leaved variety occurs in the Philippines. It is a popular garden plant and is readily available in the region due to commercial propagation. In cultivation, the 4 cm flowers are produced throughout the year.

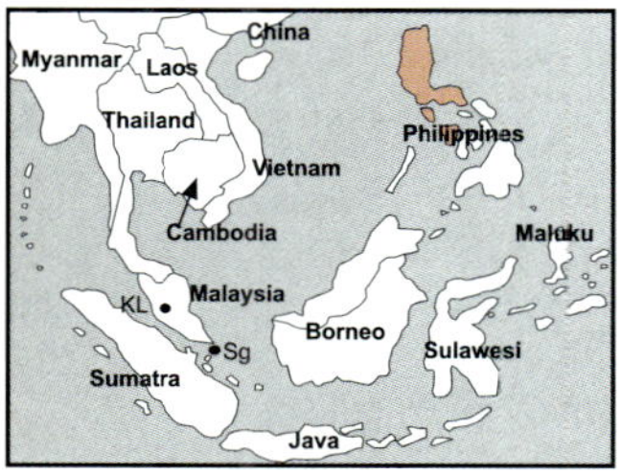


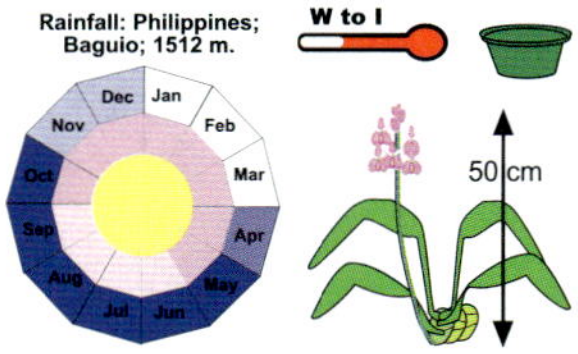


### ***Spathoglottis x parsonsii*** Ames & Quisumbing

This naturally-occurring hybrid was first found near Baguio where the parent species, *S. plicata* and *S. vanoverberghii* are both common. It has the flower shape of *S. plicata* but the downy inflorescence of *S. vanoverberghii*. Flower colour can be yellow, salmon, pink, or dark purple.

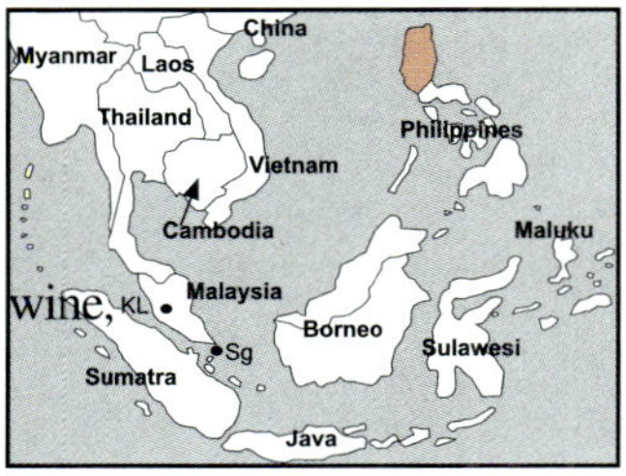


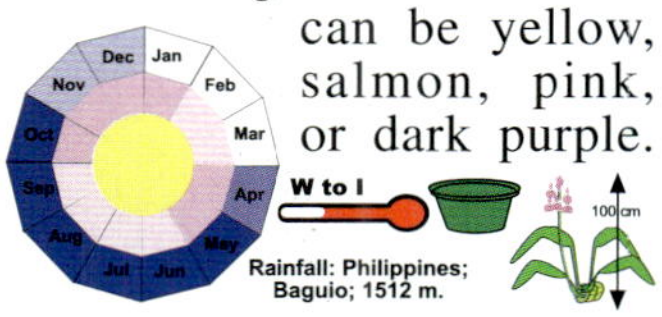

# S

### *Spathoglottis plicata* Bl.

This species is virtually ubiquitous across much of S. E. Asia. It rapidly colonises roadside cuttings and grassland from sea-level to 1600 metres altitude and a self-pollinating form has become a garden weed in several urban areas. It is highly variable in colour; in addition to the illustrated forms, an *alba* form is quite common.

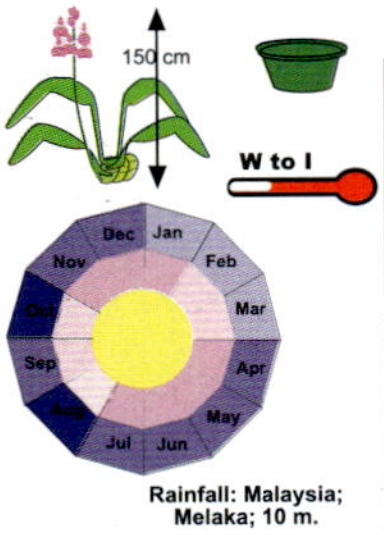


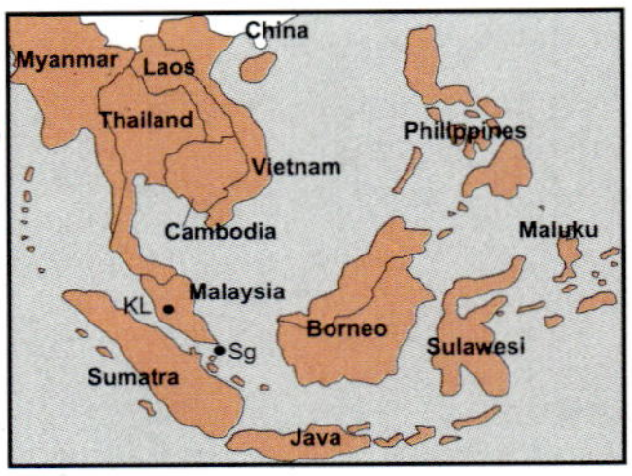

## Staurochilus

This genus of about 13 long-stemmed climbing epiphytic species is distributed from India and Myanmar to Borneo and the Philippines. The Philippines has the most species. Many of these species started off being placed in the closely-related *Trichoglottis* or *Vandopsis*, but were moved to *Staurochilus* on account of their branching inflorescences and larger flowers. Despite the fact that most of the species are easy to grow, and have long inflorescences with large, long-lasting showy flowers, they are not often seen in orchid collections outside the Philippines.

### *Staurochilus guibertii* (Linden & Rchb. f.) E.A. Christenson

This tall-growing lowland species is found on limestone rocks and trees in Quezon National Park. It has leathery 20-30 cm leaves and a rather loose-flowered inflorescence of about the same length. The 4-5 cm wide rather fleshy flowers are fragrant and remain open for up to 8 weeks. The lip is covered in short hairs and has a single keel on the midlobe.

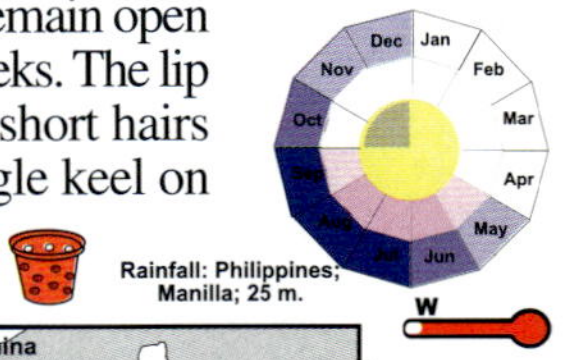

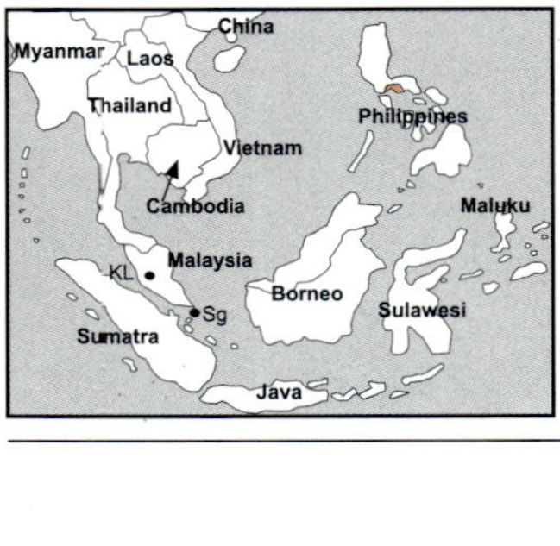

# T

## *Taeniophyllum*

This genus probably contains more than 200 species, but the number is hard to assess since so many species remain undescribed. The genus is distributed from India across S. E. Asia to Australia and the Pacific Islands. The epiphytic plants are leafless and virtually stemless, and look like a sprawl of green roots with inflorescences sprouting from the centre. They are common and widespread, but are easily overlooked and often not recognised as orchids, even when in flower. The inflorescences remain active for a long time, producing a steady succession of minute, short-lived flowers. They are hard to grow unless left attached to the original tree branch, and are not usually cultivated.

### *Taeniophyllum obtusum* Bl.

This species is common from sea-level to 500 metres altitude, growing on the twigs of any suitable tree in a range of habitats. It forms an extensive tangle of roots, with many 2-3 cm long rough brown inflorescences at the centre.

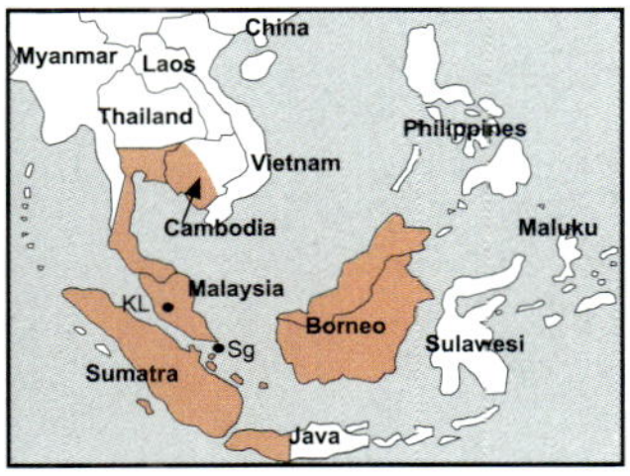


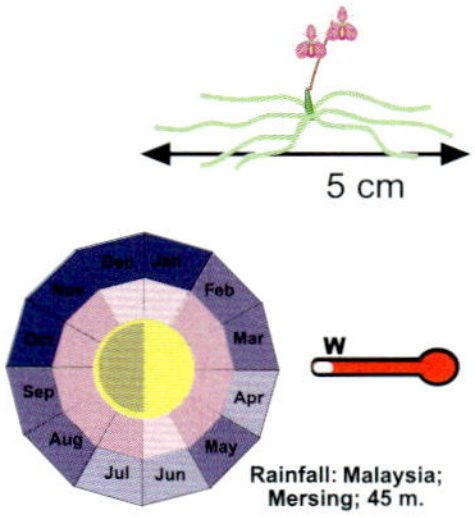


## *Tainia*

This genus contains 14 terrestrial species and is distributed from India and Sri Lanka across S. E. Asia to China, Japan, Australia and the Pacific Islands. These species have an underground rhizome with erect shoots at intervals. There are two sorts of shoot; they either have a slender pseudobulb and single leaf, or else they carry a tall inflorescence which often emerges from a rudimentary pseudobulb. The flowers are medium-size or large, and some species are showy. Despite this, they remain virtually unknown to orchid enthusiasts, and are seldom seen in collections.

### *Tainia speciosa* Bl.

This species is quite common between 250-1300 metres altitude in the moister rainforests of the region, growing in thick humus on the forest floor. It has 4-angled pseudobulbs and a 15 cm pleated leaf on a tall petiole. The 4-8 flowers are large and showy. Flower colour is variable.

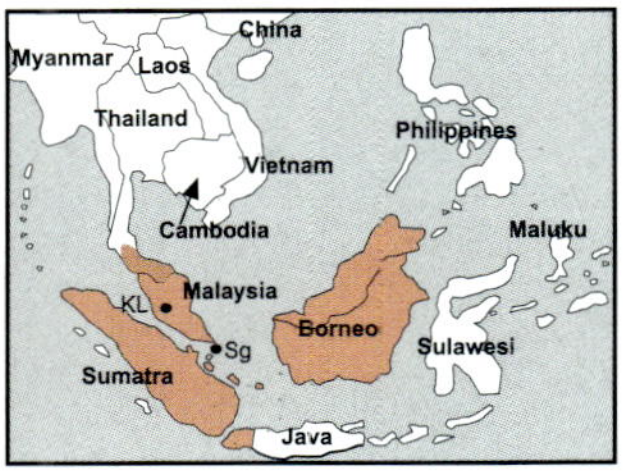


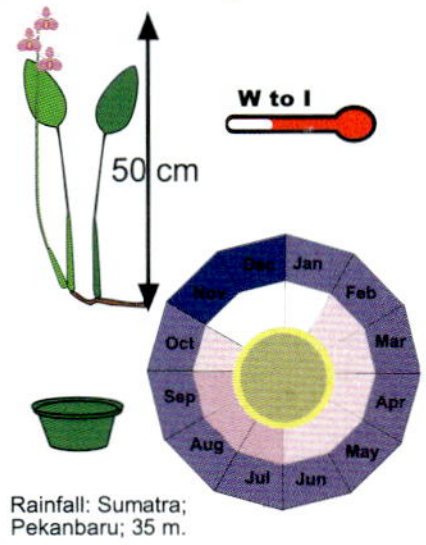

# T

## Thecopus

This genus contains 2-3 epiphytic species and is found from Thailand to Borneo. The strange flower structure means that they are not closely related to any other genus except *Thecostele*. The plants have an egg-shaped pseudobulb with a single leaf. The column foot is hollow with an entrance near the base of the lip. The lip has a complicated structure. Since the flowers are relatively large and showy, these species are quite often seen in collections in the region.

### *Thecopus secunda* (Ridl.) Seidenf.

This attractive epiphytic species is fairly rare, growing only in forests in hilly country at about 300-600 metres altitude. It has a cluster of wrinkled egg-shaped pseudobulbs with fairly broad oblong leaves. Several stiffly descending inflorescences 15 cm long are produced, with up to 15 flowers that are 3 cm wide and all open at the same time.

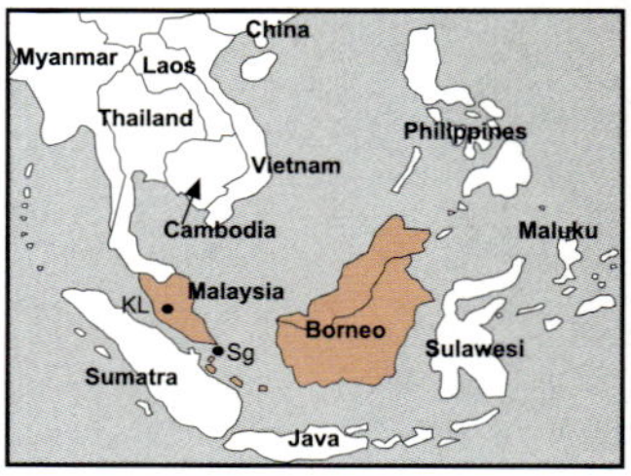


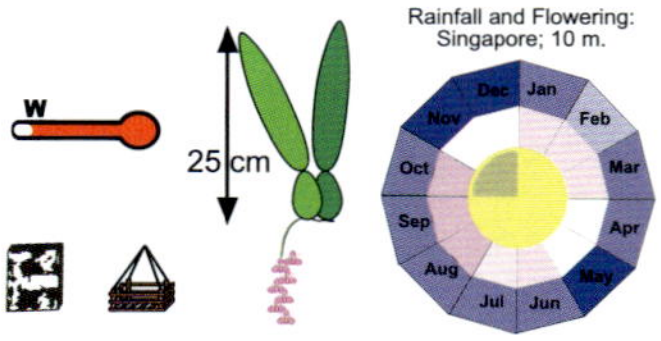


## Thecostele

This genus contains only one species. It is closely related to *Thecopus*, but can be distinguished because the column is bent into an S-shape, with a hollow tube running up from the basal arm of the 'S'. The species is commonly seen in orchid collections in the region, although it is not as attractive or showy as *Thecopus* species.

### *Thecostele alata* (Roxb.) Par. & Reichb.f.

This epiphytic species is common from sea-level to 500 metres altitude, and can be found up to 1000 metres in some regions. The cluster of wrinkled egg-shaped pseudobulbs sit in a "nest" of erect catch-roots. Several inflorescences hang down and extend slowly for several months, eventually becoming up to 50 cm long. Only a few 15 mm wide flowers are open at a time. The colour is rather variable.

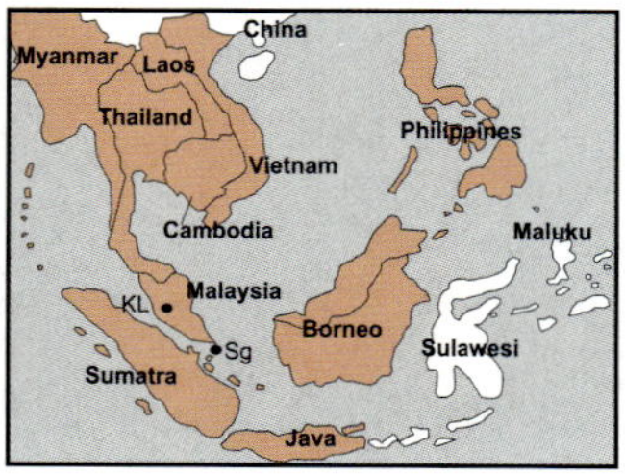


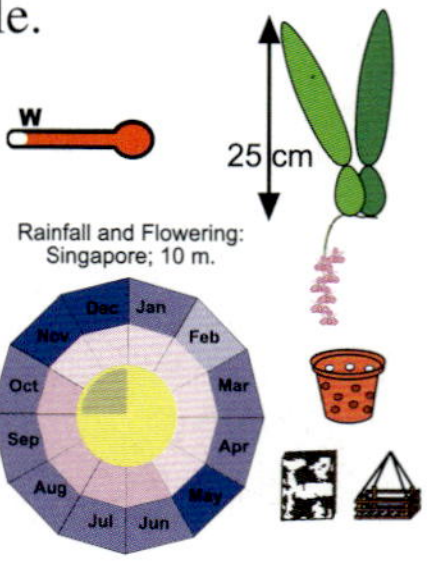

# T

## Thelasis

This genus of some 20 epiphytic species, which is distributed from eastern India to New Guinea and the Pacific Islands, contains some of the least attractive of all S. E. Asian orchids. The plants either have pseudobulbs and 1-2 main leaves per shoot, or a short stem topped by a fan of leaves. The inflorescences have many tiny flowers that hardly open. The species are of no commercial value and are seldom seen in collections.

### *Thelasis carinata* Bl.

This species grows on tree trunks or rocks in bright situations from 0-800 metres altitude, often forming large masses. It has a fan of 4-5 short leaves and usually forms two inflorescences from the lower leaf axils.

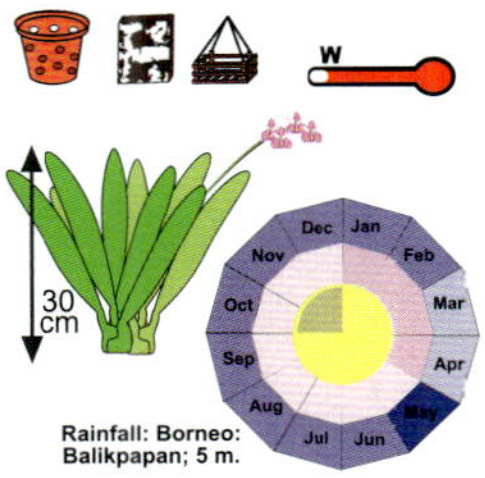


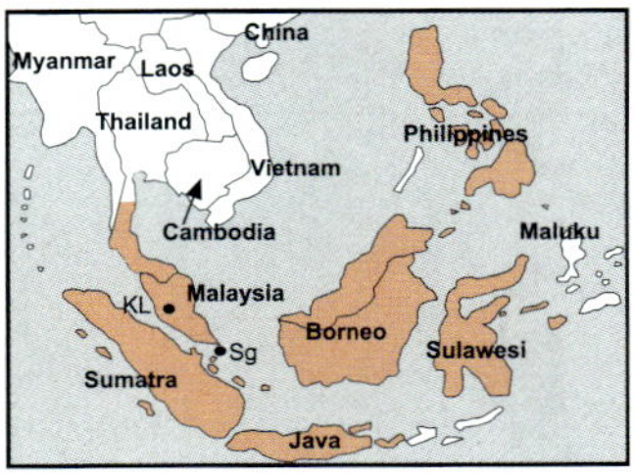


### *Thelasis micrantha* (Brogn.) J.J.Sm.

This is a common lowland species in many parts of the region, though it can occasionally be found up to 1600 metres altitude. It has a short stem and a fan of 3-4 short leaves. The 15 cm long pink-brown inflorescences emerge from the axils of the lower leaves and usually arch abruptly towards the tip. There are up to 10-12 flowers, most of which are open at once.

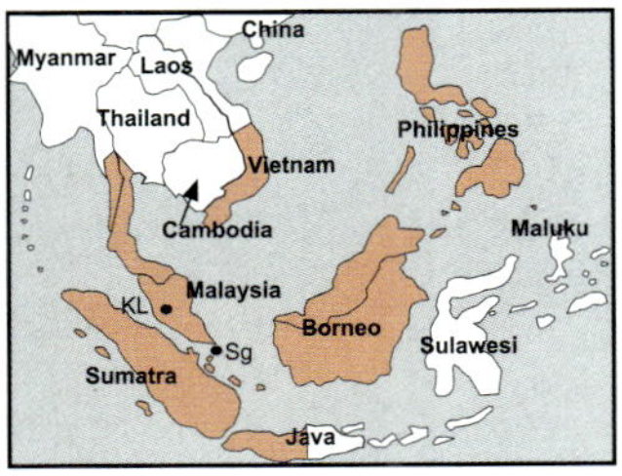


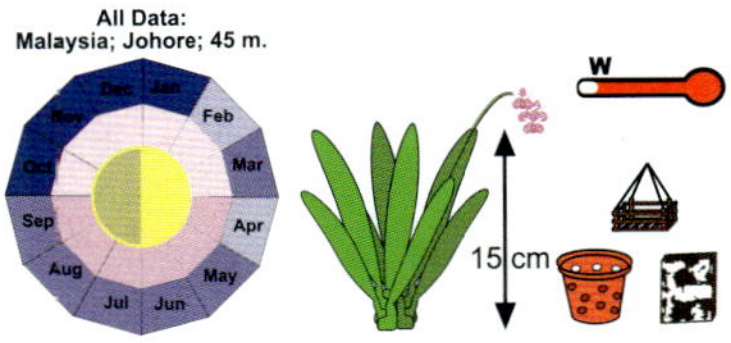


### *Thelasis obtusa* Bl.

Although it occurs as low as 300 metres altitude, this species is usually seen between 1000-1500 metres, growing on tree trunks or partly-decayed logs on the forest floor. It has a 2 cm pseudobulb with one large and two small leaves. The 20 cm tall inflorescences emerge at the base of the pseudobulb and carry a large number of densely packed small flowers, only a few of which are open at a time.

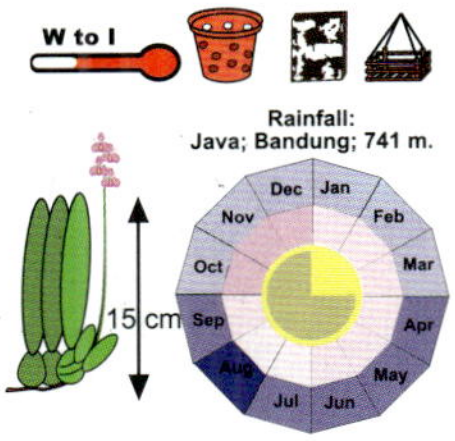


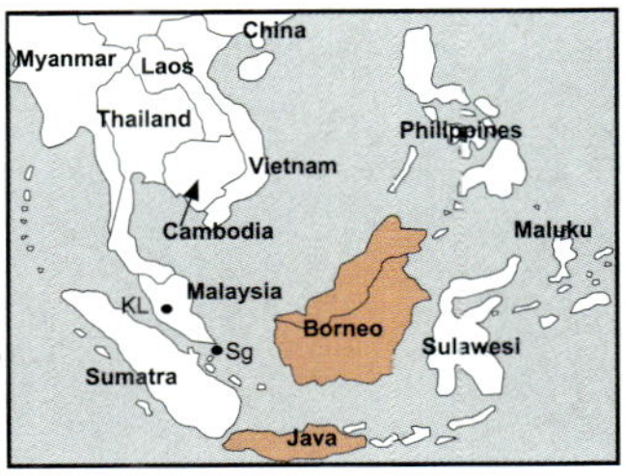

# T

## Thrixspermum

This large monopodial genus is distributed from Sri Lanka and India across S. E. Asia to the Pacific Islands. There could be as many as 150 species, but this number is highly uncertain because several large species complexes are known to exist. The plants are rather variable in size, habit and flower shape. Although several species have large attractive flowers, they are of no commercial value because the flowers are ephemeral, usually opening fully for only half a day.

### *Thrixspermum annamense* (Guill.) Garay

This small epiphytic mountain species occurs in a few widely scattered locations from Malaysia to Taiwan. It has very short stems with thick leaves that are often suffused purple. The long inflorescence has up to 20 small flowers that open one at a time at irregular intervals. The lip has a tuft of conspicuous white hairs on the midlobe.

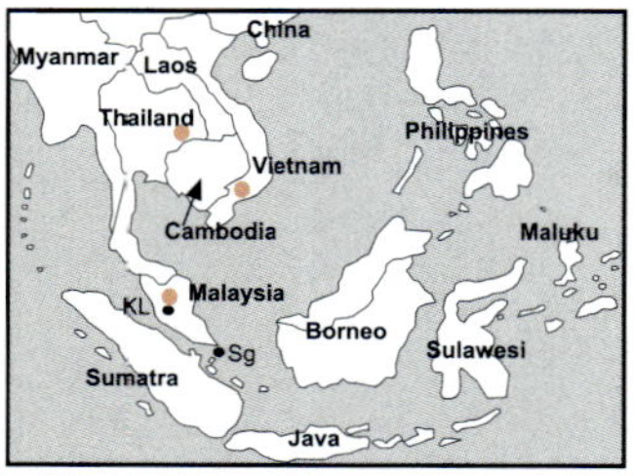


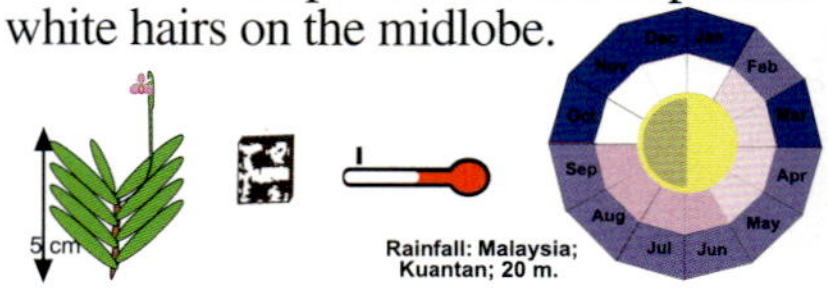


### *Thrixspermum calceolus* (Lindl.) Rchb.f.

This large lowland species climbs and scrambles through tree branches, over rocks, or even across sandy beaches, often in full sun. It can form very large colonies with stems up to 1 metre long. It has short inflorescences that usually have 1-2 flowers open simultaneously. The 5-7 cm wide fragrant flowers have a thick fleshy midlobe.

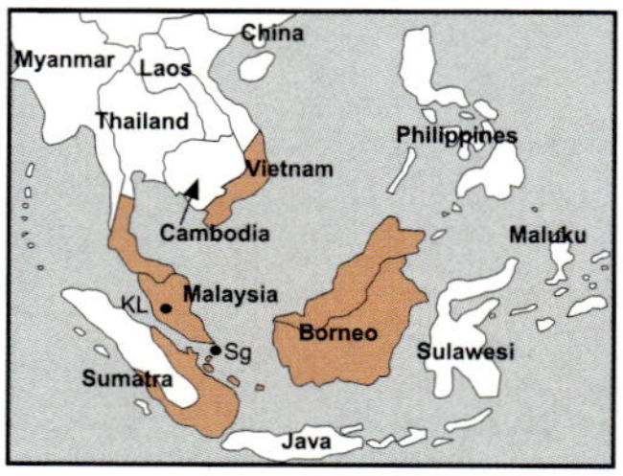


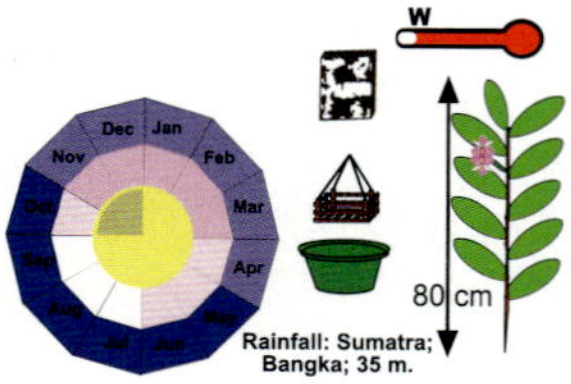


### *Thrixspermum centipeda* Lour.

It is now thought that more than 12 old 'species' are actually forms of *T. centipeda*, a highly variable species that can be found from sea-level to at least 2000 metres altitude. Some forms are better known by the old name *T. arachnites*. These untidy plants flower prolifically throughout the year.

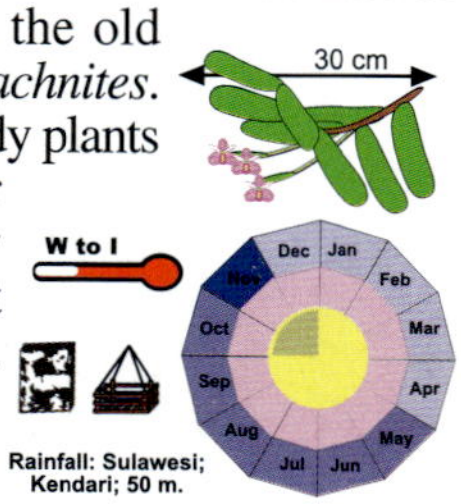


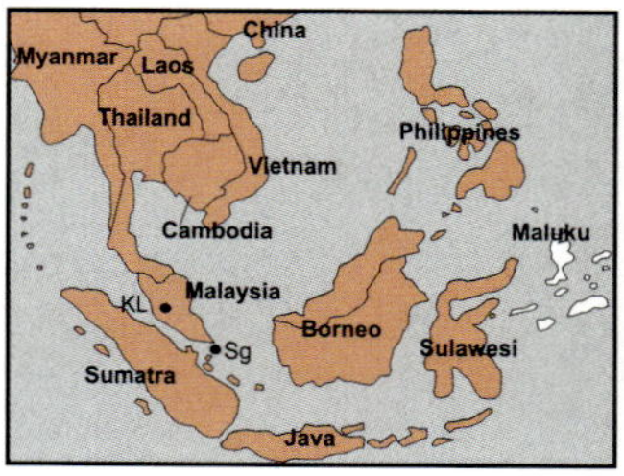

# T

## ***Thrixspermum ligulatum*** L.O. Williams

This uncommon epiphytic species has rather stout stems with many thick broad leaves. The inflorescence produces 1-2 large waxy flowers at irregular intervals.

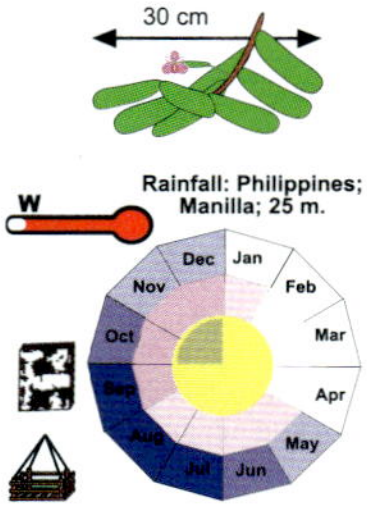


## ***Thrixspermum subulatum*** (Bl.) Rchb.f.

It is possible that the name "*T. subulatum*" is being used for more than one epiphytic species. In southern Java this is a coastal species, while in Sumatra and Sulawesi it is only found in the mountains. It has pendulous stems with dark green or green-grey leaves that have narrowed tips. The inflorescence produces a few 1-2 cm wide flowers at irregular intervals. There seems to be some variation in flower colour.

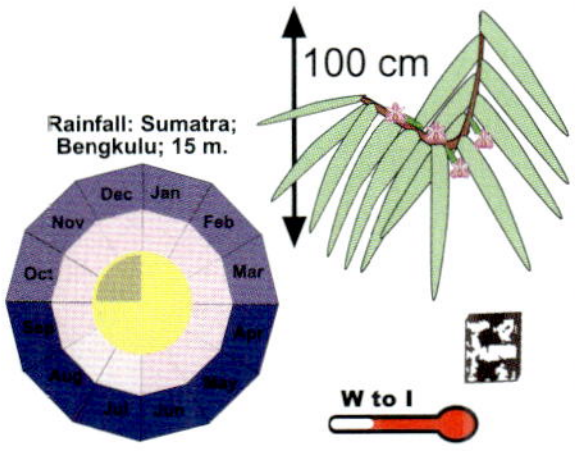


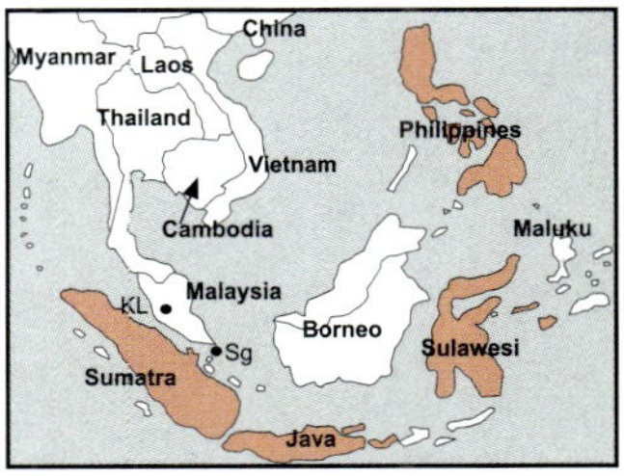

## ***Thrixspermum trichoglottis*** (Hk.f.) Kuntze

This twig epiphyte can be found in many lowland locations, but also occurs in the mountains up to 1500 metres altitude. It likes to grow on trees in exposed situations and is quite common in some city parks and gardens. The inflorescence produces 1-4 small flowers at irregular intervals. The front half of the lip is covered in small club-shaped hairs.

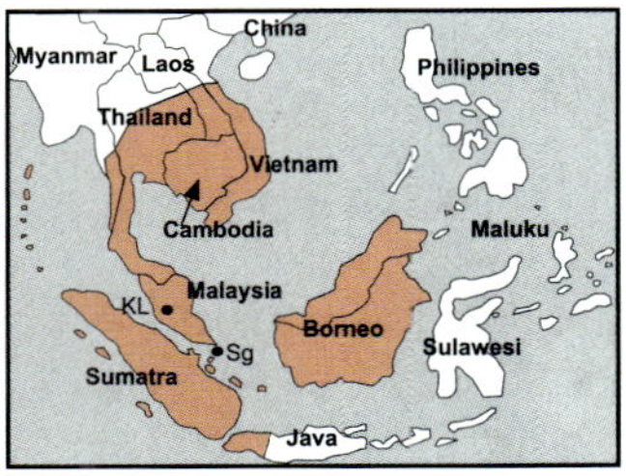


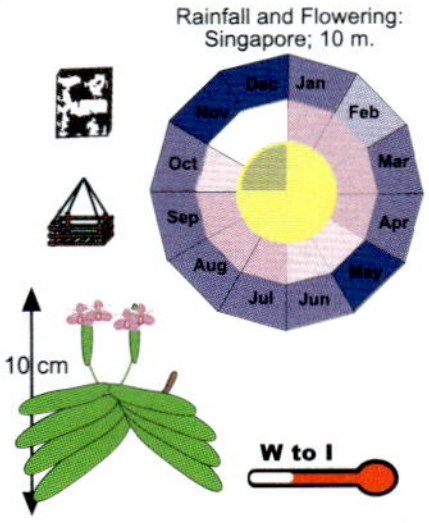

# T

## *Trichoglottis*

This monopodial genus of about 60 species is distributed from Sri Lanka and the Nicobar Islands to New Guinea, Australia and the Solomon Islands. The Philippines has the largest number of species. Most species are epiphytes with climbing or pendulous stems and narrow leaves. The inflorescences are short and usually bear 1-2 rather small flowers. Several species are quite common in cultivation in S. E. Asia, and other species are occasionally seen in collections. Some of the showier species have been used to make hybrids with species from other genera.

### ***Trichoglottis adnata*** J.J.Sm.

This little-known species is quite common in northern Sumatra between 600-1500 metres altitude, climbing up through the high branches of trees in the forest or in villages. It has slender stems with distant narrow leaves. Nodes on the stem produce 1-3 single-flowered inflorescences which emerge through a scar on the leaf sheath. The flowers are about 2 cm high.

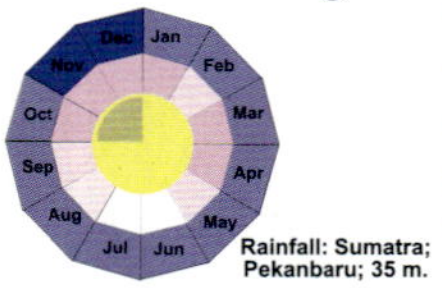

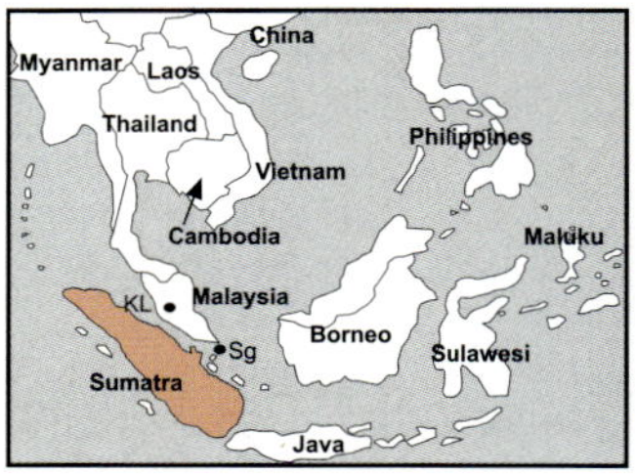


### ***Trichoglottis bataanensis*** Ames

This attractive species is quite common from sea-level to at least 500 metres altitude, growing on trees and shrubs in moderate shade. It has long pendulous stems with large close-set leaves. The flowers are 1 cm wide.

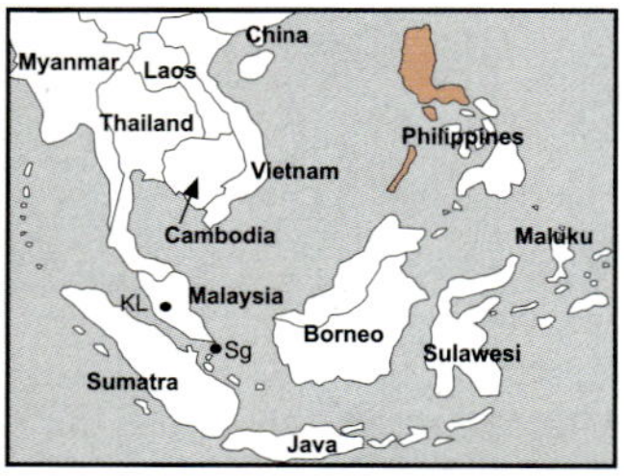


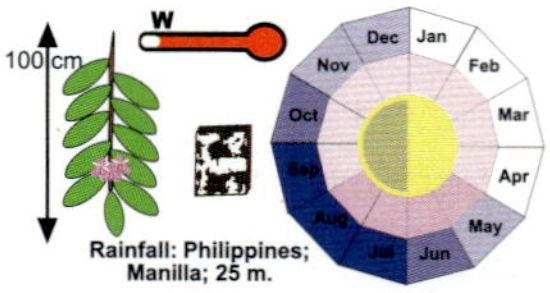


### ***Trichoglottis geminata*** (T.& B.) J.J.Sm.

The "Zebra Orchid" is common across much of its range, growing as an epiphytic climber or a lithophytic scrambler from sea-level to at least 1100 metres altitude. It has slender stems that branch freely, narrow leaves, and produces numerous 1-3 flowered inflorescences which all open at the same time. The 2 cm wide flowers are highly fragrant.

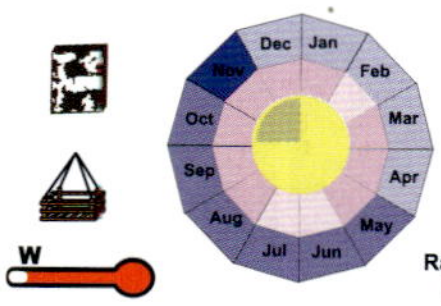

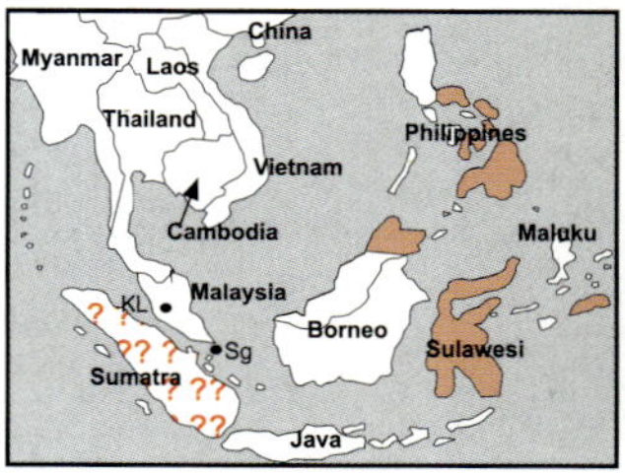

# T

## *Trichoglottis philippinensis* (Lindl.) Rchb.f. **var. *brachiata*** Ames

This large-flowered variety is found climbing up tree trunks at low altitudes. It has rather stiff stems and fairly broad 6-8 cm long leaves. Several single-flowered inflorescences are produced at a time; the flowers are about 4 cm wide. Some people call this plant *T. brachiata*. The normal form of *T. philippinensis* is found across the Philippines and has gold-brown flowers with a white lip that is suffused pale lilac.

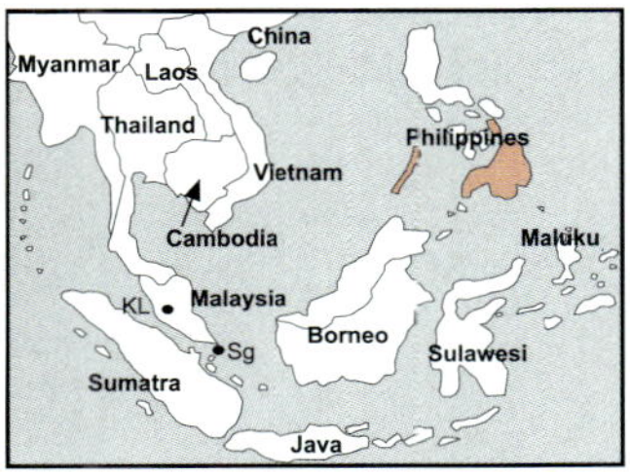


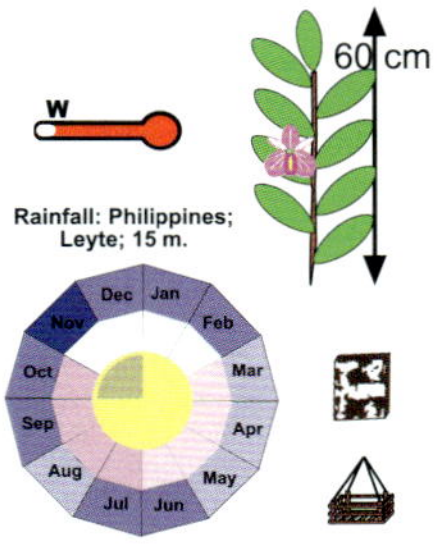


## *Trichoglottis retusa* Bl.

This widespread epiphytic climbing species grows from sea-level to about 700 metres altitude. It has slender branched stems with rather distant narrow 10-12 cm leaves. Several single-flowered inflorescences are formed from each node on the upper stem; the 1 cm high flowers have a hairy white lip.

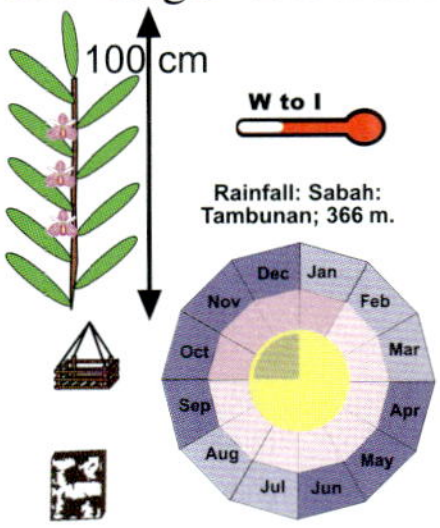


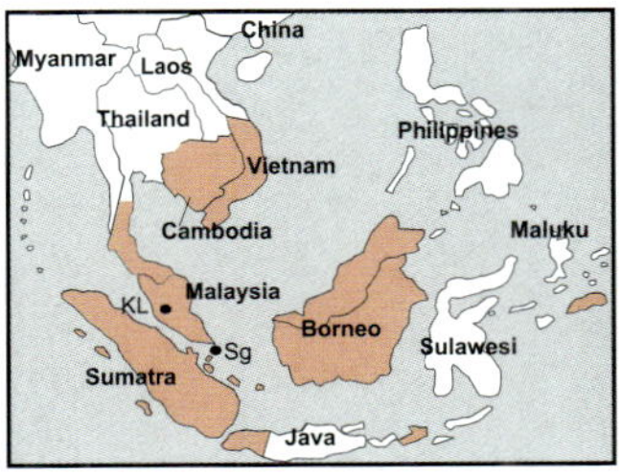

## *Trichoglottis uexkulliana* J.J.Sm.

This rare epiphytic lowland species is probably the most beautiful member of the genus. It has slender pendulous stems with rather distant broad 10 cm leaves. 1-2 single-flowered inflorescences are formed from each node on the upper stem; many of the 25 mm flowers are open at the same time.

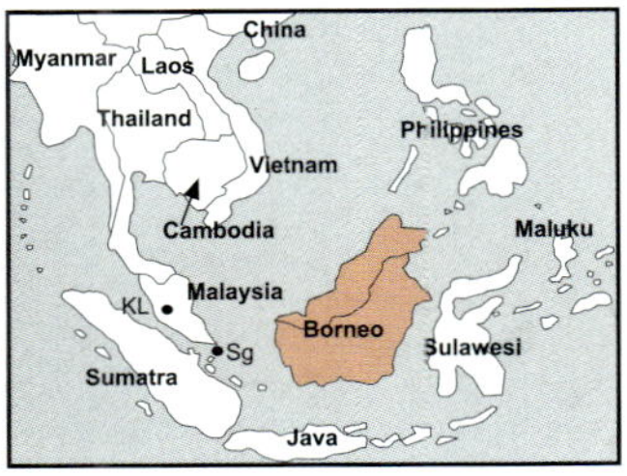


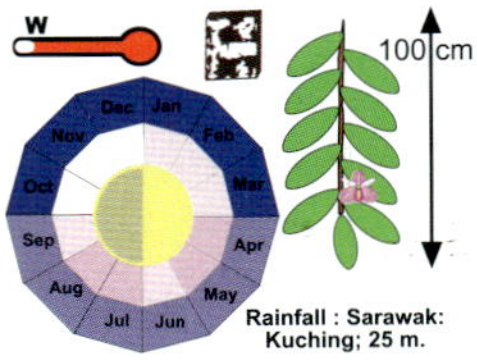

# T

## *Trichotosia*

This genus of some 50-60 species is closely related to *Eria*; some authors still consider it to be a subgenus of *Eria*. The species are distributed from China and the Himalayas to New Guinea and the Pacific Islands. They are easy to identify as their stems, leaves, and inflorescences are densely covered in short red-brown (rarely black or white) hairs. The inflorescences are usually rather short with a few flowers that do not open widely. They are not often seen in collections, even though some species are quite attractive.

### ***Trichotosia aporina*** (Hk.f.) Krzl.

This small epiphytic species has a rather limited distribution but is quite common in some montane forests from 1000-1700 metres altitude. It has short stems with 1-2 cm long leaves and produces several single-flowered inflorescences.

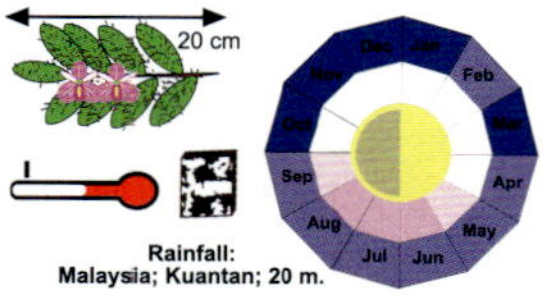


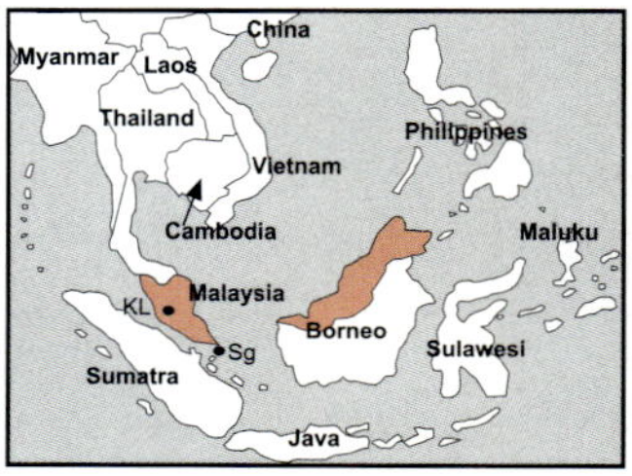

### ***Trichotosia ferox*** Bl.

This imposingly large species is quite common in mossy forests from 1000-3000 metres altitude, usually growing on tree branches or mossy rocks in fairly exposed locations. The stems are 1-3 metres long, the leaves about 12-15 cm long, and the many-flowered inflorescences are up to 10 cm long.

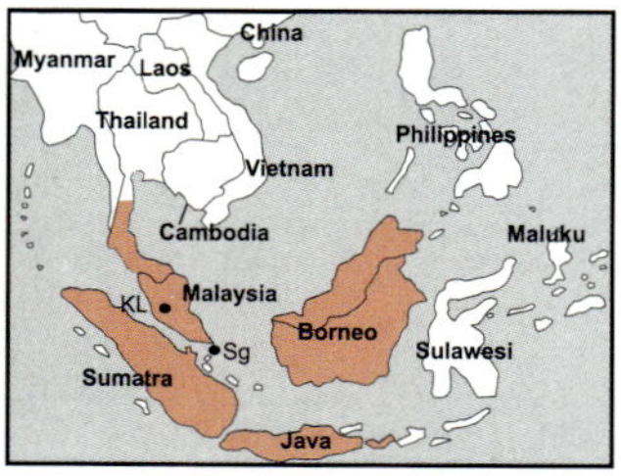


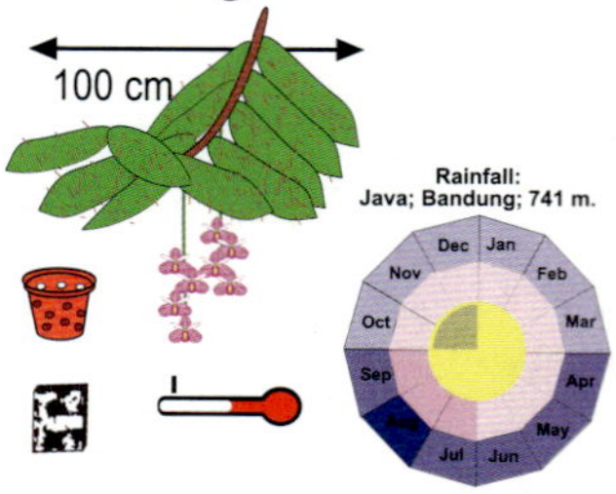


### ***Trichotosia gracilis*** Lindl.

This epiphytic species is fairly common in lowland forests, where it grows in forks on tree trunks and branches. The short stems have 5 cm long leaves and produce several inflorescences of 1-3 flowers, although usually only one is open at a time.

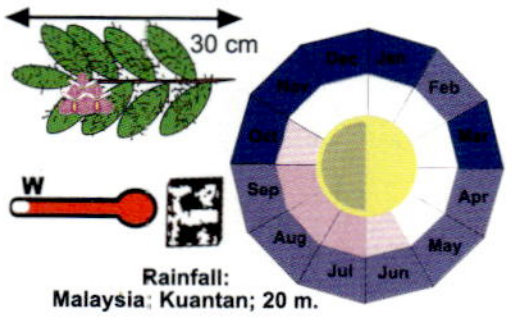


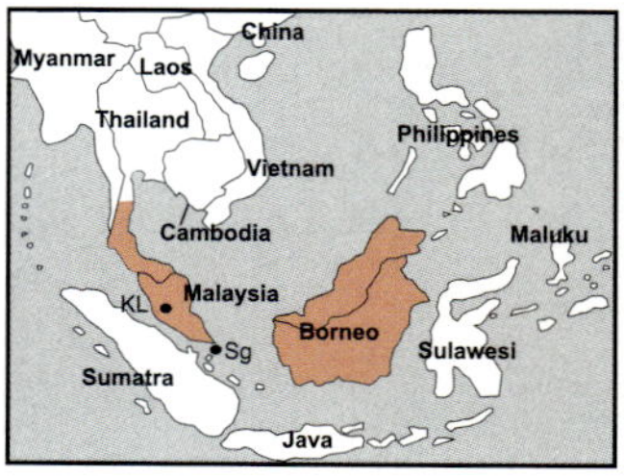

# V

## Vanda

This commercially important genus of 40 to 50 species is distributed from Sri Lanka to northern India, China, the Philippines, Australia and the Solomon Islands. *Vanda* species are grown in collections around the world, and their hybrids (often with species from other genera) form the basis of a cut-flower industry which is an important component of several S. E. Asian economies. Historically the most important species has been *Vanda sanderiana* (more correctly called *Euanthe sanderiana*) which has given rise to a seemingly endless succession of hybrids with large colourful round flat flowers.

### ***Vanda coerulea*** Cogniaux

This species grows at around 1200 m. altitude in the mountains of northern India, Myanmar, Thailand and southwest China. It is rather variable in colour, ranging from purple through shades of lilac and blue to white; in some plants the flowers are beautifully tessellated. This species has been extensively used in hybridization programs as the blue colour carries through to the offspring. Since *V. coerulea* is highly endangered, all international trade in this species is strictly controlled.

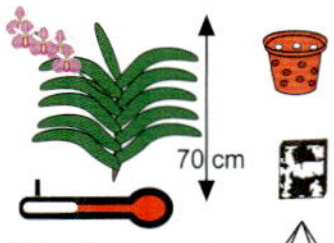


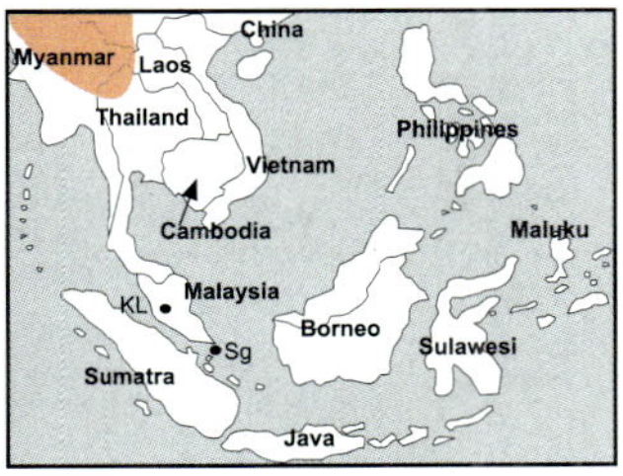


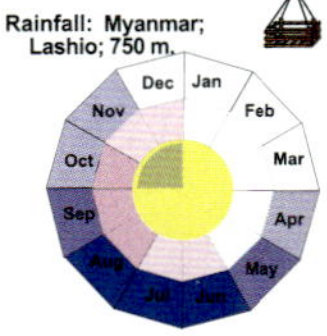


### ***Vanda dearei*** Rchb.f.

This lowland species has highly fragrant flowers that vary in colour from cream to yellow to green to brown.

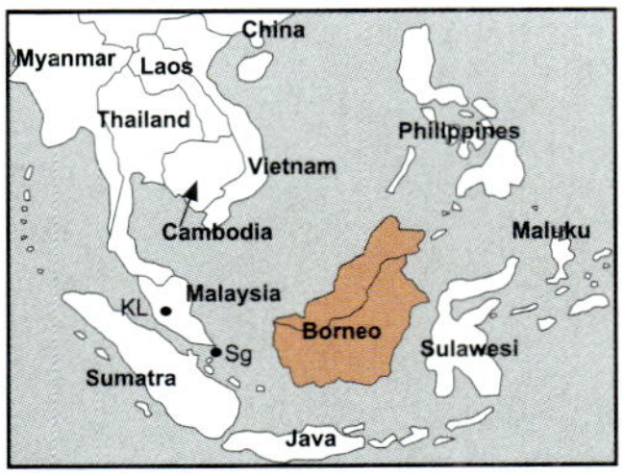

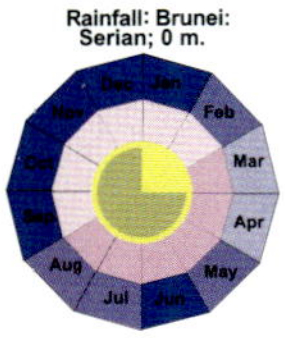


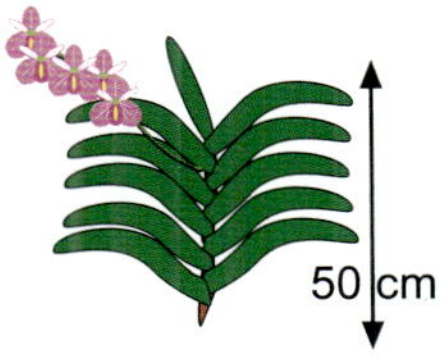

# V

## *Vanda lamellata* Lindl.

This small-flowered lightly-fragrant species is quite common in parts of the Philippines. Several colour varieties have been identified; var. *remediosae* (illustrated) has brighter colours than the normal form.

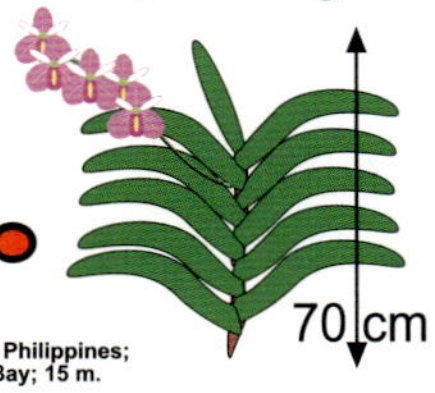


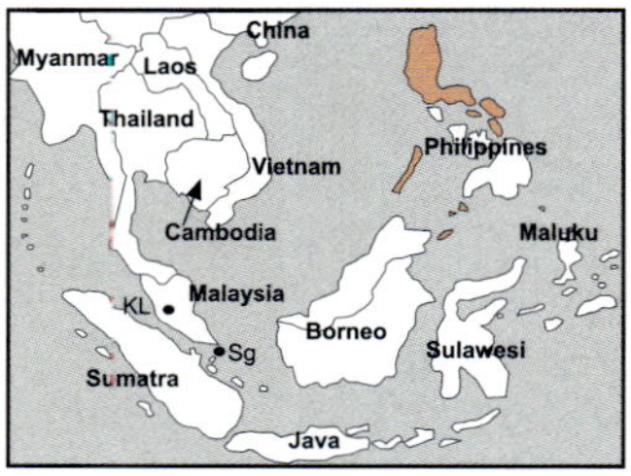

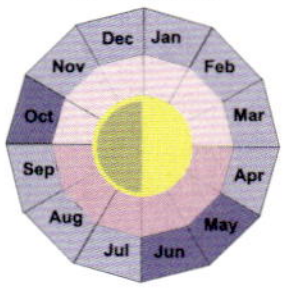


## *Vanda limbata* Bl.

There are two separate populations of this species. The Javanese form, which has yellow sepals and petals that are heavily blotched red on the front side, is still common on the island of Madura. The colours of the Philippine form (illustrated) are variable but usually have less yellow on the flower.

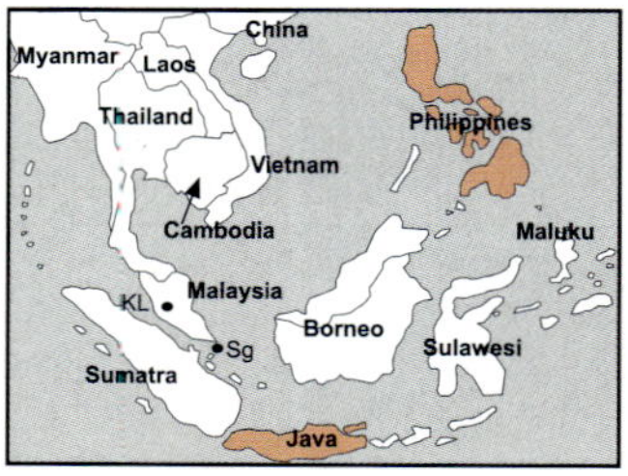


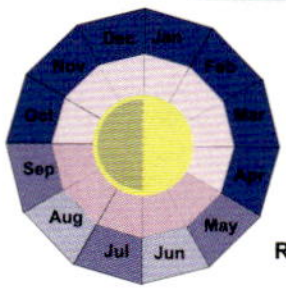


Rainfall: Philippines; Mindanao; 22 m.

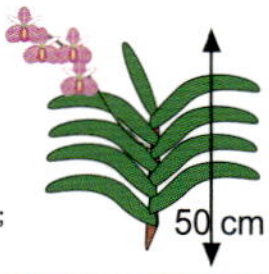


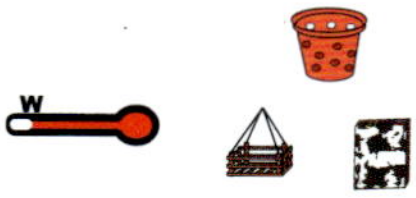


## *Vanda lindeni* Rchb.f.

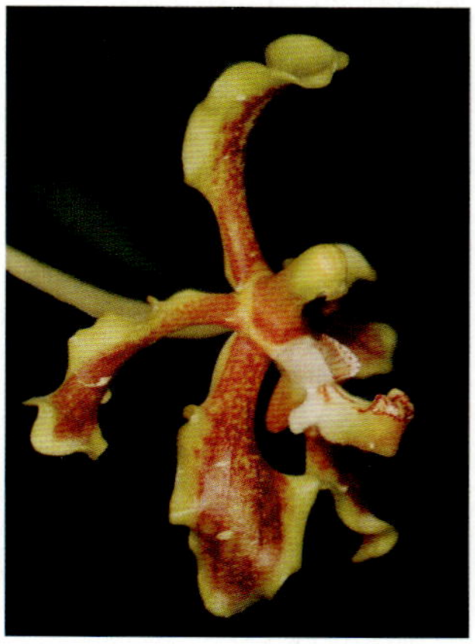

This little-known tall-growing species is closely related to the Bornean *V. scandens*, but has a different midlobe apex. It produces a medium to long inflorescence with up to 12 flowers, each of which is about 5 cm wide.

Rainfall : Maluku; Ternate; 25 m.

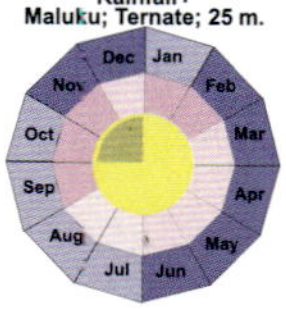


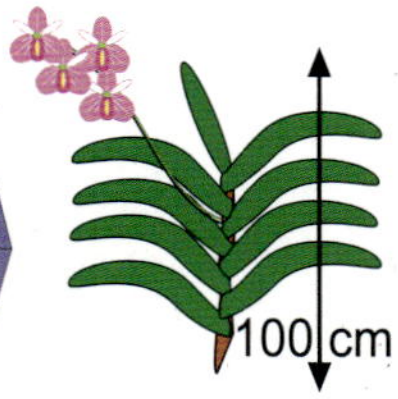


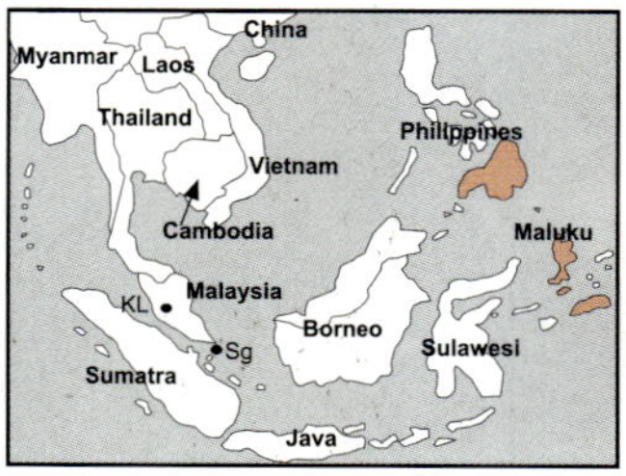

# V

## *Vanda luzonica* Loher

This beautiful highly-fragrant species is closely related to *V. tricolor*. The amount of spotting on the flower can vary considerably.

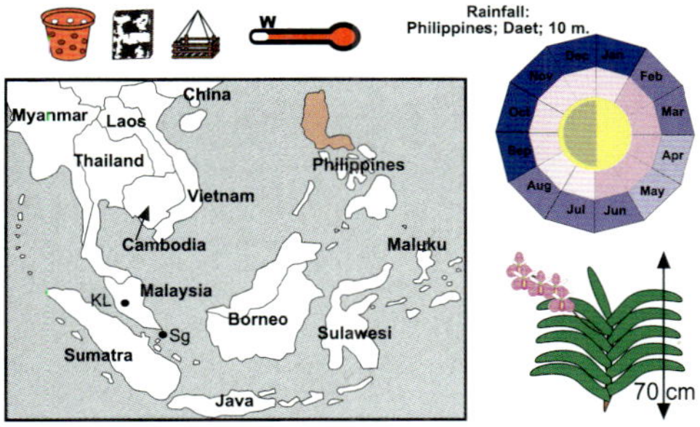

## *Vanda tessellata* (Roxb.) Hk.f. ex G.Don

This highly fragrant free-flowering species has a wide range of colour forms. It is distributed from Sri Lanka to Nepal and Myanmar.

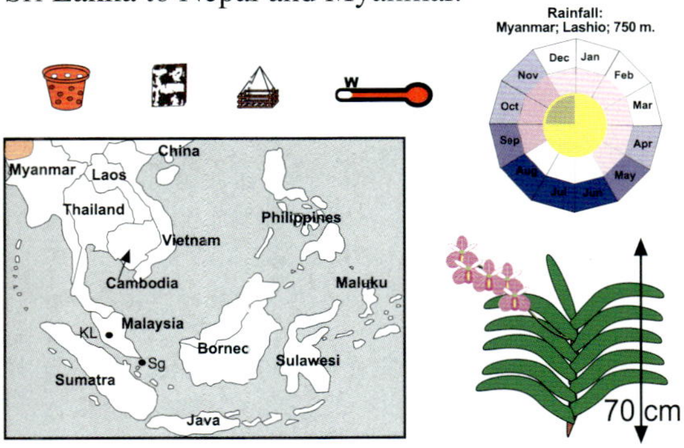


## *Vanda tricolor* Lindl.

This beautiful sweetly-scented species is still common in parts of Java and Bali, where it is often seen in village gardens. The white to pale yellow tepals are spotted red-brown or yellow on the front surface, although sometimes the tepals are almost covered in heavy red blotches. Var. *suavis* (illustrated) has longer narrower tepals than the normal type.

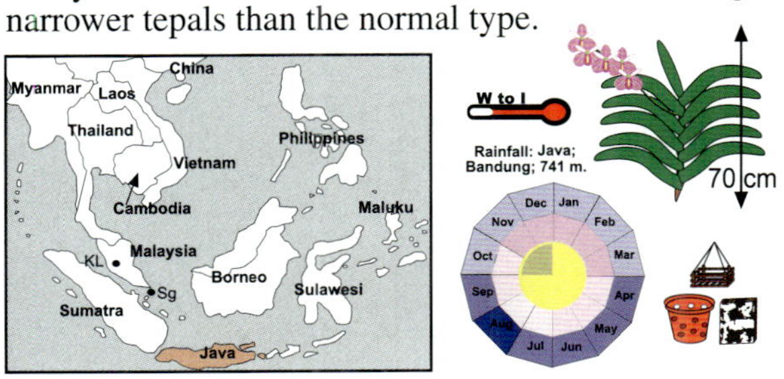

# V

## Vandopsis

This genus of only 5 species occurs from Myanmar and China through S. E. Asia to Indonesia and the Philippines. It contains some of the largest species in cultivation and has been used as a parent in hybridization programs to increase the flower size of Vanda-type hybrids. *V. gigantea,* the best-known species, occurs from Myanmar to China to Malaysia and has large dull yellow flowers with red blotches.

### *Vandopsis lissochiloides* (Gaud). Pfitz.

This showy giant species is quite common in some regions at about 1000 metres altitude, where it usually grows on cliffs and large rocks. It also grows on large mangroves at sea level. Although it is extremely slow-growing, this species is a common garden plant in villages in parts of Indonesia and the Philippines, where it is often treated as a terrestrial and grown in beds.

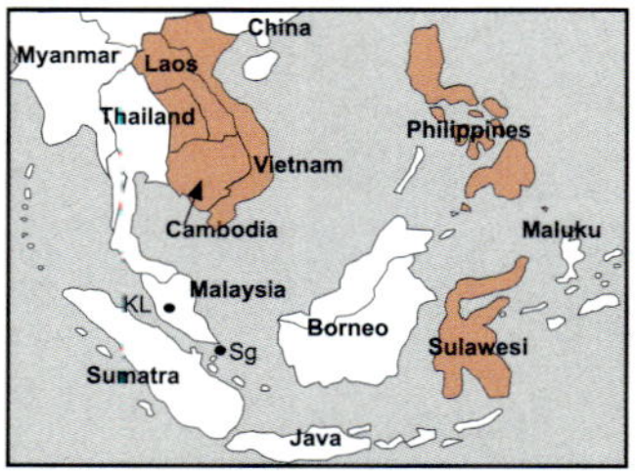


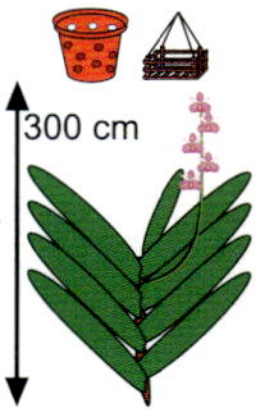


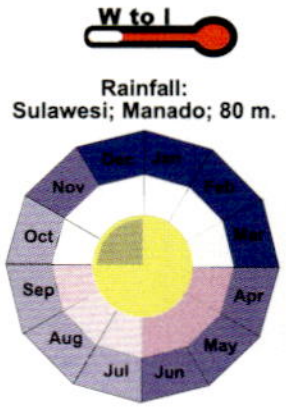


## Vanilla

This genus of about 100 species occurs in tropical regions around the world. The plants have a vine-like growth habit and start growing at ground level, then climb or scramble up the nearest tree until they reach the sunny canopy where they flower. The vines have a single leaf and root at each node; although the leaf is usually large and fleshy, in some species it is reduced to a small scale. Vanilla extract, used as a food flavour, is produced from the fermented pods of certain American species such as *V. fragrans (V. planifolia)*, which is grown commercially in several Asian countries including Malaysia and Indonesia. **Cultivation**: Vanilla species are not suitable for indoor cultivation. In tropical regions a short length of vine can be rooted in rich loam and the plants allowed to climb up a tree or climbing frame. The vines will only flower when they are several metres long.

### *Vanilla diabolica* O'Byrne

This recently-discovered species differs from other *Vanillas* in having an elongated inflorescence. Only one flower opens at a time, and each flower only lasts for one day.

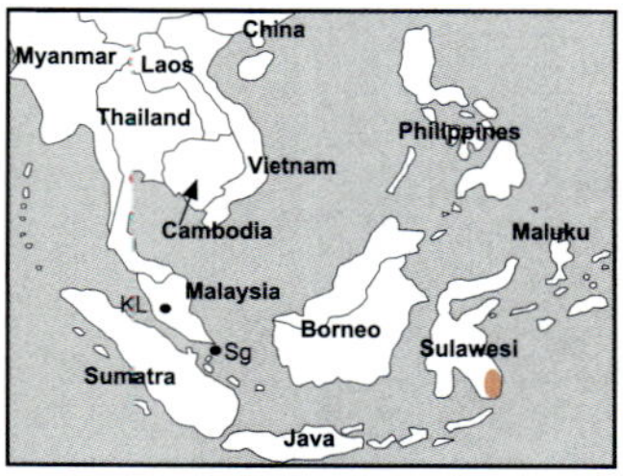


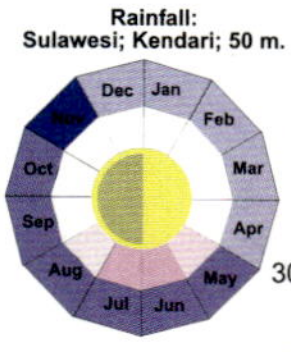


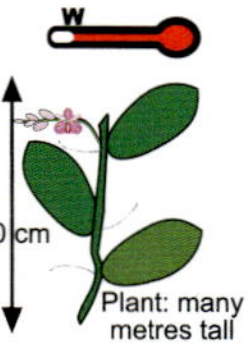

# V

### *Vanilla griffithii* Rchb.f.

This species is quite common in the lowlands of Malaysia, although the flowers are very seldom seen. Only one or two flowers open at a time and they last for only one day. This species has a particularly elaborate lip, with the margins folded back to reveal large tufts of white hairs in the throat.

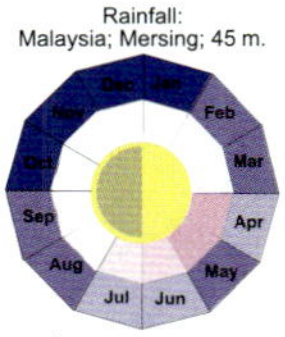


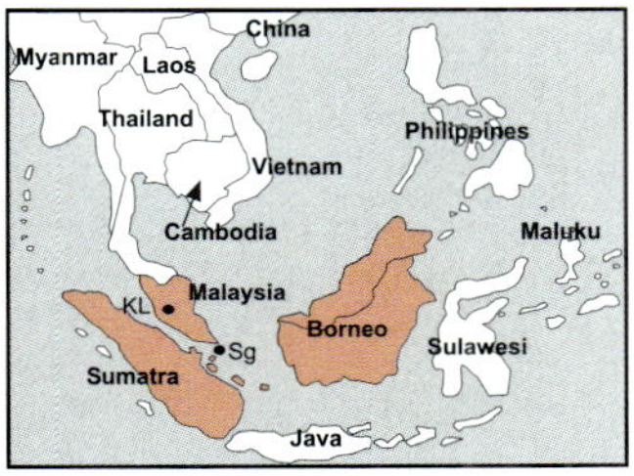


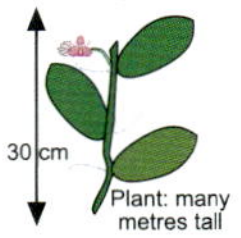

## Ventricularia

There is only one species in this genus.

### *Ventricularia tenuicaulis* (Hk.f.) Garay

This slender pendulous species is quite common in forested limestone areas. It can grow to over 40 cm long but flowers at a much smaller size. The flowers are easily overlooked as they are only about 8 mm in diameter.

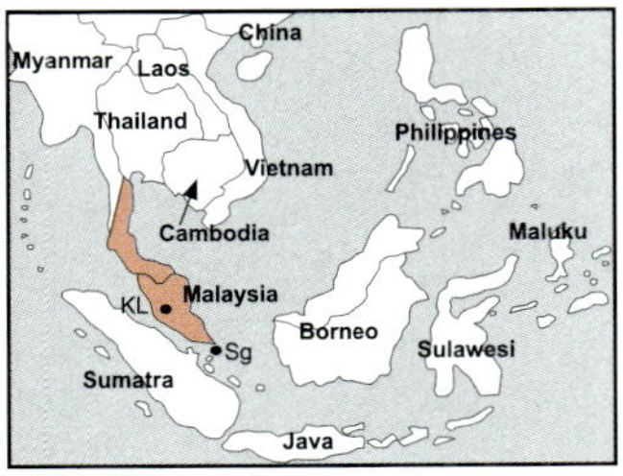


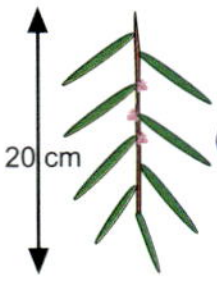


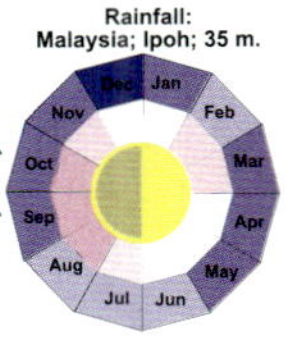

# W to Z:

There are no S. E. Asian genera beginning with W or X.

Y stands for *Yoania*, a genus of 3 strange rootless and leafless terrestrial species that have underground rhizomes. *Y. prainii* was recently discovered in north Vietnam.

Z stands for Zeuxine, a genus of 40-50 terrestrial species distributed throughout the tropical regions of Africa, Asia and the Pacific Islands. There are 10 species described from Malaysia, but only 2 of them appear to be at all widespread. Most of the species are superficially like *Anoectochilus* or *Macodes*, with weak fleshy stems that bear few to several leaves in a rosette; the leaves are not usually brightly coloured. The terminal inflorescence has few to several small flowers that do not open widely.

# *Index of Plant Names*

Bold text indicates a genus or species featured in the book. Names of species shown in roman are mentioned only in the text. Synonyms are shown in italics.

# The icons used in this book

Orchid enthusiasts always want to know how to grow their plants. Unfortunately, growing conditions in different parts of the world vary so much that it is impossible to give meaningful cultivation advice. However, it is possible to give information about the conditions that exist in the orchid's native habitat; the enthusiast can then try to replicate these conditions to get the best results from his or her plants.

In this book, this data is given in the form of pictorial icons rather than the more usual form of charts and graphs. This has been done in order to make it easier to understand the information, and also to save space. One consequence of this is that the information has become very generalised. These pictograms are, at best, a very rough guide.

## Distribution maps

Since little is known about the distributions of most orchids, these maps are, at best, a rough guide to the range of the species.

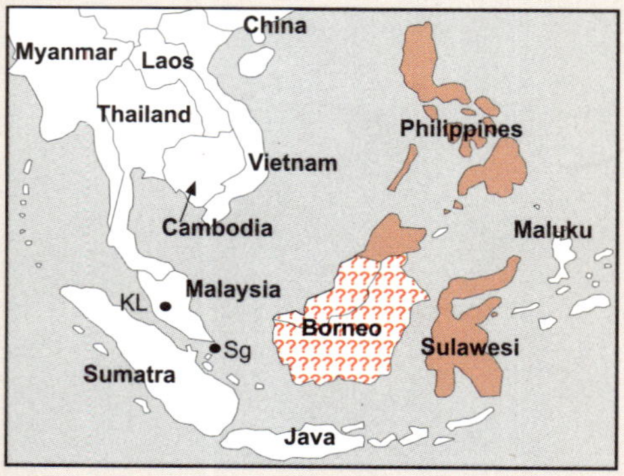


The red colour indicates the area from which the species has been recorded. This does not mean that the species is found throughout the indicated area, but that the species may be found in suitable habitats in the indicated region.

The red question mark indicates a region where similar plants have been reported, but it is uncertain if they are the same species.

## Plant habit guide

The diagram shows whether the plant is sympodial or monopodial, the arrangement and approximate shape and size of pseudobulbs or stems, the approximate arrangement and shape of the leaves, and the approximate location, size and direction of the inflorescence.

The two diagrams above show sympodial plants, which grow by producing a series of shoots from a basal rhizome. In most species the next shoot is produced after the current one has finished growing. Each shoot has the potential to grow to a maximum size (indicated), although they may fail to reach this size if conditions are unfavorable. Different colours have been used to show the different shoots.

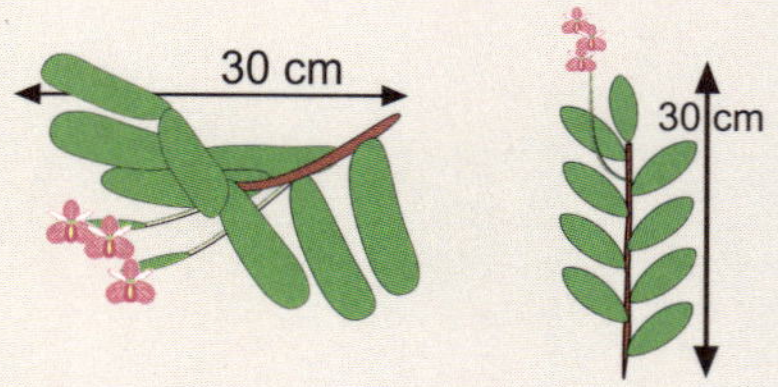


The two diagrams above show monopodial plants; the stem grows continuously from the apex. In most species, once the stem has reached a certain size it produces new shoots from the base which eventually grow into new plants. The stated size indicates how large the leafy section of the plant may become if given favorable conditions.